PHARMACEUTICAL EXCIPIENTS

PHARMACEUTICAL EXCIPIENTS

Properties, Functionality, and Applications in Research and Industry

Edited by

OTILIA M. Y. KOO

WILEY

Published by John Wiley & Sons, Inc., Hoboken, New Jersey
Published simultaneously in Canada

Library of Congress Cataloging-in-Publication Data:

Names: Koo, Otilia M. Y. (Otilia May Yue), 1974- editor.
Title: Pharmaceutical excipients : properties, functionality, and
applications in research and industry / edited by Otilia M Y Koo.
Description: Hoboken, New Jersey : John Wiley & Sons, Inc., [2017] | Includes
bibliographical references and index.
Identifiers: LCCN 2016022475| ISBN 9781118145647 (cloth) | ISBN 9781118992418
(PDF) | ISBN 9781118992425 (ePub)
Subjects: | MESH: Excipients | Dosage Forms | Technology,
Pharmaceutical–methods
Classification: LCC RS201.E87 | NLM QV 800 | DDC 615.1/9–dc23 LC record available at
https://lccn.loc.gov/2016022475

Set in 10/12pt, TimesLTStd by SPi Global, Chennai, India.

Printed in the United States of America

10 9 8 7 6 5 4 3 2 1

CONTENTS

List of Contributors **vii**

1 **Excipient Characterization** **1**
David Good and Yongmei Wu

2 **Excipients for Conventional Oral Solid Dosage Forms** **51**
Umesh Kestur and Divyakant Desai

3 **Excipients and their Functionality for Enabling Technologies in Oral Dosage Forms** **97**
Kunnal Sharma, Shreya Thakkar, Sakshi Khurana, and Arvind Kumar Bansal

4 **Excipients used in Biotechnology Products** **145**
Eva Y. Chi

5 **Excipient Standards and Harmonization** **199**
R.C. Moreton

6 **Regulatory Information for Excipients** **241**
Christopher C. DeMerlis, Alexa Smith, and David R. Schoneker

7 Development of New Excipients **269**
Karl Kolter and Felicitas Guth

8 PATability of Excipients **303**
Claudia Corredor

Index **329**

LIST OF CONTRIBUTORS

Arvind Kumar Bansal
Department of Pharmaceutics, National Institute of Pharmaceutical Education and Research (NIPER), Punjab, India

Eva Y. Chi
Department of Chemical and Biological Engineering and Center for Biomedical Engineering, University of New Mexico, Albuquerque, NM, USA

Claudia Corredor
Pharmaceutical Development, Bristol-Myers Squibb Company, New Brunswick, NJ, USA

Christopher C. DeMerlis
Global Regulatory Affairs, Colorcon, Inc., West Point, PA, USA

Divyakant Desai
Drug Product Science and Technology, Bristol-Myers Squibb Company, New Brunswick, NJ, USA

David Good
Drug Product Science and Technology, Bristol-Myers Squibb Company, New Jersey, USA

Felicitas Guth
BASF SE, Global Research & Formulation Nutrition & Health, Ludwigshafen, Germany

Umesh Kestur
Drug Product Science and Technology, Bristol-Myers Squibb Company, New Brunswick, NJ, USA

Sakshi Khurana
Department of Pharmaceutics, National Institute of Pharmaceutical Education and Research (NIPER), Punjab, India

Karl Kolter
BASF SE, Global Research & Formulation Nutrition & Health, Ludwigshafen, Germany

R. Christian Moreton
FinnBrit Consulting, Waltham, MA, USA

David R. Schoneker
Global Regulatory Affairs, Colorcon, Inc., West Point, PA, USA

Kunnal Sharma
Department of Pharmaceutical Technology (Formulations), National Institute of Pharmaceutical Education and Research (NIPER), Punjab, India

Alexa Smith
Global Regulatory Affairs, Colorcon, Inc., West Point, PA, USA

Shreya Thakkar
Department of Pharmaceutical Technology (Formulations), National Institute of Pharmaceutical Education and Research (NIPER), Punjab, India

Yongmei Wu
Drug Product Science and Technology, Bristol-Myers Squibb Company, New Jersey, USA

1

EXCIPIENT CHARACTERIZATION

David Good and Yongmei Wu

Drug Product Science and Technology, Bristol-Myers Squibb Company, New Jersey, USA

1.1 INTRODUCTION

A comprehensive understanding of the chemical and physical properties of common pharmaceutical excipients is essential to the design of high-quality drug products that provide consistent performance. In many pharmaceutical formulations, the drug substance can be susceptible to chemical and physical changes induced by the properties of the bulk excipients [1]. This is often more pronounced for drug products where the ratio of excipient content to drug is very high (i.e., low drug loading formulations). In recent years, the regular advancement of highly potent and selective drug candidates has led to more formulations that are predominately comprised of excipients and incorporate lower levels of the active pharmaceutical ingredient (API). In addition, potent drug candidates often exhibit low aqueous solubility and can require enabling formulation technologies, which include unique excipients and/or processing steps, to provide the desired clinical exposure at some stage during the clinical development program [2]. These trends in drug substance properties as well as the implementation of quality by design (QbD) product development strategies place an increased emphasis on detailed characterization of excipients to achieve robust formulations and processes.

This chapter focuses on a fundamental description of the chemical and physical properties of excipients, the associated characterization methods, and implications for formulation and processing of drug products. Numerous publications such as USP monographs provide an introduction to basic compendial excipient test methods and

Pharmaceutical Excipients: Properties, Functionality, and Applications in Research and Industry, First Edition. Edited by Otilia M.Y. Koo.

properties. These compendial descriptions and methodologies serve the basis for classification and release testing of materials; however, additional characterization is often required in the selection and processing of excipients. The content presented in this chapter provides the reader an introduction to the current methodologies and excipient properties that are most significant for the development of a commercial drug product. Included in this chapter are detailed descriptions of excipient stability and impurities as well as material variability that can influence drug product performance. These considerations are essential to the successful preparation of dosage forms for preclinical and clinical development programs. As such, this material is valuable to all scientists and students involved in pharmaceutical research from the discovery to commercial formulation and manufacture stages.

1.2 CHEMICAL AND PHYSICAL PROPERTIES

There is extensive diversity in the chemical structural elements and physical properties of pharmaceutical excipients. Excipients can be categorized in common chemical classifications including inorganics (e.g., iron oxide as pigments, calcium phosphate as filler), small molecule organics and their salts (e.g., mannitol diluent/sweetener, sodium citrate alkalizing agent), as well as polymeric excipients that can be fully synthetic or naturally derived (e.g., hypromellose, starch). The diversity is further expanded by an abundance of natural product derivatives where feedstock variability (raw materials), isolation, and chemical processing can impact the purity and structural attributes. Table 1.1 provides an overview of several common functional and chemical classifications of excipients with USP monographs. In total, there are 230 excipient monographs available to formulators with published monographs in the *Handbook of Pharmaceutical Excipients*. Each monograph can represent numerous material grades (i.e., polymer molecular weight, degree of substitution, particle size distribution, morphology) and be available from multiple manufacturers. Alternate manufacturers often employ different synthetic schemes or isolation techniques that can result in slight differences in physical properties (i.e., melt temperature, crystallinity, loss on drying, particle size) and chemical profile (i.e., trace impurities). The methods of manufacture of excipients are often proprietary trade secrets and therefore it is incumbent on formulators to identify essential material property profiles of key excipients, which is reviewed later in this chapter. To generate this knowledge formulation scientists rely on numerous compendial excipient characterization methods and develop novel methods to analyze key quality materials attributes of a formulation.

Together these USP general chapters on test methods (Figure 1.1) cover elements of the chemical and physical properties at the molecular level (e.g., NMR, IR, NIR, and UV spectrophotometry) as well as that of particulates (e.g., distribution of particle sizes, optical microscopy) and bulk material (e.g., viscosity, loss on drying, thermal analysis). While these monographs and methods provide the core testing protocols for routine certification of materials for release specifications, certificate of analysis, and compendial compliance, it is routine for manufacturers and formulation

TABLE 1.1 Examples of Excipients with Indication of Chemical Classification, Key Common Compendia Tests, and Other Specific Tests That Can Be Utilized by Manufacturers and Formulation Scientists

		Organic (Synthetic, *Natural*)		Key Characterization Methods	
Excipient function	Inorganic	Small molecule	Polymeric	Compendial (Figure 1.1)	Others
Diluent	$CaHPO_4$	*Lactose,* mannitol	*Starch, cellulose (powdered or microcrystalline)*	PSD: <429> LDS, sieving <786>, density, bulk, tap, true <616, 699>; <616>crystallinity; <731> LOD; water content <921>; SSA <846>; flow <1174>; Particle shape Particle size distribution <429> Surface area Thermal properties Polymorphic form Crystallinity <616> Amorphous content	Water activity, GPC, XRD, ssNMR, SEM, AFM, TEM, FBRM,
Binder	$CaCO_3$	Dextrates, maltitol	Copovidone, hydroxypropyl cellulose		
Disintegrant	*Magnesium aluminum silicate*	–	Crospovidone, sodium starch glycolate, croscarmellose sodium, polacrilin potassium		
Lubricant	*Talc*	Magnesium stearate, sodium stearyl fumarate, stearic acid	–		
Glidant	SiO_2 colloidal, MgO, *talc*	–	–		
Colorant	Iron oxides, TiO_2	Tartrazine, sunset yellow FCF indigo carmine, etc.	–		
Coating and film-forming agents	ZnO	Sucrose	Cellulose acetate phthalate, polyvinyl acetate phthalate, polyethylene oxide, hypromellose, polyvinyl alcohol, etc.		Film strength and adhesion

TABLE 1.1 (*Continued*)

		Organic (Synthetic, *Natural*)		Key Characterization Methods	
Plasticizer	–	*α-Tocopherol,* butyl stearate, benzyl benzoate, etc.	Propylene glycol, polyethylene glycol	911, 881, 891	Rheology, thermal mechanical analysis (TMA)
Flavor/sweetening/ fragrance	–	Vanillin, menthol, fructose, ethyl lactate, monosodium glutamate, etc.	–		Electronic tongue
Antioxidant		*α-Tocopherol,* butylated hydroxytoluene (BHT), ascorbic acid	–		UV HIL exposure

Source: Rowe [3]. Reproduced with permission of RPS Publishing.

Universal Tests

I. Description
<1091> Labeling of Inactive Ingredients

II. Identification
<181> Identification---Organic Nitrogenous Bases
<191> Identification Tests---General
<197> Spectrophotometric Identification Tests
<201> Thin-Layer Chromatographic Identification Test
<401> Fats and Fixed Oils
<621> Chromatography
<731> Loss on Drying
<736> Mass Spectrometry
<781> Optical Rotation
<851> Spectrophotometry and Light-Scattering
<941> Characterization of Crystalline Solids by XRPD
<1119> Near-Infrared Spectrophotometry

III. Assay
<311> Alginates Assay
<345> Assay for Citric Acid/ Citrate and Phosphate
<425> Iodometric Assay---Antibiotics
<431> Methoxy Determination
<541> Titrimetry
<621> Chromatography
<801> Polarography
<851> Spectrophotometry and Light-Scattering

IV. Impurities
(Primary importance for rational testing to achieve chemical stability of the drug substance in formulation)

i. Organic
<226> 4-Epianhydrotetracycline
<461> Nitrogen Determination
<466> Ordinary Impurities
<621> Chromatography *(routine)*
<781> Optical Rotation
<801> Polarography
<851> Spectrophotometry and Light-Scattering
<1086> Impurities in Drug Substances and Drug Products

ii. Inorganic
(Copper not routinely specified, but can catalyze many drug substance oxidation rxns)
<206> Aluminum
<211> Arsenic
<221> Chloride and Sulfate
<231> Heavy Metals *(being phased out)*
<232> Elemental Impurities---Limits *(replace <231>)*
<233> Elemental Impurities---Procedures *(replace <231>)*
<241> Iron *(importance for oxidation rxns Fe+3)*
<251> Lead
<261> Mercury
<281> Residue on Ignition
<291> Selenium
<471> Oxygen Flask Combustion
<730> Plasma Spectrochemistry
<733> Loss on Ignition

iii. Residual Solvents
<228> Ethylene Oxide and Dioxane
<467> Residual Solvents
<621> Chromatography
<731> Loss on Drying

Specific Tests
(Additional characterization for defining compendial properties of certain excipients)

I. Physicochemical Characterization
<429> Light Diffraction Measurement of Particle Size
<616> Bulk Density and Tapped Density
<631> Color and Achromicity
<641> Completeness of Solution
<651> Congealing Temperature
<695> Crystallinity
<699> Density of Solids
<721> Distilling Range
<731> Loss on Drying
<741> Melting Range or Temperature
<761> Nuclear Magnetic Resonance
<776> Optical Microscopy
<781> Optical Rotation
<785> Osmolality and Osmolarity
<786> Particle Size Distribution Estimation by Analytical Sieving ***(lesser resolution than <429>, most frequently for materials with wide distributions, e.g. granulations)***
<791> pH
<811> Powder Fineness
<821> Radioactivity
<831> Refractive Index
<841> Specific Gravity
<846> Specific Surface Area
<881> Tensile Strength
<911> Viscosity---Capillary Viscometer Methods
<912> Rotational Rheometer Methods
<913> Rolling Ball Viscometer Method
<941> Characterization of Crystalline Solids by XRPD
<1045> Biotechnology-Derived Articles
<1119> Near-Infrared Spectrophotometry
<1174> Powder Flow
<1761> Applications of Nuclear Magnetic Resonance Spectroscopy

II. Pharmaceutical Water
(Importance for physical and chemical stability of drug substance and formulation processing)
<541> Titrimetry
<643> Total Organic Carbon
<645> Water Conductivity
<791> pH
<891> Thermal Analysis
<921> Water Determination
<1230> Water for Health Applications
<1231> Water for Pharmaceutical Purposes
See Microbiology (Chart 10)
<1644> Electrical Conductivity Measurements of Solutions

III. Functionality/Safety/GMPs
<301> Acid-Neutralizing Capacity
<1059> Excipient Performance
<1074> Excipient Biological Safety Evaluation Guidelines
<1078> GMPs for Bulk Pharmaceutical Excipients
<1080> Bulk Pharmaceutical Excipients---Certificate of Analysis
<1081> Gel Strength of Gelatin
<1097> Bulk Powder Sampling Procedures
<1174> Powder Flow
<1195> Significant Change Guide for Bulk Pharmaceutical Excipients
<1197> Good Distribution Practices for Bulk Pharmaceutical Excipients

Figure 1.1 Classification of USP/NF compendial testing methods specified for excipients. Universal tests cover required testing of ID, assay, and impurities. Specific tests are additional methods to better describe and control excipient chemical and physical properties. Most commonly utilized methods for excipients intended for oral solid formulations are noted in bold text.

scientists to conduct extensive supplemental testing to ensure the quality and consistency of excipient properties. While it is of great interest to formulators to conduct additional noncompendial functional testing regarding the distinct critical material properties of a developmental product, there is routine attention given to the core information recorded in compendial tests. This is exemplified by publications that demonstrate the compilation and statistical analysis of reported CoA data to identify material properties that are unique to a manufacture location or period of time [4]. This type of analysis is commonly pursued by quality groups that track results of certified testing and can be valuable to formulators seeking to identify critical quality attributes by incorporating excipient lots that most represent the material diversity in the early screening and development stages. In addition, the Excipient Consortium (NIPTE – Advanced Pharmaceutical Materials Knowledge Center) and other similar groups provide extensive testing and make data and materials available to membership composed of universities, manufacturers and pharmaceutical companies. Searchable databases of material records and supplemental functional testing (e.g., shear cell and compaction testing) greatly improve the ability of formulators to project potential variability to critical material attributes and design robust formulations to accommodate the typical range of material properties.

Further expansion of the library of pharmaceutical excipients to include new chemical entities is a challenging endeavor with regulatory requirements that involve significant investment and time [5]. These requirements include extensive safety and toxicology studies for the introduction of new excipient chemistries and create an incentive to develop unique innovative physical material properties from the existing library of chemicals. Materials are often engineered to meet compendia specifications for existing excipient monographs; however, they often employ unique processing methods or combinations of primary excipients (coprocessing) to provide innovative properties and eliminate or lessen the regulatory burden for acceptance.

1.3 COMPENDIAL CHARACTERIZATION METHODS AND EXCIPIENT PERFORMANCE

Compendial test methods contained in detailed pharmacopeia monographs are readily available to formulation scientists. These monographs serve the basis for core techniques in chemical and physical analyses to identify excipients and to ensure quality through routine analysis. Quality specifications regarding the purity and stability of excipients rely on these compendial test methods (Figure 1.1). USP/NF monographs contain both general tests and specific tests that are applied to characterize excipients. USP/NF monographs are stability indicating and contain a suitable assay method or an accompanying procedure to identify impurities that can demonstrate stability.

Common elements of excipient monographs include name and description, identification test, assay and impurities method(s), packaging and storage conditions as well as any specific tests needed to better describe and control an excipient (e.g. microbial limit test, pH, etc.). The functionality of excipients are mostly dictated by an individual formulation (i.e., formulation quality attribute) and the processing technologies utilized to manufacture a dosage form. Therefore, incorporation of functional tests and acceptance criteria are limited in monographs to cases where routine test are not sufficient to support the majority applications for a material.

1.3.1 Pharmacopoeial Harmonization

Excipients that have established worldwide acceptance in compendial testing and specification are given considerable preference as they could be universally integrated into a drug product. This universal compendial designation greatly simplifies the ability to demonstrate quality and equivalence of a formulation filed with numerous regulatory health authorities. Efforts on global harmonization of the international pharmacopeia landscape (United States Pharmacopeia–National Formulary (USP/NF), EP, Japanese Pharmacopoeia (JP)) have sought to enact standards that enable consistent quality of excipients and minimize the need for regional test methods and repeat testing, therefore facilitating drug products to be rapidly introduced to international markets. Harmonization of general chapters including analytical methods as well as excipient monographs is coordinated by one pharmacopeia (USP, EP, or JP) during a staged working procedure by a Pharmacopeial Discussion Group (PDG). There are six stages to reach a harmonized monograph, which include identification, investigation, expert committee review, official inquiry, consensus, and implementation. A listing of harmonization activity for excipients and analytical methods is included in Figure 1.2, which demonstrates a majority of monographs identified for harmonization have reached the implementation stage. The PDG reports 58% of excipients at stage six as of July 2013. Similarly, the majority of general chapters related to characterization methods have reached completion of stage six harmonization status.

1.3.2 Monograph Revisions

USP–NF monographs are subject to routine review and the USP provides guidance for revisions to allow for changes to testing methods and excipient specifications. A recent example of interest is the revision of the monograph for characterization of heavy metals <231>, which has been replaced by elemental limits <232> and test procedure <233>. Under monograph <231> it is incumbent on excipient manufacturers to certify the control of inorganic materials of potential harm are below toxic levels. In addition, it is the responsibility of excipient users to substantiate the absence of impurities before incorporating into drug products.

A wet chemistry colorimetric test method specified in <231> has been in routine use for decades; however, this test relies on subjective visual inspection for precipitation of metal sulfides. A colored precipitate of sulfide-forming elements is visually compared to a 10 ppm Pb standard to determine compliance with the heavy metal limit. Resolution of individual elements is not viable with USP <231>. Experiences with <231> have demonstrated poor resolution and quantification that has resulted in lower than actual amounts for numerous heavy metals known to be toxic (lead, arsenic, mercury, and cadmium). In particular, the required 600 °C ignition temperature prevents the <231> method from resolving mercury and other volatile analytes. The revisions incorporated in <232> account for a wider range of metals with potential to impact quality and define individual limits according to known toxicity (Table 1.2). Included in the new limits are catalysts that were not previously able to be resolved. Multielement ICP-MS and ICP-OES techniques have been established in <233> to simultaneously detect a great number of metals of interest with high specificity and sensitivity. Of particular interest to the stability of drug products is the high resolution of copper, which is often linked to the catalysis of oxidative reaction in drug products. However, identification of speciation (oxidation state), which is important in reactivity of the metal impurity with other formulation components, is not covered by the new testing. Toxicity associated with the defined limits assumes that the entire amount of metal recorded is present in the oxidation state that demonstrated the greatest toxicity.

Chapter title	Coordination Pharmacopeia	Harmonization Stage
Amino Acid Determination	USP	6
Bacterial Endotoxins (Rev 1)	JP	4 rev
Bulk Density and Tapped Density	EP	5A2
Conductivity	EP	2
Color (Instrumental Method)	EP	3
Density of Solids	EP	5B
Disintegration	USP	6
Dissolution (Rev 1)	USP	6
Capillary Electrophoresis	EP	6
Polyacrylamide Gel Electrophoresis	EP	6
Extractable Volume (Rev 1)	EP	6
Heavy Metals	USP	3
Inhalation	EP	4
Isoelectric Focusing	EP	6
Laser Diffraction Measure of Particle Size	EP	4
Limits for Nonsterile Products	EP	6
Microbial Contamination	EP	6
Tests for Specified Microorganisms	EP	6
Microbial Enumeration	EP	6
Optical Microscopy	USP	6
Particle Size Distribution Estimation by Analytical Sieving (Rev 1)	USP	5B
Particulate Contamination (Rev 1)	EP	6
Peptide Mapping	USP	6
Porosimetry by Mercury Instrusion	EP	4
Powder Fineness	USP	5A
Powder Flow	USP	6
Protein Determination	USP	6
Residue on Ignition (Rev 2)	JP	6
Specific Surface Area	EP	6
Sterility Tests	EP	6
Tablet Friability	USP	6
Thermal Behavior of Powders	EP	3
Uniformity of Content /Mass	USP	6
Uniformity of Delivered Dose of Inhalations	EP	2
Water–Solid Interaction	EP	3
X-Ray Powder Diffraction	EP	4

(a)

Figure 1.2 Harmonization status of general compendial analytical testing methods and excipient monographs. Listing includes the stage of review and publication (1–6) and the agency leading the harmonization process. Status indicated reflects public announcement for July 2013.

Excipient Name	Coordinating Pharmacopeia	Harmonization Stage
Alcohol (Rev 2)	EP	4
Benzyl Alcohol (Rev 1)	EP	6
Dehydrated Alcohol (Rev 2)	EP	4
Butylparaben	EP	6
Calcium Carbonate	USP	4
Calcium Disodium Edetate	JP	6
Calcium Phosphate Dibasic (and anhydrous)	JP	6
Carmellose Calcium (Rev 1)	USP	6
CarmelloseSodium	USP	4
Carmellose	JP	4
Cellulose Acetate (Rev 1)	USP	6
Cellulose Acetate Phthalate	USP	6
Microcrystalline Cellulose (Rev 1)	USP	6
Cellulose, Powdered (Rev 1)	USP	6
Citric Acid, Anhydrous (Rev 1)	EP	6
Citric Acid, Monohydrate (Rev 1)	EP	6
Copovidone	JP	4
Croscarmellulose Sodium	USP	6
Crospovidone	EP	4
Ethylcellulose	EP	6
Ethylparaben	EP	6
Gelatin	EP	3
Glucose Monohydrate/Anhydrous	EP	3
Glycerin	USP	3
Glyceryl Monostearate	USP	2
Hydroxyethyl Cellulose	EP	4/2
Hydroxypropyl Cellulose	USP	4
Hydroxypropyl Cellulose, Low Substituted	USP	4
Hydroxypropylmethyl Cellulose	JP	6

Excipient Name	Coordinating Pharmacopeia	Harmonization Stage
Hypromellose Phthalate	USP	6
Lactose, Anhydrous (Rev 2)	USP	6
Lactose, Monohydrate	USP	6
Magnesium Stearate	USP	5A
Mannitol	EP	3
Methylcellulose	JP	6
Methylparaben	EP	6
Petrolatum	USP	4
Petrolatum, White	USP	4
Polyethylene Glycol	USP	4
Polysorbate 80	EP	4 rev
Povidone	JP	6
Propylene Glycol	EP	4
Propylparaben	EP	6
Saccharin	USP	6
Saccharin, Calcium	USP	6
Saccharin, Sodium (Rev 1)	USP	6
Silicon Dioxide	JP	4 rev
Silicon Dioxide, Colloidal	JP	4 rev
Sodium Chloride (Rev 2)	EP	6
Sodium Lauryl Sulfate	USP	3
Sodium Starch Glycolate (Rev 1)	USP	6
Starch, Corn (Rev 2)	USP	6
Starch Potato	EP	6
Starch, Pregelatinized	JP	3
Starch, Rice	EP	6
Starch, Wheat	EP	6
Stearic Acid	EP	5B
Sucrose	EP	4
Sterile Water for Injection in Containers	USP	3
Talc	EP	6
Titanium Dioxide	JP	5A2

(b)

Figure 1.2 *Continued*

The combined experiences of excipient manufacturers and end users as well as the evolution of analytical technologies are considered when revisions are proposed to existing monographs. Typical justification for revision includes public safety and health reasons, insufficient supply of pharmacopoeial quality material, poor availability of specified reagents, new reagents or methods of preparation, and advances to analytical procedures (more appropriate, accurate, or precise). Additional monographs proposed for revision in 2013 include <41> balances, <659> packaging and storage requirements, high fructose corn syrup, and <1092> dissolution procedure: development and validation.

1.4 NOVEL CHARACTERIZATION TECHNIQUES

The critical material properties of excipients are unique to every drug formulation. These properties are impacted by the chemical and physical nature of the drug substance as well as other excipients, required route of administration dosage form, formulation processing methods, and the intended storage and handling of the final product or intermediates. Critical material properties defined by the needs of a particular formulation are typically not entirely elucidated by the common characterization techniques described in USP/NF monographs.

TABLE 1.2 Revised Default Concentration Limits for Heavy Metal Impurities in Excipients and Drug Substances in Monograph <232>

	Concentration Limits (μg/g)		
Element	Oral Drug Products Maximum Daily Dose of ≤ 10 g/day	Parenteral Drug Products Maximum Daily Dose of ≤ 10 g/day	Inhalational Drug Products Maximum Daily Dose of ≤ 10 g/day
Cadmium	2.5	0.25	0.15
Lead	0.5	0.5	0.5
Inorganic arsenic	0.15	0.15	0.15
Inorganic mercury	1.5	0.15	0.15
Iridium	10	1.0	0.15
Osmium	10	1.0	0.15
Palladium	10	1.0	0.15
Platinum	10	1.0	0.15
Rhodium	10	1.0	0.15
Ruthenium	10	1.0	0.15
Chromium	–	–	2.5
Molybdenum	10	1.0	1.0
Nickel	50	5.0	0.15
Vanadium	10	1.0	3.0
Copper	100	10	10

In addition to toxicity, the numerous heavy metals present at these controlled levels can catalyze reactions with drug substance and negatively impact the stability and impurity profile.

Analytical testing in monographs cover many techniques suitable for routine materials characterization; however, formulation scientists frequently need to employ specialized equipment and methods that are tailored to needs for identifying specific issues related to the performance and quality of a particular formulation. It is important to note that monographs and associated tests or specification ranges for a particular excipient are not suitable to indicate exact equivalence in performance or composition. Typically, a significant variety of material properties exist for a group of excipients that all meet a common compendia standard. A pharmaceutical scientist needs to understand material differences of excipients with identical compendial classification and identify where there could be potential to influence drug product performance or quality.

Examples provided in this section serve to demonstrate novel excipient characterization methods that are created by pharmaceutical scientists to develop high-quality robust processing and performance attributes of new formulations. These few contributions demonstrate how material properties and variability (lot–lot or manufacturer) can be identified and related to formulation performance and process development.

1.4.1 Chemical Imaging

Application of chemical imaging throughout the drug product manufacture process enables more comprehensive identification and understanding of critical material attributes by resolving how excipients respond to applied process conditions (compaction, milling, temperature, moisture, etc.) and affect downstream performance properties (disintegration/dissolution, tablet hardness, chemical stability, etc.). Raman, FTIR, NIR, and other chemical imaging methods are strong examples of specific functional testing of excipients and their interaction(s) with other formulation additives to identify key material attributes. These methods often require extensive development to tune resolution and sensitivity to the materials of interest and to apply for measurement of drug products or various drug product intermediates. In addition, these techniques are typically paired with a chemometric processing tool such as partial least squares (PLS), principal component analysis (PCA), multivariate curve regression (MCR), or other suitable means to treat and analyze the acquired data. Often, the complete variability of excipients cannot be fully anticipated and this provides challenges to calibrating methods and extrapolating data outside prior experiences [6]. However, a working method can still provide mechanistic insight into the attributes of functional excipients and aid in the design of robust drug product processes and selection of high-quality materials.

Basic IR spectroscopy is described in general compendial test methods, but advances in the application of NIR methods for release testing of tablet potency and uniformity have recently been demonstrated in regulatory documents and the pharmaceutical literature [7]. This type of characterization method is product specific and requires extensive method development and validation. The value for developing these novel techniques is the ability to have rapid and extensive testing of tablets that can better track the robustness of a formulation and process. NIR chemical imaging (NIR-CI) techniques, which are often applied to API for potency and uniformity determination, can also readily be employed to track excipient performance in a dosage form. The analysis of functional excipients (disintegrants, binders, lubricants, etc.) can be performed with numerous commercially available NIR-imaging systems capable of spatial and chemical resolution for analysis of intact tablets or drug product intermediates. NIR can determine content uniformity, moisture content, particle size/distribution of all the sample components, contaminants, as well as polymorph distributions (e.g., lactose α vs β) [7a, 8]. A powerful example of the utility of combined chemical and spatial information is the ability to localize the drug substance degradation products and overlay information regarding the excipient composition and moisture of the immediate region to elucidate drug product degradation mechanisms and the impact of specific formulation components.

The localization of excipients in drug products or intermediates can be important to specific performance or quality attributes including the chemical stability of the drug substance. One example is the routine use of excipients as pH modifiers that alter the solubility and dissolution rate of ionizable pharmaceutical compounds through influence of the local pH. If chemical stability of the drug substance is also sensitive to pH, the formation of impurities can be accelerated by additives intended to impact solubility and dissolution. Figure 1.3 shows a tablet that was stored at accelerated stability condition (40 °C and 40%RH) with an overall dark color and distinct localized spots. Raman and IR imaging confirmed the visual intensity of spots was related to regions that were rich with particles of API and the acidic modifying excipient. In cases where impurities from the excipient or drug

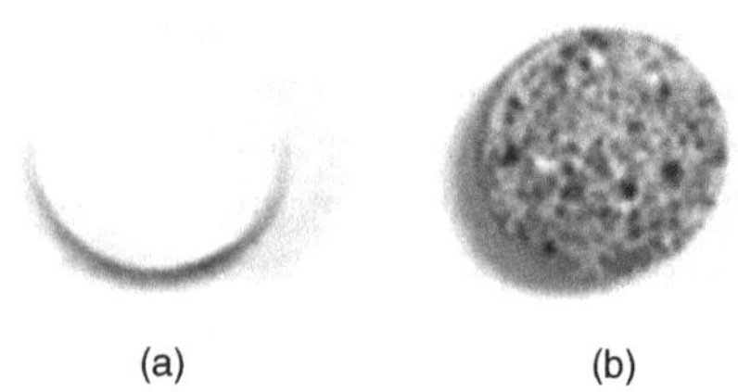

Figure 1.3 (a) Initial and (b) aged tablets containing drug substance sensitive to local pH environment provided by a minor excipient (<10 wt%). The aged tablet was exposed to 40 °C and 40% RH for 2 weeks.

substance that form during storage do not present a vivid color, the localization of components can be facilitated by chemical imaging techniques (Raman, NIR, FTIR, etc.). Figure 1.4a demonstrates one case from the literature where surface-enhanced Raman chemical imaging localized a degradation product of acetaminophen in tablets containing PVP as excipient [9]. Similar work has looked at furosemide tablets chemical degradation using NIR imaging and a PLS model generated from pure component data spectral to derive the contribution and distribution

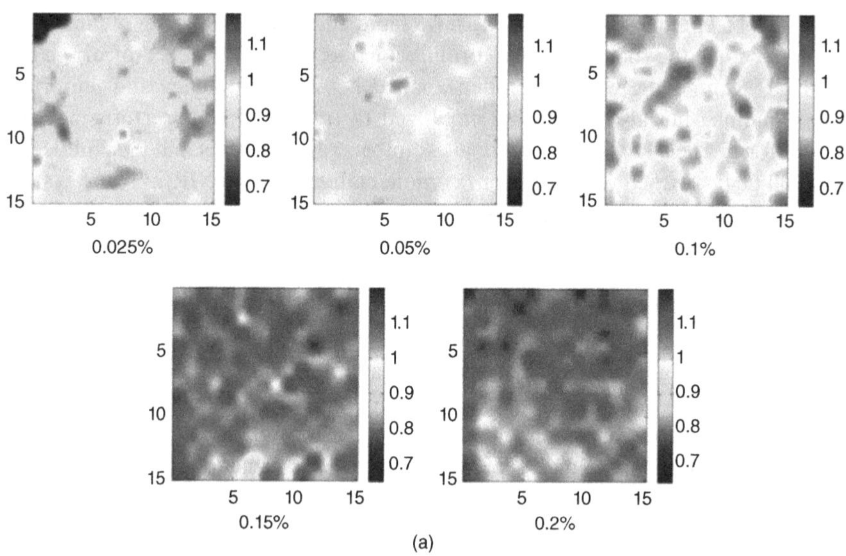

Figure 1.4 Chemical imaging of drug product stability showing (a) surface-enhanced Raman chemical imaging of between 0.025% and 0.2% 4-aminophenol (degradant/impurity) versus the pixel position in tablets of acetaminophen and PVP. Images were obtained from plotting the median intensity of the principal band of 4-aminophenol normalized butanethiol peak. Source: De Bleye [9]. Reproduced with permission of Elsevier. (b) Identification of prodrug (top) to parent (bottom) conversion in a prototype BMS tablet formulation. Images and data courtesy of Boyong Wan and Christopher Levins (Bristol-Myers Squibb, 2015). *See color plate section for color representation of this figure.*

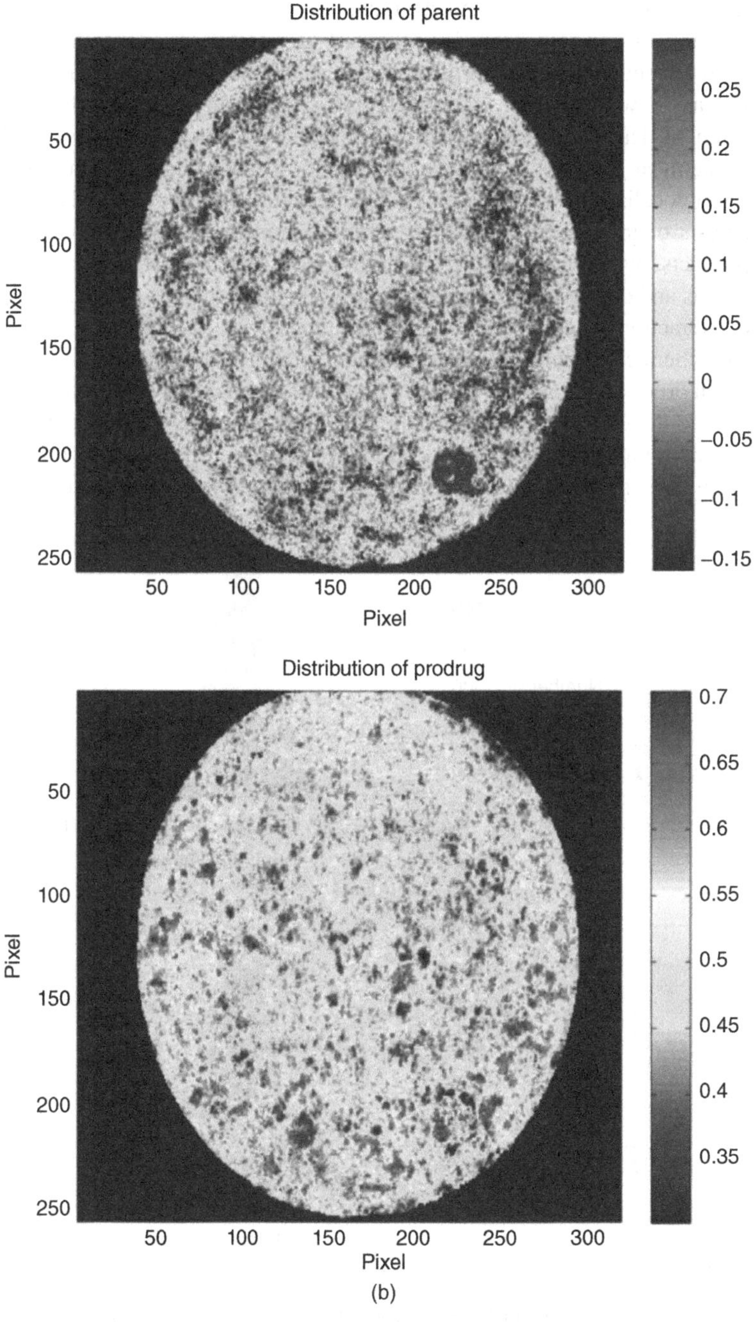

(b)

Figure 1.4 *Continued*

of excipients, drug substance, and degradation products [10]. The utility of NIR imaging analysis has also been demonstrated for a BMS developmental prodrug compound to understand mechanisms that lead to the formation of parent drug in tablets as shown in Figure 1.4b. These types of localized degradants can often be difficult to detect in mean spectrum from the bulk samples; however, NIR or Raman chemical imaging provides high-resolution spatial data that improves the detection of localized minor components. Chemical imaging allows rapid acquisition and analysis of trace materials resulting from excipients and their interactions in drug products, which in turn provides improved fundamental understanding of mechanisms and degradants to support the design of high-quality products.

The interaction of excipients and formulations with moisture can go beyond impacting chemical stability to induce changes in physical properties and response to common processing conditions. *In situ* chemical imaging has demonstrated utility to determine the densification behavior of excipient and drug mixtures exposed to different environmental conditions and stresses [11]. Figure 1.5 shows FTIR images with HPMC absorbance bands from mixtures with ibuprofen under two compaction

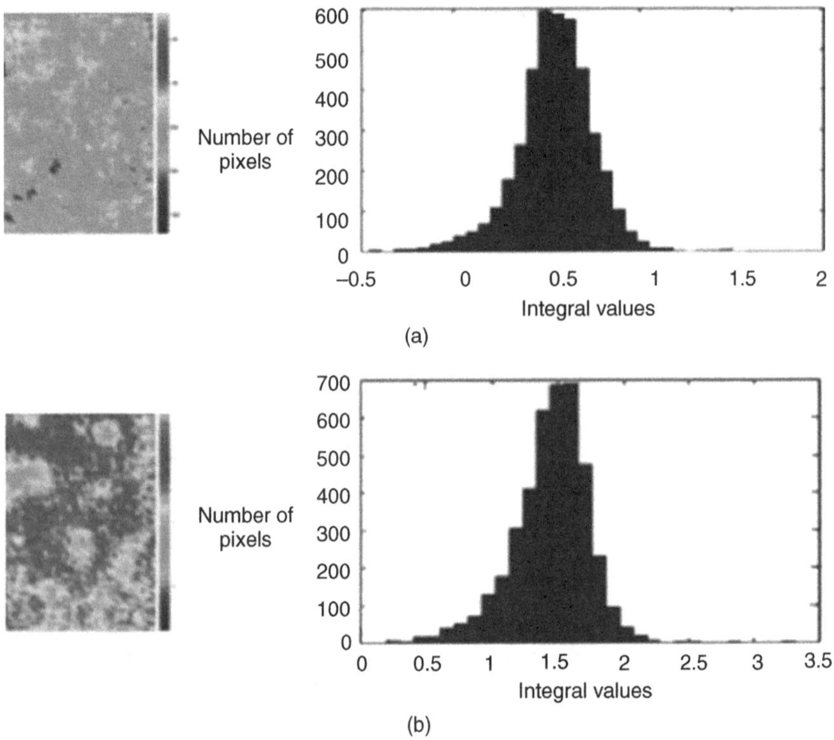

Figure 1.5 FTIR images and histograms of HPMC ibuprofen tablets using blends stored at two RH conditions and compressed at two forces: (a) 60% RH blend compressed at 80 cN m; (b) 80% RH blend compressed at 120 cN m. Source: Elkhider [11]. Reproduced with permission of Elsevier. *See color plate section for color representation of this figure.*

pressures and relative humidity conditions. It is clear from the images and associated histograms that greater densification (dark pixels) is achieved from higher moisture content and compaction force. This technique enables the developers to study the impact of moisture on multiple components during compaction process and the potential to tailor composition and conditions to provide robust tablet processing and performance. The same group has also demonstrated complementary use of X-ray microtomography techniques, which provide greater penetration of tablet samples compared to *in situ* FTIR images that are restricted to resolving surface attributes [12].

The distribution of magnesium stearate lubricant is often critical to the processing attributes as it alleviates sticking of powder/compacts to machine surfaces. In addition, downstream performance characteristics, including dissolution/disintegration behavior of the dosage form, can be influenced by the physical and chemical characteristics of magnesium stearate such as particle size and morphology as well as ratio of stearic to palmitic content. The dispersion of magnesium stearate in powder blends is of specific interest to formulators since these materials are intended for activity at the interface between particles and the surfaces of processing equipment. The resolution of magnesium stearate dispersion is not readily elucidated by any specified compendial testing methods and requires unique instruments and methods.

Raman mapping is one chemical imaging technique which has been used to quantify the blendability of a lubricant. Raman analysis has been applied to increase processability and determine the appropriate blend time and level of shear. Additionally, Raman imaging data can be correlated to the wetting or dissolution of dosage forms where negative performance has been demonstrated if a lubricant provides too much coverage of particle surfaces (from either overblending or amount of lubricant). This can also be applied when changing equipment and on scale-up. Figure 1.6 demonstrates the localization of magnesium stearate with regard to the surface of a tablet comprised mostly of API and other excipient particles [13]. Lubricant particles are bright intensity areas, while the API and other excipients are represented in lower intensity (dark/black). Samples representing prolonged mixing time (lower panels of Figure 1.6) exhibited less pixels/domains associated with high (>15%) lubricant concentration and a greater number of domains with low (~1–4%) lubricant concentrations. High lubricant concentration domains are associated with aggregated lubricant particles. When concentrated lubricant domains are broken, a greater abundance of low lubricant concentration domains are formed, which is consistent with more uniformly distributed lubricant particles. The images in Figure 1.6 suggest that extended bin blending or larger scale blending operations that increase total shear can improve lubricant uniformity on the surface of tablets. This example demonstrated the potential to resolve excipients and analyze interactions with material properties (surface area, particle size, etc.) and process conditions (blend time, scale, speed, etc.). This type of technique complements bulk analysis and downstream process evaluation of blend performance (tabletability, hardness, friability, segregation, uniformity, etc.) and provides insights that can save time and material through detailed characterization of small blends in early stages of formulation or process development.

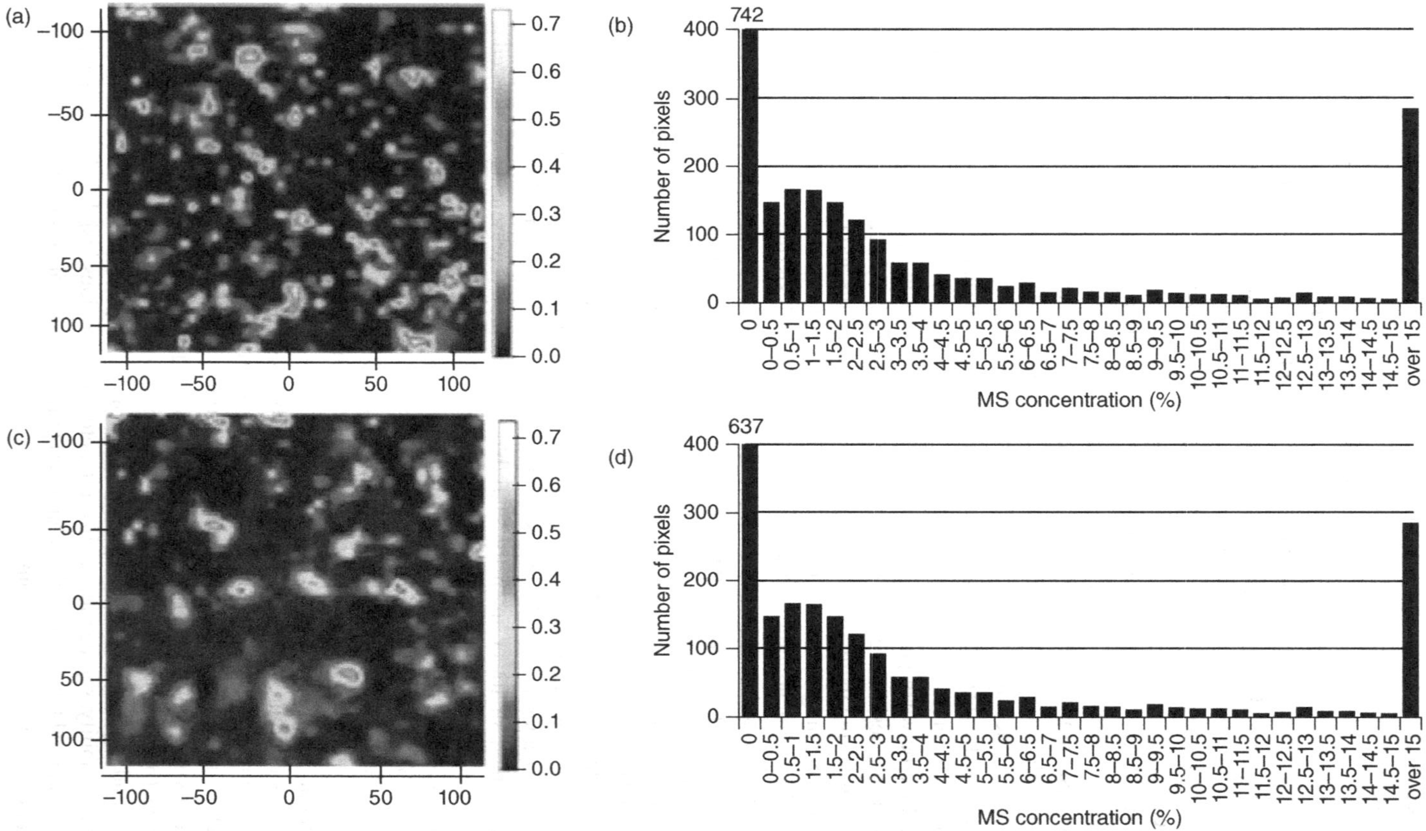

Figure 1.6 Dispersion of magnesium stearate (MS) lubricant particles in physical blends analyzed by Raman chemical imaging. Quantification of domain size, number, and localization is provided. Blending time increases from 2 to 60 minutes from the top to bottom tablet images. Source: Lakio [13]. Reproduced with permission of Springer. *See color plate section for color representation of this figure.*

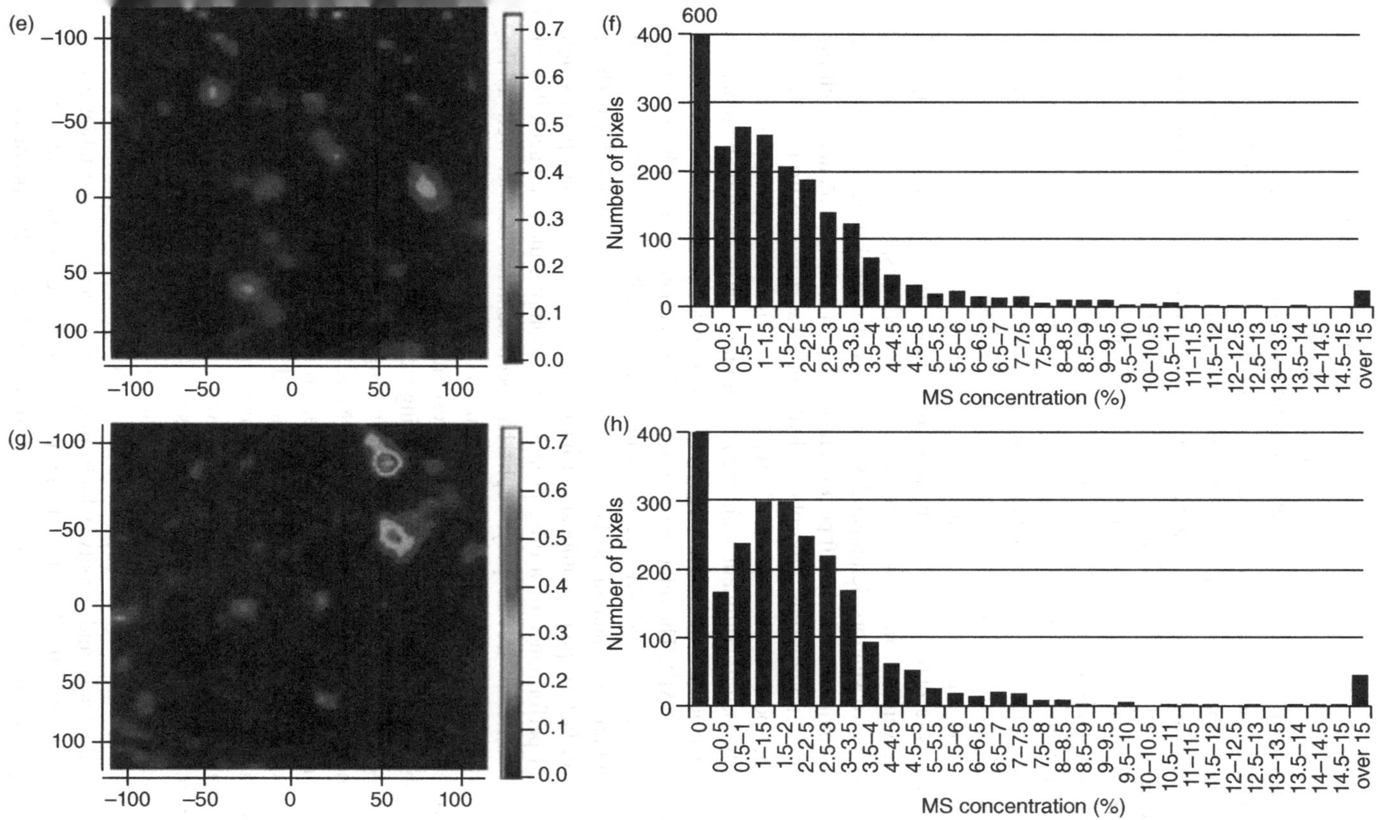

Figure 1.6 *Continued*

1.4.2 Advanced NMR Techniques

The increased availability and use of solid-state NMR (ssNMR) is one example where advanced analytical techniques facilitate a greater fundamental understanding of excipient properties that can impact formulation. Drug product formulations must consider and account for the variability of excipient properties that are in many cases attributed to proprietary sourcing and production methods of multiple vendors. To ensure uninterrupted supply of medicine to patients, the qualification of multiple excipient sources for a drug product is routinely sought. The equivalence of excipient performance from multiple vendors has to be determined by the formulation scientists. The application of ssNMR makes it possible to identify many unique characteristics of some excipients that could be associated with a specific manufacturing processes or the material supply chain. If differences in chemical or physical properties exist for a critical excipient, suitable analytical methods and controls must be established to maintain product quality.

ssNMR was recently used to study structural characteristics of lactose acquired from multiple vendors. In Figure 1.7a, the carbon-13 NMR spectrum of lactose as received from two vendors demonstrates numerous structural differences that are evident from unique chemical shift peaks. The resolution and assignment of multiple physical phases (polymorphs, amorphous, hydrated) was achieved from a detailed analysis of the spectrum as shown for Kerry sourced lactose in Figure 1.7b. Lactose is a commonly used filler/diluent available from multiple large vendors in numerous grades and each of these materials comprises of a complex mixture of multiple phases as resolved by ssNMR. The relative quantitative phase compositions of lactose from vendors in Figure 1.7a ssNMR are listed in Table 1.3. The largest differences in phase content are for α-anhydrous lactose (0–13%) and β lactose (50–75%). These differences in lactose phase composition have the potential to impact processing, stability, and performance of a drug product. For example, the compactability of tablets with different lactose polymorphs has been demonstrated as well as the sensitivity of certain drug substances to the transfer of moisture from excipients [14]. Similarly, examples of ssNMR have shown resolution of bulk lactose polymorphs and amorphous phases from common processing techniques such as spray drying of aqueous suspensions. Other reports have shown a high degree of structural similarity across numerous microcrystalline cellulose (MCC) grades from a single supplier [15]. Specialized techniques such as ssNMR can provide fundamental data to support the design of robust high-quality formulations and processes when considering the selection, substitution, or processing of excipient grades from numerous suppliers.

NMR has also been used in recent years for direct performance indicating analysis of excipients through imaging. One example is the swelling and erosion of extended release matrixes such as the HPMC tablet matrix shown in Figure 1.8. This type of functional test can distinguish attributes relevant to the release mechanism(s) such as gel layer thickness and density, which impact the selection of the tablet excipients. Here a round HPMC matrix tablet containing a weakly basic drug and an acid-modifying excipient demonstrates a consistent and faster disappearance of the dry core when exposed to an aqueous neutral buffer solution. The dry tablet core

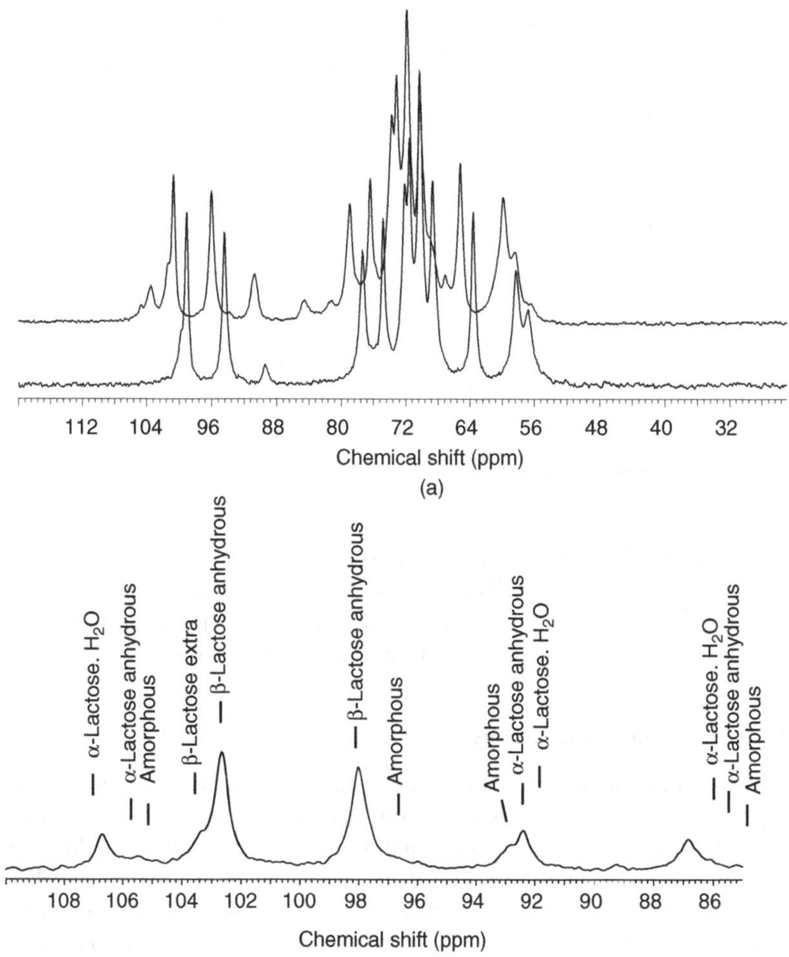

Figure 1.7 ^1H–^{13}C CPMAS solid-state NMR spectra of lactose (a) from different vendors; Kerry-1320016404 (top) and DFE Pharma-42312-7356 [587] (bottom). (b) Zoomed region of Kerry-1320016404 showing five different phases. Figure courtesy of Anuji Abraham and George Crull (Bristol-Myers Squibb, 2014).

is shown in dark contrast and the hydrated HPMC gel layer is resolved in lighter (white/grey) area surrounding the core. The rate of disappearance of the dry tablet core can be analyzed by integrating the area of the dark core in these images and plotting the time course (bottom plot in Figure 1.8). At all time points, the HPMC matrix with the acid-modifying excipient had a lower dry core area and this difference became more pronounced at later time points. This method allows dynamic measurement of the contribution of erosion and diffusion of drug through an HPMC polymer matrix to rationally design the target release rate. In addition, it is possible to design

TABLE 1.3 Molar Phase Composition of Lactose Batches from Vendors Kerry and DFE Pharma

Vendor Name	Kerry	DFE
Lactose batch no	1320016404	42312-7356 (587)
Molar composition	%	%
α-Lactose, H_2O	5	0
β-Lactose	50	75
α-Lactose anhydrous	13	0
Unknown	25	19
Amorphous	7	5.4

Source: Bristol-Myers Squibb, 2014. Reproduced with permission of Bristol-Myers Squibb Company.

Phase composition was determined by calculating the area under the peaks (of anomeric carbon atoms) of ^{1}H–^{13}C CPMAS NMR spectra after deconvolution of the spectra using ACD (version 12) software.

more elaborate methods that can also capture chemical information by quantification of elements of interest such as ^{19}F, ^{35}Cl, and ^{31}P contained in numerous excipients (e.g., impact of residual salts or degree of phosphate cross-linking in super disintegrant swelling kinetics) or to track the diffusion of a labeled drug substance.

1.4.3 Atomic Force Microscopy

Atomic force microscopy (AFM) cantilevers can be functionalized with excipients or drug particles and used as probes to investigate the effect of surface chemistry on the interaction with another material in a drug product. Functionalized AFM probe tips can be constructed to provide a localized solid–solid interface between pharmaceutically relevant materials. This microscopic interface provides for high-resolution contact that can identify specific physical and chemical interactions such as studying solid-state decomposition reactions between excipients and drug substances. The use of a force–displacement mode also provides direct quantification of adhesion forces from controlled interactions between two material surfaces.

An example of a novel AFM technique was recently published where the authors determined the impact of dicalcium phosphate dihydrate (DCP) toward the solid-state hydrolysis of aspirin [16]. In this study, the anisotropic surface chemistry of the crystals, which present different reactive functional groups on various crystal faces, were determined to contribute to the reactivity of aspirin in contact with DCP. A strong interaction was identified between DCP and the aspirin (100) surface at 75% RH leading to formation of local pits. These pits were also associated with formation of needle shaped crystals normal to the surface in Figure 1.9a and hypothesized (absent chemical data) to be consistent with the growth of crystalline salicylic acid due to aspirin hydrolysis. Furthermore, the interaction was highly dependent on the formation of a water layer on the aspirin (100) surface above 40% RH marked by a significant shift in the force-displacement profile (Figure 1.9b) resulting from chemical, electrostatic,

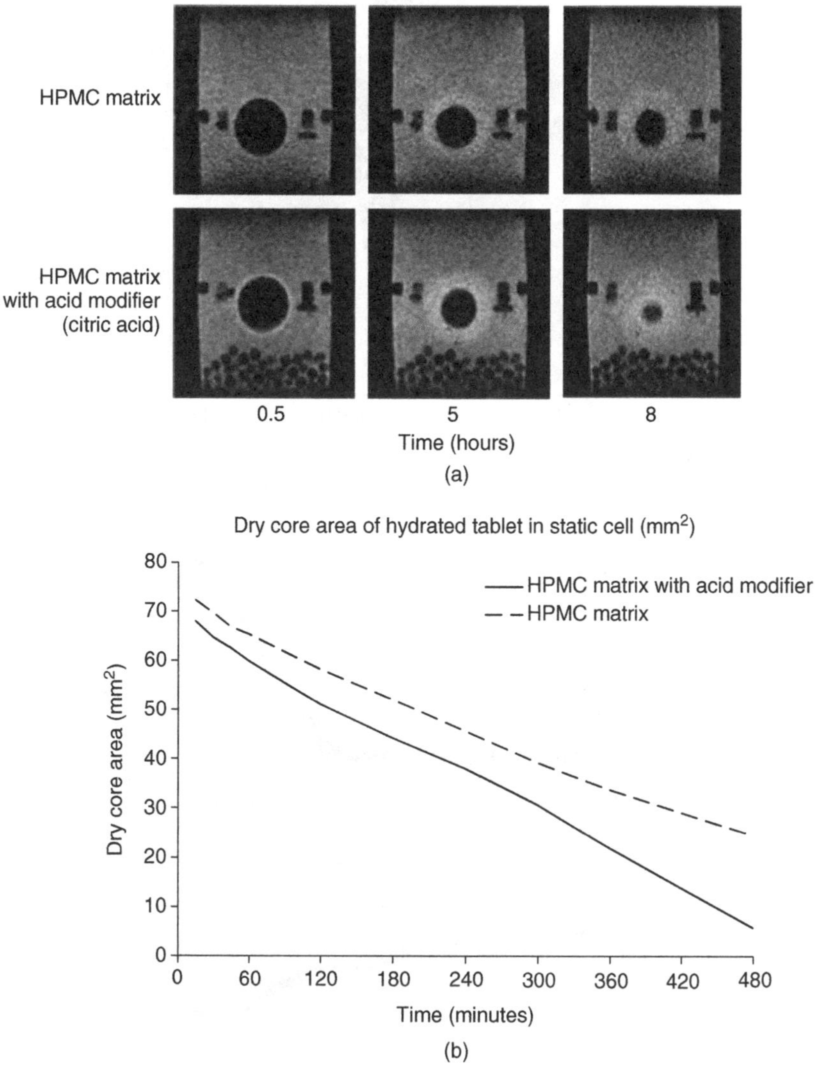

Figure 1.8 (a) Low-field NMR analysis of HPMC matrix tablets providing time course images of the darker dry tablet core and lighter gel layer in a static aqueous buffer solution (SEMS, $T_r = 1800$ ms, NS = 2, $T_e = 6$ ms) and (b) corresponding area of the dry tablet core over 8 hours in aqueous media. Images courtesy of Sarah Hanley and Jonathan Brown (Bristol-Myers Squibb, 2012).

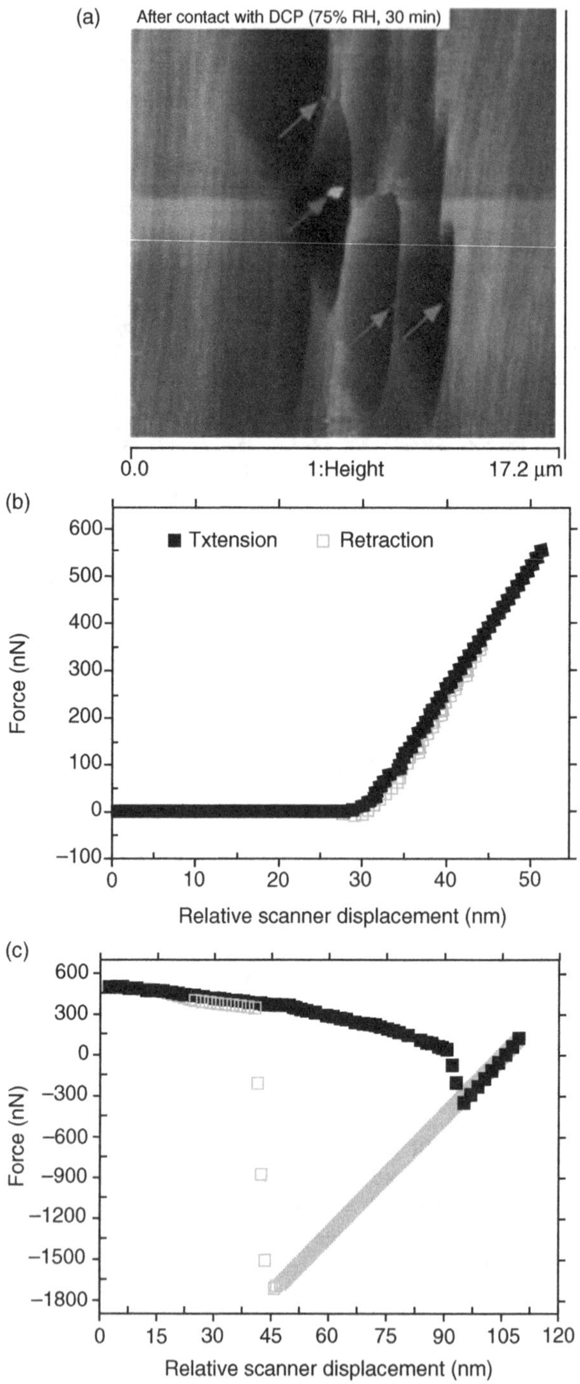

Figure 1.9 AFM measurements of dicalcium phosphate dihydrate in contact with aspirin (100) surface (a) image 30 minutes after contact demonstrating pits and new crystalline grown on aspirin surface and (b) and (c) force–displacement curves for low and high RH condition, respectively. Source: Cassidy [16]. Reproduced with permission of Elsevier.

and meniscus contributions. This elegant study demonstrates how highly specific tests involving unique design elements such as probe tip fabrication can greatly advance the fundamental understanding of material properties and the solid-state reactivity of excipients. This type of characterization does require highly experienced analysts for the diligent conduct and analysis of detailed data sets, which can limit routine use in development programs. However, AFM is well suited to addressing complex chemical and physical behaviors in a wide variety of pharmaceutical systems.

1.4.4 Process Analytical Technologies (PAT)

Many established analytical measurement systems are being utilized with novel integration strategies to provide real-time data on pharmaceutical processing of drug product intermediates and excipients. These efforts have been encouraged by global regulatory agencies and the pharmaceutical industry to monitor and control critical process parameters that are linked to important performance and material attributes. Excipients are widely used to impart function and/or processability to drug products and are therefore critical components of any effort to monitor drug processes in real time. Extensive process analytical technology (PAT) examples and reviews are available in the literature and also described in more detail in Chapter 9, which demonstrates the combined efforts to increase the use throughout development and commercial manufacture. A simple examination of the literature demonstrates a large increase for the number of publications focused on PAT in the last decade as shown in Figure 1.10. A strong focus in this area by regulators, academic institutions, and the pharmaceutical industry is also apparent in a listing of top contributors from US institutions.

One representative example of PAT implementation is for the fluid bed drying of ibuprofen granulation. In this study, NIR was utilized to directly provide a continuous measurement of moisture content for the drug product intermediate [8b]. Since most drug products have chemical and physical stability that is sensitive to temperature and moisture content during processing, it is critical to monitor and optimize drying conditions and parameters. Therefore, the propensity of excipients to sequester moisture and the associated thermodynamics and kinetics of moisture transfer is important to determine when screening formulations and processes. Excipients that provide strong associations with water (bound water) can lead to formulations with a dominant diffuse phase of drying and very small evaporative phases. This type of excipient behavior can cause lengthy drying processes and a predominant exponential region of the drying curve. However, excipients and formulations with linear evaporative cooling behavior are amenable to accelerated drying conditions. This behavior is shown for ibuprofen–starch granulation in Figure 1.11 where fast drying was suitable to achieve an approximately 50% time reduction over the normal process. This type of advanced analytical monitoring facilitates a mechanistic understanding of the drying process and identification of any excipient or material constraints to assure high-quality robust operation parameters are selected.

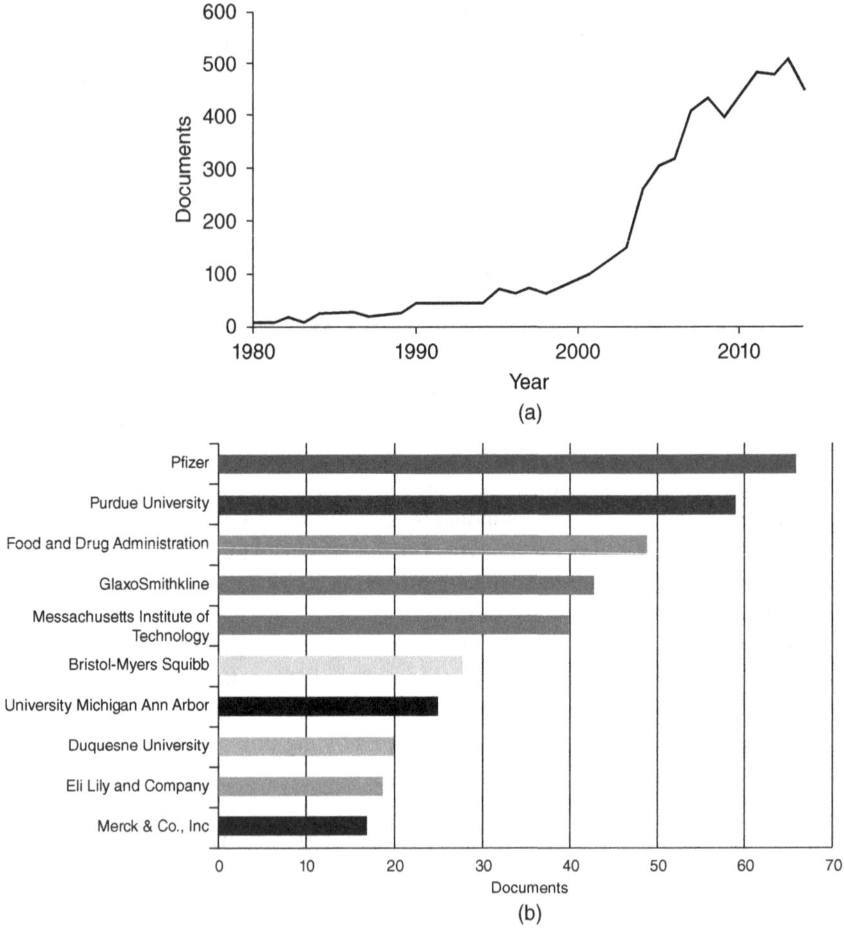

Figure 1.10 Number of documents containing the keyword process analytical technology in pharmacy and engineering journals between 1980 and 2014 in Scopus database. The number of documents per year is indicated in (a) and the top contributing US institutions are listed in (b).

1.5 EXCIPIENT IMPURITIES AND IMPLICATIONS TO DRUG PRODUCT STABILITY (DRUG–EXCIPIENT INTERACTIONS)

Pharmaceutical excipients have been studied extensively to obtain a detailed understanding of the properties and functionalities they exhibit in solid dosage formulations. Excipients can play different functions in a formulation, such as: to attribute proper mechanical property for a formulation to enhance drug product manufacturability; to ensure drug product performance by governing the mode and rate of drug release from a dosage form for immediate or extended drug delivery; or to improve drug product stability.

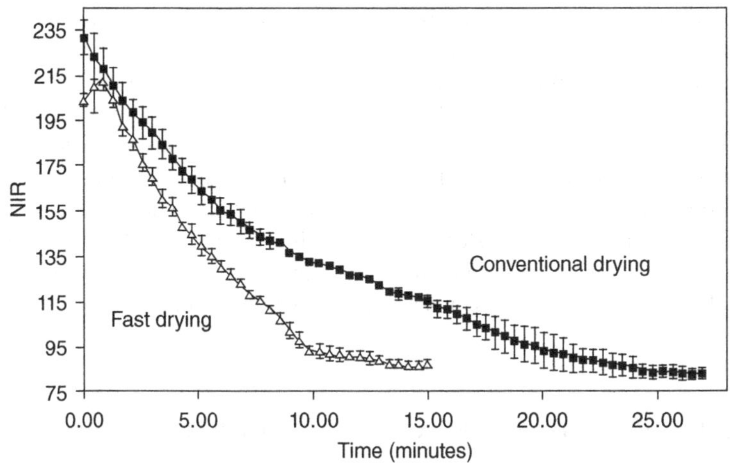

Figure 1.11 Fluid bed drying curves for ibuprofen granulation monitored by in-line NIR measurement of the moisture content. Source: Wildfong [8b]. Reproduced with permission of Elsevier.

TABLE 1.4 A Sample of Drug Incompatibility with Excipient Impurities

Drug	Impurity	Excipient	Drug Loading (w/w)
BMS-203452	Formaldehyde	PEG 300 or Tween 80	1%
Fluoxetine HCl	Reducing sugars	Lactose	10%
Org-30659	Lactose phosphate	Lactose	0.10%
A	Peroxides	Povidone/copovidone	2–3%
B	Peroxides	Povidone/copovidone	2–3%
Raloxifene	Peroxides	Povidone/copovidone	12.50%
CP448187	Free radicals/peroxides	Microcrystalline cellulose	0.50%
BMS-A	Free radicals/peroxide/ reducing sugars	Microcrystalline cellulose	0.83%
Vigabatrin	Reducing sugars, aldehydes	Microcrystalline cellulose	Not available
Irbesartan	Formaldehyde	PEG in film coating	Low strength
Haloperidol	Furfuraldehyde	Lactose	0.575%
Varenicline	Formic acid/ formaldehyde/ acetic acid	PEG or acetate	0.68%
Hydralazine	Aldose	Starch	10%

Excipients and impurities contain reactive functional groups. A survey of current literature indicates that the majority of the drug product stability issues are due to interactions of drugs with excipients and/or reactive impurities contained in the excipients (Table 1.4). Impurities are introduced into excipients as residues from the manufacturing process and raw materials or as degradants from excipient aging.

Even though the level of impurities is usually low, reactive species such as peroxides and aldehydes can interact with drugs directly and cause significant drug product stability issues. In a formulation with low drug loading (1%), 1 ppm of formaldehyde could be equivalent to 0.1% of API in molar to molar ratio (assuming the API has a 10× MW to HCHO). Trace amount of metals (Cu, Fe) can catalyze a series of chemical reactions and degrade the quality, shelf life, and value of drug products. Some of the common reactive excipient impurities include peroxides, reducing sugars, aldehydes, organic acids and esters, heavy metals and trace metals, nitrates/nitrites, and free radicals. In this section, we discuss the sources of impurities, potential chemical interactions with APIs, and analytical methods to measure these impurities at trace levels.

1.6 EXCIPIENT IMPURITIES AND THEIR SOURCES

1.6.1 Peroxides

Peroxides, in general, can be either organoperoxides (ROOR′) or hydroperoxides (ROOH) [17]. Peroxides are generated through a free-radical mechanism involving homolytic cleavage of the C—H bond next to a heteroatom, followed by the addition of oxygen that leads to peroxy radical formation. The peroxy radical can then participate in an autocatalytic cycle by abstraction of hydrogen radical from another reactant to form a hydroperoxide, while generating another carbon free radical [18].

Many pharmaceutical excipients contain trace levels of hydroperoxide impurities, especially polymeric excipients that are commonly generated through radical reactions leaving trace peroxides as a by-product. Polyvinylpyrrolidones (povidone, PVP), hydroxypropylcellulose, crospovidone, polyethylene glycol (PEG), polyethylene oxide, and polysorbate are known to autoxidize to hydroperoxides that subsequently degrade to short-chain aldehydes and carboxylic acids.

Peroxides could be introduced into an excipient during the manufacturing process. In the chemical synthesis of povidone, peroxides were used to initiate the polymerization reaction, and it is difficult to completely eliminate them from the final product [19]. Synthesis of crospovidone from PVP polymerization was carried out in the presence of cross-linking reagents *N,N*-divinyl imidazolidone and radical initiators (AIBN, organic peroxides). Tallon et al. reported that cross-linker Type B, *N,N*-divinyl imidazolidinone has twice the number of oxidation sites compared with cross-linker Type A, which is ethylidene vinylpyrrolidinone. Peroxide accumulation on storage is much greater in Type B crospovidone [20].

Cellulosic excipients utilize either peroxide or a hypochlorite bleaching agent that can remain in the excipients at trace levels and become a source of undesirable oxidation of drug substance. For example, one excipient vendor indicated that hydrogen peroxide was used in the bleaching process of MCC. Polyethylene oxides are formed by generating high-molecular-weight material and oxidizing the material to the desired molecular weight range, leaving residual peroxides as a potential trace impurity.

Peroxide levels increase in polymeric excipients with aging at dry conditions. At high temperature and humidity, peroxide formation decreases. Chemical reactivity of peroxide is high – classic nucleophilic substitution reaction could occur with electropositive oxygen atom of hydroperoxide; secondary amines could react with hydroperoxides to form hydroxyl amines, which could further react to form imines or nitrones, while a tertiary amine reacts to form *N*-oxides. Sulfides react with peroxides to form sulfoxides or sulfones. Detailed examples and case studies of peroxide reaction with APIs are provided in the excipient guidance section.

1.6.2 Reducing Sugars

Reducing sugar impurities are of particular concern if the drug has primary amine and secondary amine in its structure, as Maillard reaction could occur causing discoloration of drug product or even further degradation of the API [21]. Reducing sugar impurities could be generated during the manufacturing processes of the excipients where acid hydrolysis and milling are used. They can also be generated as degradation products of the polysaccharide excipients during long-term exposure to heat and moisture. For example, MCC is manufactured by controlled hydrolysis of α-cellulose from plant with dilute mineral acid [22]. Trace levels of glucose (40–80 ppm) were reported in some lots of MCC [23]. Starch is a mixture of amylose and amylopectin, the ratio of which differs depending on the source of plants. It is prepared from plant seeds or roots such as corn, wheat, potato, and tapioca through coarse milling, water washing, wet sieving, and centrifugal separations. Degradation of starch into reducing sugars during the isolation and fractionation steps can hardly be avoided. Mannitol is produced by catalytic or electrolytic reduction of monosaccharides such as mannose and glucose. Trace levels of reducing sugar from mannitol were reported to cause oxidative degradation of a cyclic heptapeptide from a lyophilized formulation [24]. Lactose is a natural disaccharide consisting of galactose and glucose. Lactose and its monosaccharide components can undergo Maillard reaction with primary amines and secondary amine drugs as reported in the literature [25].

1.6.3 Aldehydes

Formaldehyde, acetaldehyde, and furfuraldehyde are common aldehyde impurities in excipients. Formaldehyde could be formed from the breakdown of the polymeric chain of PEG and polysorbates (Figure 1.12) [26]. Nassar et al. reported 2–165 ppm of formaldehyde in various lots of PEG 300 and polysorbate 80 [27]. Trace level of formaldehyde (8 ppm) was sufficient to generate 1% degradation product for BMS-204352 (Figure 1.13). Similarly, film-coated tablets of Avapro™ (irbesartan) were found to degrade to a hydroxymethyl derivative of the drug substance during long-term stability studies of the low dose (Figure 1.14) [28]. The formaldehyde adduct formation was attributed to the formaldehyde impurity from PEG used in the tablet coating material – Opadry II white. Eliminating PEG from the blend of Opadry prevented the formation of the degradant. Formaldehyde could also be formed as a degradation product of the drug. In the case of hydrochlorothiazide (HCTZ)

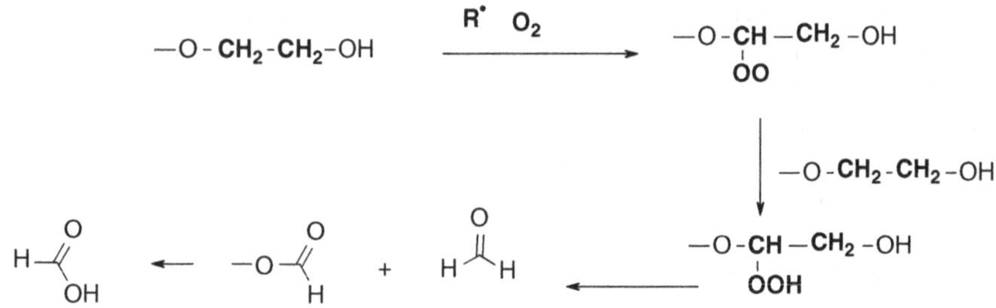

Figure 1.12 Formaldehyde/formic acid formation from oxidation and breakdown of polyethylene glycol and polysorbates.

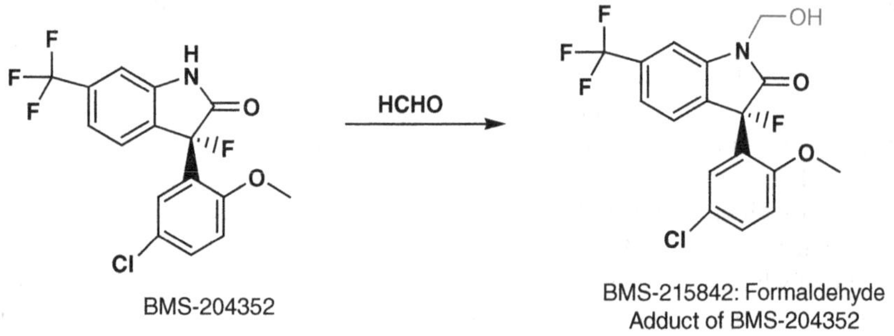

Figure 1.13 Chemical interaction between BMS-204352 and formaldehyde.

bead formulation, formation of a trace amount of formaldehyde was observed due to hydrolysis of HCTZ under high humidity conditions. It subsequently reacted with sodium starch glycolate, decreasing its functionality as a disintegrant, and therefore retarding dissolution of the formulation [29]. Formaldehyde reacts with amine drugs to form *N*-formyl adducts (hemiaminals) that can react further to form dimer(s). Formaldehyde is also known to cross-link gelatin capsule shells leading to dissolution slowdown and incomplete drug release. The cross-linking is a result of formaldehyde interaction with amino groups in gelatin to form insoluble protein. Formaldehyde is susceptible to air oxidation and could be partially converted into formic acid. Therefore, excipients having residual formaldehyde are expected to contain some formic acid impurity as well.

Furfuraldehyde, an aromatic aldehyde, can be formed during the manufacturing process of those excipients that are sourced from plants. Many plant materials contain hemicellulose, a polymer of sugars containing five carbon atoms. When heated with sulfuric acid, hemicellulose undergoes hydrolysis to yield xylose and other five carbon sugars, which may undergo dehydration to form furfuraldehyde [30]. 5-Hydroxymethyl-2-furfuraldehyde (HMF) can also be formed as a result of heat sterilization of parenteral solutions containing hexoses. It is reported that spray-dried lactose contains furfuraldehyde [31]. Presence of HMF was found to correlate with discoloration of lactose. The reaction between HMF and primary

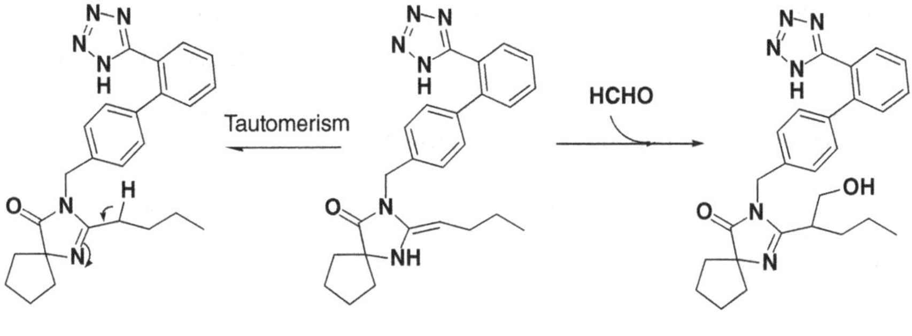

Figure 1.14 Proposed mechanism of degradation of Irbesartan by formaldehyde.

amine drugs could lead to the formation of Schiff bases [1f, 32]. The "browning reaction" is base-catalyzed and may therefore be accelerated in solutions with high pH or in solid dosage forms containing alkaline lubricants (e.g., magnesium stearate).

1.6.4 Metals

Metals are ubiquitous in pharmaceutical excipients at very trace levels and can catalyze oxidation of pharmaceuticals. ICP-AES (atomic emission spectroscopy) is the common analytical method to determine trace metals in excipients. The sensitivity of the method can be as low as ppb to ppm levels. The reaction of molecular oxygen with most organic molecules is thermodynamically favored; however, the triplet state of molecular oxygen (the ground state) represents a kinetic barrier. Trace metal impurities can react with triplet oxygen reducing the molecule to more kinetically favored oxidizing agents such as superoxide [33].

Another common mode of transition-metal-mediated oxidation involves the Fenton-like reactions where the oxidized or reduced form of a catalytic transition metal such as Fe(III) or Fe(II) reacts with hydrogen peroxide to produce several more reactive species. Hydrogen peroxide can be reduced to a hydroxyl radical and hydroxide or oxidized to a peroxy radical and a proton [34].

1.6.5 Organic Acids

Formic acid and its esters, acetic acid and monochloro acetic acid, are trace organic acid impurities that may be present in pharmaceutical excipients (Table 1.5) [35]. Residual organic solvents from the synthesis and purification of excipients may go through further degradation to form organic acids.

Oxidation of PEGs at high temperatures at a central carbon followed by chain scission could generate formaldehyde. Air oxidation of formaldehyde at temperatures used for accelerated stability testing could lead to the formation of formic acid, which would then react with alcohols to form esters. Formic acid could also be formed through oxidation of terminal groups in PEG and polysorbates (Figure 1.12) [36].

TABLE 1.5 Reported Trace Organic Acids Impurities in Pharmaceutical Excipients

Impurity	Excipient	Reported Levels (ppm)
Formic acid	Polyethylene glycol	10–1000 s (MW dependent)
(including formyl esters)	Hydroxypropyl methylcellulose	10–100
	Povidone	1000
	Polyvinyl alcohol	30–40
Acetic acid	Polyvinyl alcohol Cellulose acetate	100 s
Monochloro acetic acid	Sodium starch glycolate	0–14
	Croscarmellose sodium	22–53

Polyvinyl alcohol (PVA) is frequently used as a film-forming polymer in tablet coating systems. Hydrolysis of polyvinyl acetate is a common method for manufacture of PVA, which leads to the presence of polyvinyl acetate, acetic acid, and also some methyl acetate in the commercially available PVA. Also, relatively high levels (20–60 ppm) of formic acid can be present in the PVA as an impurity. We observed that formic acid level does not increase in PVA (solid powder) on storage, while it increases significantly in a mixture of PVA with PEG (both as solid powder). The rate of increase of formic acid in the mixture of PVA and PEG was greater than that in PEG alone [37].

1.6.6 Monochloroacetate

Monochloroacetate is a potentially reactive impurity in croscarmellose sodium and sodium starch glycolate. Croscarmellose sodium is produced from the reaction of sodium monochloroacetate with primary alcohols on the cellulose backbone. The levels of monochloroacetate in selected batches of croscarmellose sodium and sodium starch glycolate are presented in Table 1.6.

1.7 GUIDANCE ON EXCIPIENT IMPURITY AND INTERACTIONS

Compatibility studies are usually the first step to assess the potential chemical interaction between drug and excipients and excipient impurities; however, the compatibility with each and every lot of excipients that will be used in the drug product's lifetime cannot be guaranteed. In fact, the levels of reactive impurities in excipients are variable between lots and vendors.

Pharmacopoeia monographs of excipients do not contain test for the reactive impurities in excipients as the tolerance of the reactive impurities varies widely among drug products. A reactive impurity that may be problematic for a particular drug product may not be a problem for many other drugs. It would be useful for a formulator to know the typical levels of potential reactive impurities in excipients to assess the risk of selecting an excipient in the formulation. The drug degradation would be

Excipients	Lot Number	Impurity (ppm)							
		Glucose	HCHO	Hydrogen Peroxide	NO_2	NO_3	Monochloro Acetate	Heavy Metals	Trace Metals
Microcrystalline cellulose, PH102	R3846	79.6	4.8	<2	N/A	N/A	N/A	<10	<5 Mg, Mn; <10 Al, Cr, Cu, Fe, Ni, Zn; 10 Ca
	4H72517	59.5	5.1	<2	9.4	23.0	0.9	N/A	N/A
	3D69565	40.7	4.1	ND	N/A	N/A	N/A	N/A	N/A
Lactose fast flo	4K93184	ND	N/A	<2	10.4	12.4	12.0	<10	<5 Cr, Cu, Fe, Mg, Mn, Ni, Zn; <10 Al; 15 Ca
Lactose monohydrate	4B80671	ND	1.4	<2	5.1	9.1	1.0	<10	<5 Cr, Cu, Fe, Mg, Mn, Ni, Zn; <10 Al, Ca
	4F77552	ND	ND	<2	5.5	8.0	0.9	<10	<5 Cr, Cu, Fe, Mg, Mn, Ni, Zn; <10 Al, Ca
Lactose anhydrous	4B73319	ND	7.4	<2	5.4	4.3	0.6	<10	<5 Cr, Cu, Fe, Mg, Mn, Ni, Zn; <10 Al, 37 Ca
	1D36857	ND	3.6	<2	3.7	6.0	0.6	<10	<5 Cr, Cu, Fe, Mg, Mn, Ni, Zn; <10 Al, 32 Ca
Pregelatinized starch	2B49405	ND	14.7	<2	14.5	29.2	4.4	<10	<10 Cr, Cu, Fe, Mg, Mn, Ni, Zn; <20 Al, Ca
	4B76327	ND	10.9	<2	11.8	22.9	2.3	<10	<10 Cr, Cu, Fe, Mg, Mn, Ni, Zn; <20 Al, 21 Ca
	9J23860	ND	11.1	N/A	N/A	N/A	N/A	N/A	N/A
Povidone	4K86262	INC	INC	37	2.2	13.6	ND	<10	<5 Cr, Cu, Fe, Mg, Mn, Ni, Zn; <10 Al, Ca
	2D51392	INC	INC	72	1.6	13.1	ND	<10	<5 Cr, Cu, Fe, Mg, Mn, Ni, Zn; <10 Al, Ca
Crospovidone	2F59164	ND	40.8	66	17.2	52.4	ND	N/A	<5 Mn; <10 Al, Cr, Cu, Fe, Ni, Zn; 5 Mg; 10 Ca
	5A09915	ND	8.5	69	10.5	30.4	ND	N/A	<5 Mg, Mn; <10 Al, Ca, Cr, Cu, Fe, Ni, Zn
Sodium starch Glycolate	2B49270	–	4.6	<2	279.2	183.1	ND	<10	<5 Cr, Cu, Fe, Mn, Ni, Zn; <10 Al, 79 Ca; 9 Mg
	4G81568	-	1.5	<2	285.6	117.3	135.8	<10	<5 Cr, Cu, Fe, Mn, Ni, Zn; <10 Al, 75 Ca; 8 Mg
Croscarmellose Na	2H57677	ND	6.5	<2	2.4	23.8	52.2	N/A	N/A
	5D09176	ND	6.6	<2	1.4	10.3	21.6	<10	<10 Cr, Cu, Fe, Mg, Mn, Ni, Zn; <20 Al, 42 Ca
Magnesium stearate	2C54809	ND	3.8	<2	2.1	6.0	ND	<10	<5 Mn; <10 Al, Ca, Cr, Cu, Fe, Ni, Zn
	4K93152	ND	3.7	<2	5.3	12.5	0.7	N/A	N/A
Stearic acid	4K87827	ND	3.1	<2	3.5	6.6	ND	ND	<5 Mn; <10 Al, Ca, Cr, Cu, Fe, Ni, Zn; 30 Mg
Hydroxypropyl cellulose	3H66588	ND	11.4	13	N/A	N/A	N/A	N/A	N/A
	4H72633	ND	9.4	13	0.9	3.5	ND	<10	<5 Cr, Cu, Fe, Mg, Mn, Ni, Zn; <10 Al, 23 Ca
Silicon dioxide	4B80431	ND	6.1	<2	5.8	12.5	ND	N/A	7 Mg; <5 Mn; <10 Al, Ca, Cr, Cu, Fe, Ni, Zn
	5E02038	N/A	N/A	<2	1.5	8.7	ND	N/A	200 Al; 480 Ca; 30 Fe; 130 Mg; <5 Mn, <10 Cr, Cu, Ni, Zn

ND: not detectable; N/A: not available; INC: incompatible

dependent on the availability/molecular mobility of the drug and reactive impurity, particle size of API, drug to excipient ratio, water activity/amount in the formulation, microenvironmental pH, other excipients in the formulation, and environmental factors such as temperature and humidity. This section provides specific examples of potential chemical interactions between drug–excipient/excipient impurity and a general guidance in choosing the suitable excipient for a robust formulation.

1.7.1 Lactose

Major reactive impurities in lactose include glucose, galactose, acetic acid, formic acid, furfuraldehyde, and potentially other aldehydes. Potential chemical reactions between API and lactose are Maillard reactions, Claisen–Schmidt condensation reaction and hydrolysis and catalytic reactions facilitated by lactose.

Lactose, a disaccharide of glucose and galactose, is a reducing sugar. Therefore, lactose and its monosaccharide components can undergo Maillard reaction with primary and secondary amines [38]. These reactions are complicated and can generate a complex variety of degradation products. The first degradation product in a Maillard reaction is thought to be N-glycosamine, which can undergo Amadori rearrangements to form a wide variety of products [25, 39]. Some of other degradation products reported in the literature are N-formylated, N-acetylated condensation products with furfuraldehyde [25]. Furfuraldehyde is reported to be present in spray-dried lactose. Presence of 5-hydroxymethyl-2-furfuraldehyde (HMF) was found to correlate with discoloration of lactose. The reaction between 5-hydroxymethyl-2-furfuraldehyde and primary amine drugs could lead to the formation of Schiff bases. The "browning reaction" is base-catalyzed and may therefore be accelerated if alkaline lubricants are used [1f, 31, 40]. The amorphous content of lactose, equilibrium moisture content, microenvironmental pH, and salt/free base form of the drug can contribute to the extent of Maillard reaction.

1.7.1.1 Claisen–Schmidt Condensation Reaction The impurity 5-hydroxymethyl -2-furfuraldehyde can react with the carbonyl (ketone) to form a condensed product. Janicki and Almond showed that Haloperidol reacted with HMF to form the condensed product as shown in Figure 1.15 [40, 41] Interestingly, Haloperidol is a tertiary amine.

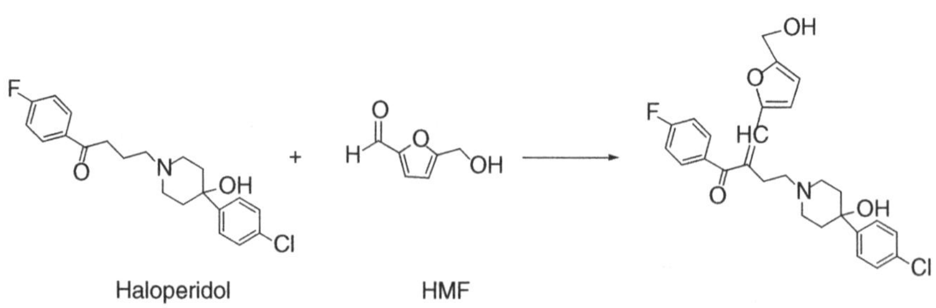

Figure 1.15 Reaction of haloperidol with HMF to form a condensation product.

1.7.1.2 Hydrolysis and Catalytic Effects of Lactose Hydrolytically labile drugs can undergo degradation in the presence of lactose (especially amorphous, spray-dried grade) due to the moisture associated with the excipient itself. Hydrolysis of certain esters was shown to be catalyzed by sugars. Badawy et al. showed that hydrolysis of DMP 754 was catalyzed by lactose in solution as well as solid state [42].

1.7.2 Microcrystalline Cellulose

The major reactive impurity in MCC is glucose. It may also contain trace formaldehyde, nitrates, and nitrites.

MCC is a stable, though hygroscopic, material. The moisture content of MCC is specified to be not more than 6.0% w/w. The sorption and penetration of water into formulations containing MCC has been studied. The moisture content of MCC and the availability of sorbed moisture were found to be responsible for the instability of drugs prone to hydrolysis. There are several literature reports of incompatibility of drugs with MCC for drugs susceptible to hydrolysis [43]. Ahlneck et al. have cited accessibility of water molecules in MCC and drug diffusion into the sorbed water layer as governing factors for hydrolytic degradation of acetyl salicylic acid [1a]. The authors conclude that by controlling the relative humidity the amount of water absorbed by MCC can be controlled, thus reducing the fraction of weakly bound water in MCC. This in turn will reduce the rate of hydrolysis of drug.

George et al. reported a specific drug incompatibility associated with MCC [44]. Products of the Maillard reaction between a primary amine drug, vigabatrin, and the monosaccharide component of MCC, that is, glucose were implicated for the browning of vigabatrin tablets on aging. Amadori rearrangement could follow Maillard reaction resulting in formation of 1-amino-1-deoxyketose, which can react further to form dicarbonyl compounds and a multitude of products. Reaction of vigabatrin with 5-hydroxymethyl-2-furfuraldehyde, which is a known glucose degradation product, was also cited by the authors as a potential cause for the discoloration of tablets. In another study, MCC was found to be responsible for granulation discoloration in a capsule formulation of BMS Compound A. Approximately 40 ppm of glucose was detected in the particular lot of MCC used in the formulation. The discoloration was indicative of Maillard reaction occurring between glucose impurity from MCC and L-phenylalanine in the drug complex.

A few reports of nonspecific drug-MCC incompatibility include instability attributed to the strong hydrogen bonding capability of MCC, which catalyzed the proton abstraction step for a drug, DMP 543, which undergoes degradation by nucleophilic substitution [45].

1.7.3 Povidone and Crospovidone

Povidone and crospovidone contain significant levels of peroxides. Povidone may also contain formic acid and formaldehyde. EP, JP Compendial test for peroxides in povidone and crospovidone limits the level to within 400 ppm. This level could be too high for some formulations if the drug substance is highly susceptible to oxidation.

In addition, peroxide content can change on storage – increasing at high temperature and decreasing with humidity.

Major incompatibilities with povidone/crospovidone and impurities include (i) Oxidation – Any drug substance that is prone to oxidation can be susceptible to interactions with PVP such as formation of *N*-oxide [20, 46], oxidation of thiols. (ii) Nucleophilic addition – N-terminus of peptides and amino acids (e.g., phenylalanine) can react with carbonyl of lactam group of PVP to form an amide [47]. (iii) Hydrolysis – Hydrolytically labile drugs can undergo degradation in the presence of PVP due to the moisture associated with the excipient itself and the increased solubility of the drug in the moisture layer.

1.7.4 Hydroxypropyl Cellulose

Hydroxypropyl cellulose (HPC) is made from the reaction of cellulose with propylene oxide at elevated temperature and pressure [3a]. Cellulose is first treated with sodium hydroxide to produce a swollen alkali cellulose, which could react with propylene oxide more easily. Propylene oxide can be substituted on the cellulose through an ether linkage at the three hydroxyls present on each anhydroglucose unit of the cellulose chain. Etherification takes place at almost all secondary hydroxyls. The secondary hydroxyl present in the side chain can further react with propylene oxide, and "chaining out" may occur. More than 1 mol of hydroxypropyl substituent can be present on the side chain. Depending on the degree of substitution, HPC can be manufactured into low-substituted grade. Following the reaction, the reactant is recrystallized by neutralization, washed, and milled to obtain a white to slightly yellowish-colored powder.

Formaldehyde and hydrogen peroxide are the potential reactive impurities that are present in HPC (Table 1.6).

As HPC might contain trace hydroperoxide, any drug substance that is prone to oxidation can be susceptible to interactions with HPC and can exhibit reactions such as formation of *N*-oxide [48] (Figure 1.16), oxidation of thiols, and abstraction of benzylic hydrogen atoms [49] (Figure 1.17). Hydrogen atoms whose bond is weakened by nearby aromatic structures can delocalize the resulting radical enabling oxidation at other points in the ring system and potentially opening the ring system [47].

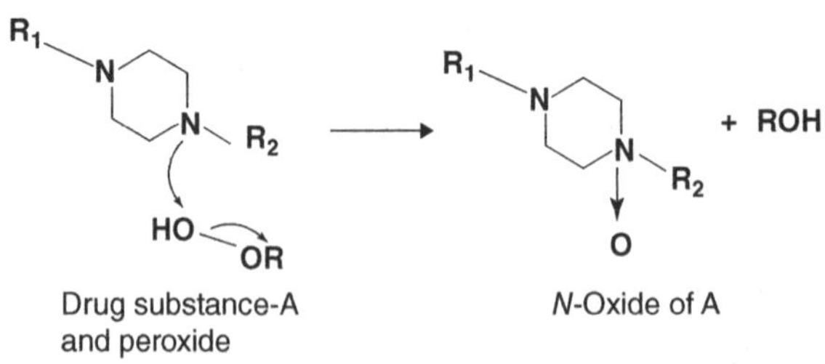

Figure 1.16 Piperazine reaction with hydrogen peroxide to form *N*-oxide. Source: Freed [48]. Reproduced with permission of Elsevier.

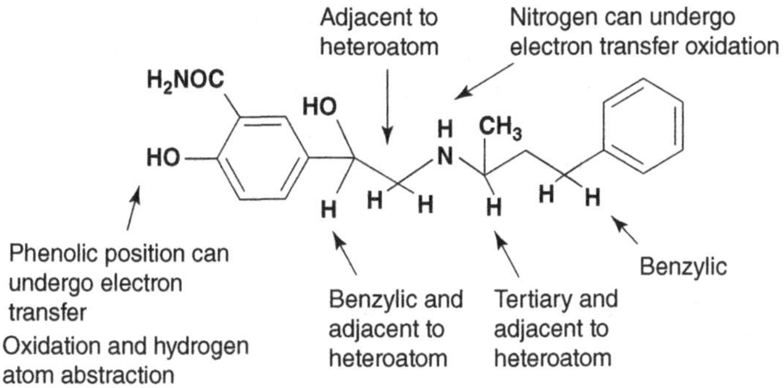

Figure 1.17 Sites susceptible for oxidation.

Reactions with Formaldehyde Impurity Formaldehyde is known to react with amine drugs to form *N*-formyl adducts (hemiaminals) that can react further to form dimer(s). Formaldehyde is also known to cross-link gelatin capsule shells leading to dissolution slowdown and incomplete drug release. The cross-linking is a result of formaldehyde interaction with amino groups in gelatin to form insoluble protein. Formaldehyde is susceptible to oxidation and could be partially converted into formic acid in contact with air. Therefore, excipients having residual formaldehyde are expected to contain some formic acid impurity as well.

1.7.5 Croscarmellose Sodium

To produce croscarmellose sodium, alkali cellulose is prepared by steeping cellulose, obtained from wood pulp or cotton fibers, in sodium hydroxide solution. The alkali cellulose is then reacted with sodium monochloroacetate to obtain carboxymethylcellulose sodium. After the substitution reaction is completed and all of the sodium hydroxide has been used, the excess sodium monochloroacetate slowly hydrolyzes to glycolic acid. The glycolic acid changes a few of the sodium carboxymethyl groups to the free acid and catalyzes the formation of cross-links to produce croscarmellose sodium. The croscarmellose sodium is then extracted with aqueous alcohol and any remaining sodium chloride or sodium glycolate removed. After purification, croscarmellose sodium of greater than 99.5% purity is obtained. Croscarmellose sodium may be milled to break the polymer fibers into shorter lengths and hence improve its flow properties [22].

Nitrate, nitrite, monochloroacetate are the major impurities found in croscarmellose sodium. However, none of these impurities are listed in the pharmacopoeial specifications.

1.7.5.1 Adsorption Some weakly basic drugs can compete with the sodium counterion, adsorbing onto the surface of the disintegrant particles. Adsorption of weakly basic drugs and salts of weakly basic drugs to croscarmellose sodium has been observed to cause incomplete *in vitro* dissolution and/or incomplete extraction [50]. Adsorption occurs in deionized aqueous and aqueous-organic media and is typically modeled by either the Freundlich or Langmuir adsorption isotherms [51]. Adsorption of the drug substance is a function of pH, ionic strength, and ionic species and can be mitigated by the addition of competing electrolytes [52]. No effect on *in vivo* absorption has been observed likely due to competition for adsorption sites by gastric electrolytes.

1.7.5.2 Solid-State Reactions A solid-state reaction has been observed under high humidity conditions that converts the freely soluble, weakly basic delavirdine mesylate salt to the very slightly soluble free base, causing incomplete dissolution [53]. The methanesulfonic acid produced from that reaction then converts sodium carboxymethyl moieties of croscarmellose sodium to the free acid. It is suggested that disintegrant performance may also be impaired due to this reaction [54].

1.7.5.3 Specific Interaction Croscarmellose sodium may have small residual amounts of monochloroacetate from manufacturing. This impurity is capable of performing an SN2 with available nucleophiles yielding an impurity that has an atomic mass of +58 amu with respect to the parent compound [55].

1.7.6 Sodium Starch Glycolate

Sodium starch glycolate is a substituted and cross-linked derivative of potato starch [22]. Starch is carboxymethylated by reacting it with sodium chloroacetate in an alkaline medium, followed by neutralization with citric or some other acid. Cross-linking may be achieved by either physical methods or chemically by using reagents such as phosphorus oxytrichloride or sodium trimetaphosphate. Monochloroacetate, nitriles, and nitrates are the potential reactive impurities that are present in sodium starch glycolate (Table 1.6).

Some weakly basic drugs can compete with the sodium counterion, adsorbing onto the surface of the disintegrant particles causing incomplete *in vitro* dissolution and/or incomplete extraction [55]. Displacement of chlorine in monochloroacetate by an amine or alcohol could lead to an impurity with MW +58 with respect to the parent compound, which is known as SN2 reaction [54b] (Figure 1.18).

1.7.7 Starch

Starch is composed of amylose and amylopectin, polymers of glucose connected by α-1,4-glycosidic linkages (in contrast to cellulose β-1,4 linkages). Amylopectin has occasional branch chains connected by α-1,6-glycosidic linkages. Starch is extracted

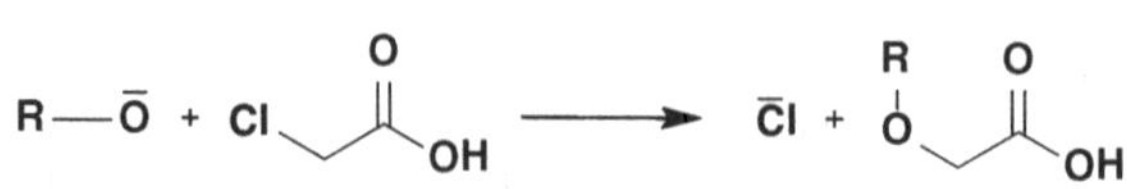

Figure 1.18 SN2 reaction between API and monochloroacetate impurity.

from plant sources through a sequence of processing steps involving coarse milling, repeated water washing, wet sieving, and centrifugal separation. The wet starch obtained from these processes is dried and milled before use in pharmaceutical formulations.

Pregelatinized starch is a starch that has been chemically and/or mechanically processed to rupture all or part of the starch granules and so render the starch flowable and directly compressible. Partially pregelatinized grades are also commercially available. Formaldehyde, nitrates, and nitrites are the major reactive impurities in starch.

The terminal aldehydes in starch have been shown to react with hydrazine moieties of hydralazine HCl [56] and form phthalazine hydrazone and/or triazolophthalazine derivative. Starch has been implicated in moisture-mediated reactions due to residual moisture content in the excipient [43c, 57]. Starch can react with formaldehyde decreasing its functionality as a disintegrant and therefore retarding dissolution of a formulation [29]. Starch also has the potential to physically adsorb drug substance onto the surface of starch particles [50a, 58] in the examples of ketotifen and oxytetracycline HCl, possibly causing drug-retention during extraction or dissolution, which can translate to incomplete physiological absorption [58].

1.7.8 Silicon Dioxide

Colloidal silicon dioxide is widely used in pharmaceuticals, cosmetics, and food products. Its small particle size and large specific surface area give it desirable flow characteristics that are exploited to improve the flow properties of dry powders in a number of processes, for example, tableting [59]. Colloidal silicon dioxide is also used to stabilize emulsions and as a thixotropic thickening and suspending agent in gels and semisolid preparations [60]. In aerosols, other than those for inhalation, colloidal silicon dioxide is used to promote particulate suspension, eliminate hard settling, and minimize the clogging of spray nozzles. Colloidal silicon dioxide is also used as a tablet disintegrant and as an adsorbent dispersing agent for liquids in powders or suppositories [61].

1.7.8.1 Moisture Scavenging Colloidal silicon dioxide is hygroscopic; it adsorbs large quantities of water without liquefying. When used in aqueous systems at a pH between 0 and 7.5, colloidal silicon dioxide is effective in increasing the viscosity of a system. However, at a pH greater than 7.5, the viscosity-increasing properties of colloidal silicon dioxide are reduced and at a pH greater than 10.7, this ability is lost entirely since the silicon dioxide dissolves to form silicates. Colloidal silicon dioxide powder should be stored in a well-closed container.

1.7.8.2 Incompatible with Diethylstilbestrol Preparations Johansen and Moller studied the dissolution behavior and diffuse reflectance of several solvent deposition of drugs on different excipients. They found that drugs solvent-deposited and on SiO_2 were so strongly adsorbed that the therapeutic effect failed to appear presumably due to incomplete release or degradation of the drug, as in the case of diethylstilbestrol. Silicon dioxide can act as Lewis acid under anhydrous conditions and promote diverse reactions. In the case of diethylstilbestrol, silicon dioxide catalyzed oxidation of the compound to produce peroxide and conjugated quinone degradation products [62].

1.7.8.3 Autoxidation of Linoleic Acid Methylester (LME) Interaction between ester carbonyl and silica silanol groups could happen in mixture of ester drugs and colloidal silicon dioxide. Tischinger-Wagner et al. reported that linoleic acid methylester (LME) adsorbed to the surface of silica through hydrogen bonds, and the porous and colloidal silicas accelerated the oxidative degradation of LME [63].

1.7.9 Stearic Acid

Stearic acid is made via hydrolysis of fat by continuous exposure to a countercurrent stream of high-temperature water and fat in a high-pressure chamber. The resulting mixture is purified by vacuum-steam distillation and the distillates are then separated using selective solvents [22].

Stearic acid may also be made via hydrogenation of cottonseed and other vegetable oils; by the hydrogenation and subsequent saponification of oleic followed by recrystallization from alcohol; and from edible fats and oils by boiling with sodium hydroxide, separating any glycerin and decomposing the resulting soap with sulfuric or hydrochloric acid. The stearic acid is then subsequently separated from any oleic acid by cold expression [1, 22].

1.7.9.1 Incompatibilities with Stearic Acid Stearic acid is incompatible with most metal hydroxides and may be incompatible with oxidizing agents. Insoluble stearates are formed with many metals; ointment bases made with stearic acid may show evidence of drying out or lumpiness due to such a reaction when compounded with zinc or calcium salts. A number of differential scanning calorimetry studies have investigated the compatibility of stearic acid with drugs. Although such laboratory studies have suggested incompatibilities, for example, naproxen, of drugs with stearic acid, they may not necessarily be applicable to formulated products. Stearic acid has been reported to cause pitting in the film coating of tablets coated using an aqueous film-coating technique; the pitting was found to be a function of the melting point of the stearic acid [64].

Stearic acid could affect the hydrolysis rate of API if the degradation is pH dependent. It could also potentially react with an API containing a primary amine to form a stearoyl derivative [4, 65].

1.7.10 Magnesium Stearate

Magnesium stearate is the most commonly used lubricant in the pharmaceutical processes. It is an essential component of a drug formulation and plays a key role in successful manufacturing of pharmaceutical solid dosage forms. In unit operations such as blending, roller compaction, tableting, and capsule-filling, lubricant helps to reduce the friction between the surfaces of manufacturing equipment and the solids to avoid sticking and to ensure the continuation of the process [66].

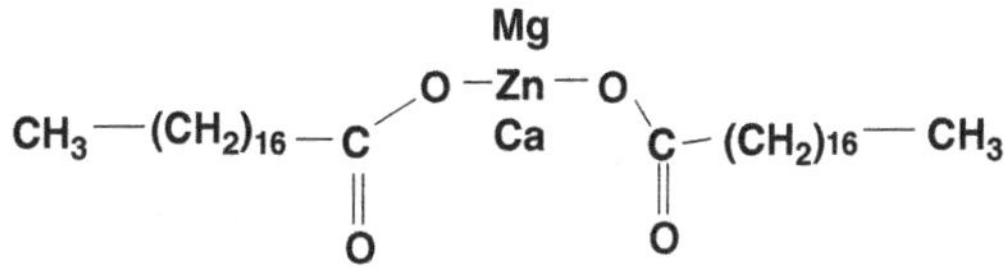

Figure 1.19 Chemical structure of magnesium stearate and other metallic salts (calcium and zinc) of stearic acids.

Magnesium stearate (Figure 1.19) is a solid, fluffy white powder that could be derived from plants as well as animal sources. It is prepared either by the chemical reaction of an aqueous solution of magnesium chloride with sodium stearate or by the reaction of magnesium oxide, hydroxide, or carbonate with stearic acid at elevated temperatures. The raw materials used in the manufacturing of magnesium stearate are refined fatty acids, a mixture of palmitic and stearic acid. Magnesium stearate might contain various impurities originated from its manufacturing process, which might lead to incompatibilities with APIs in the formulations.

Interactions between magnesium stearate and API include potential chemical interactions with the impurities (MgO), the effect of alkalinity caused by magnesium stearate, its catalytic effect, and other chemical reactions initiated and mediated by magnesium ions.

1.7.10.1 Potential Interactions with Impurities (MgO) The commercial materials of magnesium stearate contain several impurities such as magnesium oxide (MgO) and palmitic acid; these impurities often react with APIs in the solid state causing stability issues. For instance, as reported by Kararli et al., MgO reacts with ibuprofen at certain temperatures and humidity values in the solid [1g]. Specifically, when the mixture of MgO and ibuprofen was stressed at 40 °C and 75% RH, a significant amount of degradation was detected by differential scanning calorimeter (DSC), thermal gravimetric analysis (TGA), and multiple internal reflectance infrared (MIR). In another study, ketoprofen was found to form a eutectic mixture with magnesium stearate [64, 67].

1.7.10.2 Hydrolytic Degradation at Basic pH The presence of magnesium stearate in a formulation can increase the microenvironmental pH of the formulation, creating an alkaline condition and consequently accelerating the hydrolysis of some drugs. For example, the degradation rate of acetylsalicylic acid (aspirin) in a blend increased with the addition of magnesium stearate; the hydrolysis rate depended on the concentration of magnesium stearate in the blend. This is because acetylsalicylic acid is a moisture-sensitive drug, and a buffer layer around the particles of acetylsalicylic acid was formed, creating an environment that was detrimental to the chemical stability of the compound [68]. Furthermore, the presence of MgO impurity in magnesium stearate may also play a role since it could enhance the degradation by creating an alkaline pH environment. For example, Gordon et al. noticed that in the presence of magnesium stearate, ibuprofen forms a eutectic mixture that sublimates [69]. In addition, quinapril (a tetrahydroisoquinoline carboxylic acid), an angiotensin-converting enzyme (ACE) inhibitor, was also found to be incompatible with magnesium stearate due to the basicity of the lubricant; the degradation of quinapril was mediated by the availability of moisture. In addition to hydrolysis, oxidation is another reaction causing chemical instability issues associated with the presence of magnesium stearate, which is discussed in the following section.

1.7.10.3 Oxidation The presence of magnesium stearate in a formulation can also induce an oxidation reaction. For instance, the decomposition of drotaverine HCl was accelerated when magnesium stearate and talc were present in a formulation [70]. Drotaverine HCl was degraded to drotaveraldine by an oxidative degradation pathway, which can be inhibited using an antioxidant or an acidic auxiliary material. A similar catalytic action of magnesium stearate was observed with the autoxidation of 2,6,10,14-tetramethylpentadecane, where magnesium stearate catalyzed the decomposition of hydroperoxide first to boost autoxidation of the compound [71]. Aside from its effect on oxidation, the metal ions from magnesium stearate also cause chemical instability.

1.7.10.4 Metal Ion-Mediated Degradation Degradation of drugs is also mediated by the presence of magnesium ions. For example, upon an accelerated stress treatment, fosinopril sodium was degraded into a β-ketoamide (III) and a phosphoric acid (IV) in a prototype tablet formulation with magnesium stearate [72]. It was shown by further investigation that the degradation of fosinopril was mediated by magnesium metal ions, and thus a mechanism invoking metal chelation was postulated. Based on a kinetic study, it was established that the degradation was a second-order reaction between fosinopril and magnesium. Since many drugs are susceptible to ion-catalyzed degradation, it has been suggested that stearate salts should be avoided as tablet lubricants. However, by the addition of malic acid, hexamic acid, and maleic acid in a formulation, the degradative effect of alkali stearates can be inhibited due to competition for the lubricant cation between the drug and an additive acid. The incompatibility of magnesium stearate with a drug also depends on the functional groups of the drug. For example, drugs with an amine group are often very reactive, which is discussed in the following section.

1.7.10.5 Reaction with Amines Many drugs contain amine groups, and amines are typically prone to reactions with excipients and salt counterions. Specifically, the potential for a reaction with magnesium stearate or stearic acid is particularly of concern when a drug has a primary amine group. In the case of norfloxacin, after a prolonged storage at 60 °C, the formation of a stearoyl derivative was observed in the tablets containing magnesium stearate. Other drugs, found to be incompatible with magnesium stearate, include glimepiride, cephalexin, glipizide, ibuproxam, indomethacin, ketoprofen, moexipril, nalidixic acid, primaquine, promethazine hydrochloride, temazepam, glibenclamide, penicillin G, oxacillin, clopidogrel besylate, and erythromycin [73]. In summary, drugs with a primary amine group are often very unstable in formulations containing magnesium stearate.

1.8 ANALYTICAL METHODS FOR DETERMINING TRACE REACTIVE EXCIPIENT IMPURITIES

1.8.1 Reducing Sugar

Determination of reducing sugar in excipients is not required by compendia. As the reducing sugars usually exist at trace levels, HPLC determination is challenging due to the absence of a chromophore for UV detection and the difficulty of achieving adequate retention in an HPLC column. Conventional glucose assays, such as enzymatic and colorimetric assays, typically lack specificity and are not sufficiently sensitive to detect glucose at ppm concentrations. A reversed-phase HPLC method to determine trace levels of glucose and formaldehyde in pharmaceutical excipients was reported by our laboratory [23]. The method utilizes precolumn derivatization of analytes with 2,4-dinitrophenylhydrazine (DNPH) to enable UV detection of trace glucose and formaldehyde. The detection limit for glucose is as low as 1 ppm, and the limit for formaldehyde is 0.3 ppm. Several batches of excipients were analyzed for glucose, formaldehyde, and other reactive impurities, and the results are presented in Table 1.4.

1.8.2 Aldehydes

Headspace gas chromatography (GC) is the most commonly used method to determine trace volatile impurities in pharmaceutical excipients. The method involves derivatization of aldehydes with *O*-2,3,4,5,6-(penta fluorobenzyl) hydroxylamine hydrochloride (PFBHA), followed by static headspace GC of PFBHA aldehyde oximes with mass spectrometry (MS) detection. Another GC/MS method reported determines formic acid and formaldehyde in excipients simultaneously [35]. The method utilized a one-step procedure requiring dissolution or dispersion of samples in acidified ethanol to convert formic acid and its esters to ethyl formate and formaldehyde to diethoxymethane. Identification and quantification of the derivatized analytes were conducted by GC/MS. The authors claimed the detection limit for formic acid to be 0.5 ppm (range 0.5–10,000 ppm), and 0.2 ppm (range 0.2–10,000 ppm) for formaldehyde. There are several other methods for the measurement of formaldehyde in the literature, such as the colorimetric method using chromotropic acid (CTA) or acetylacetone, and the Purpald (4-amino-3-hydrazino-5-mercapto-1,2,4-triazole) method that requires oxidation of the formaldehyde-Purpald adduct for color development. The CTA (4,5-dihydroxynaphthalene-2,7-disulfonic acid) assay is a popular method for the detection of formaldehyde as it is highly specific. This colorimetric assay, however, requires lengthy heating of the sample under strong acidic conditions (e.g., 100 °C, 30 minutes). The CTA assay is not suitable for the determination of formaldehyde in starch and cellulose-based excipients as the strong acid may hydrolyze the end glucose on the polymeric chain of these excipients and release intrinsic aldehydes, causing false high levels of aldehyde readings.

The DNPH methods described under Section 1.6.2 was developed in our laboratories and was shown to be a simple and sensitive method to determine trace formaldehyde, acetaldehyde, and other aldehydes in excipients. It works at room temperature, and the reaction time is 1 hour. DNPH derivatization is an acid-catalyzed reaction at pH 1.5–2; however, starch and cellulose excipients (and their derivatives) will not be degraded as they are not in contact with the acidic reaction solution. The reactive impurities from these excipients were extracted out in 50/50 acetonitrile/water in the sample preparation step. The extracted solution was filtered through 0.45 μm membranes before adding to the DNPH solution for the derivatization reaction. The method utilized HPLC for separation and quantification and could be beneficial for laboratories without GC/MS capability.

1.8.3 Peroxides

Measurement of trace hydroperoxide (HPO) in pharmaceutical excipients has been challenging as the impurity is not stable. One of the methods is the HPLC-based HPO assay, involving triphenylphosphine (TPP) for total HPO content (ROOH and H_2O_2) [74]. Some other techniques for HPO measurement include the ferrous oxidation-xylenol orange (FOX2) method for total HPO, the liquid chromatography-based electrochemical determination of hydrogen peroxide using platinum and enzyme electrodes (detection limit > 1 ng/mL), and the enzyme-based Reflectoquant™ colorimetric test for inorganic peroxide (0.2–20 ppm) [18]. Other analytical methods such as the titanium sulfate method or the iodide titration methods reported in US Pharmacopeia (USP), British Pharmacopoeia (BP), and European Pharmacopoeia (Ph. Eur.) are for the determination of more concentrated hydrogen peroxide solutions.

1.8.4 Organic Acids

The levels of organic acids in excipients are not usually tested by the excipient manufacturer due to their low toxicity as class III solvents. GC/MS method is the most commonly used method to determine these impurities [27]. For example, the formyl species detection methods usually require derivatization with an alcohol, such as ethanol, to form an ester followed by high-performance liquid chromatography (HPLC) or GC separation and detection. These methods are nonspecific with respect to type (free acid vs ester) and relative proportion of formyl species present in the starting materials.

1.9 CONCLUSION

Characterization of pharmaceutical excipients is gaining increased attention in the pharmaceutical industry as it provides a deeper understanding of the material in terms of its properties, functionalities, and potential interactions with APIs. It plays a key role in successful drug product development and manufacturing. In this chapter, we have reviewed the characterization of excipients from compendia monographs to the novel techniques developed and used in studying excipients, and the excipient impurities, potential interactions and implications to drug product stability. Comprehensive knowledge of excipients structure and function along with an understanding of the stability "soft spots" of the drug will lead to better risk assessment and implementation of a strategy for robust drug product development. While some strategies involve implementations of novel technologies in drug product design and process, other options involve greater understanding of excipients through cooperation and collaboration between excipient manufacturers and end users of excipients.

REFERENCES

1. (a) Ahlneck, C. and Alderborn, G. (**1988**) Solid state stability of acetylsalicylic acid in binary mixtures with microcrystalline and microfine cellulose. *Acta Pharmaceutica Suecica*, **25** (1), 41–52;

   (b) Ahlneck, C. and Lundgren, P. (**1985**) Methods for the evaluation of solid state stability and compatibility between drug and excipient. *Acta Pharm. Suec.*, **22**, 303–314;

   (c) Ahlneck, C. and Zografi, G. (**1990**) The molecular basis of moisture effects on the physical and chemical stability of drugs in the solid state. *International Journal of Pharmaceutics*, **62** (2–3), 87–95;

   (d) Carstensen, J.T. (**1974**) Stability of solids and solid dosage forms. *Journal of Pharmaceutical Sciences*, **63** (1), 1–14;

   (e) Carstensen, J.T., Johnson, J.B., Spera, D.C., and Frank, M.J. (**1968**) Equilibrium phenomena in solid dosage forms. *Journal of Pharmaceutical Sciences*, **57** (1), 23–27;

   (f) Duvall, R.N., Koshy, K.T., and Pyles, J.W. (**1965**) Comparison of reactivity of amphetamine, methamphetamine, and dimethylamphetamine with lactose and related compounds. *Journal of Pharmaceutical Sciences*, **54** (4), 607–611;

   (g) Kararli, T.T., Needham, T.E., Seul, C.J., and Finnegan, P.M. (**1989**) Solid-state interaction of magnesium oxide and ibuprofen to form a salt. *Pharmaceutical Research*, **6** (9), 804–808;

   (h) Kontny, M.J. and Zografi, G. (**1985**) Moisture sorption kinetics for water-soluble substances. IV: Studies with mixtures of solids. *Journal of Pharmaceutical Sciences*, **74** (2), 124–127;(i) Narang, A.S., Rao, V.M., and Raghavan, K. (**2009**) Excipient compatibility, in *Developing Solid Oral Dosage Forms: Pharmaceutical Theory and Practice*, Elsevier, Burlington, pp. 125–146;

   (j) Wu, Y., Levons, J., Narang, A.S. *et al.* (**2011**) Reactive impurities in excipients: profiling, identification and mitigation of drug–excipient incompatibility. *AAPS PharmSciTech*, **12** (4), 1248–1263;

   (k) Yu, X., Buevich, A.V., Li, M. *et al.* (**2013**) A compatibility study of a secondary amine active pharmaceutical ingredient with starch: identification of a novel degradant formed between desloratadine and a starch impurity using LC–MS(*n*) and NMR spectroscopy. *Journal of Pharmaceutical Sciences*, **102** (2), 717–731;

   (l) Zografi, G., Grandolfi, G.P., Kontny, M.J., and Mendenhall, D.W. (**1988**) Prediction of moisture transfer in mixtures of solids: transfer via the vapor phase. *International Journal of Pharmaceutics*, **42** (1–3), 77–88;

   (m) Byrn, S.R., Xu, W., and Newman, A.W. (**2001**) Chemical reactivity in solid-state pharmaceuticals: formulation implications. *Advanced Drug Delivery Reviews*, **48** (1), 115–136.

2. (a) Qiu, Y., Chen, Y., Zhang, G.G.Z. *et al.* (**2009**) *Developing Solid Oral Dosage Forms: Pharmaceutical Theory & Practice*, Elsevier Science;

   (b) Takagi, T., Ramachandran, C., Bermejo, M. *et al.* (**2006**) A provisional biopharmaceutical classification of the top 200 oral drug products in the United States, Great Britain, Spain, and Japan. *Molecular Pharmaceutics*, **3** (6), 631–643;

   (c) Dahan, A., Miller, J.M., and Amidon, G.L. (**2009**) Prediction of solubility and permeability class membership: provisional BCS classification of the world's top oral drugs. *The AAPS Journal*, **11** (4), 740–746.
3. (a) Rowe, R.C., Sheskey, P.J., and Weller, P.J. (**2003**) *Handbook of Pharmaceutical Excipients*, Pharmaceutical Press;(b)United States Pharmacopeial, C., U.S. Pharmacopeia & National Formulary. United States Pharmacopeial Convention, Incorporated: **2006**.
4. Kushner, J.t. (**2013**) Utilizing quantitative certificate of analysis data to assess the amount of excipient lot-to-lot variability sampled during drug product development. *Pharmaceutical Development and Technology*, **18** (2), 333–342.
5. Koo, O.M.Y. (**2011**) Excipients-application challenges and examples of new excipients in advanced drug delivery systems. *American Pharmaceutical Review*, **14** (2), 60–68.
6. Igne, B., Shi, Z., Drennen, J.K. 3rd,, and Anderson, C.A. (**2014**) Effects and detection of raw material variability on the performance of near-infrared calibration models for pharmaceutical products. *Journal of Pharmaceutical Sciences*, **103** (2), 545–556.
7. (a) Cogdill, R.P., Anderson, C.A., Delgado, M. *et al.* (**2005**) Process analytical technology case study: Part II. Development and validation of quantitative near-infrared calibrations in support of a process analytical technology application for real-time release. *AAPS PharmSciTech*, **6** (2), E273–E283;

   (b) Agency, E.M. (**2012**) Assessment report – Eliquis; *European Medicines Agency*, Document EMA/641505/2012 (WC500136575), 9–10.
8. (a) Hailey, P.A., Doherty, P., Tapsell, P. *et al.* (**1996**) Automated system for the on-line monitoring of powder blending processes using near-infrared spectroscopy. Part I. System development and control. *Journal of Pharmaceutical and Biomedical Analysis*, **14** (5), 551–559;

   (b) Wildfong, P.L., Samy, A.S., Corfa, J. *et al.* (**2002**) Accelerated fluid bed drying using NIR monitoring and phenomenological modeling: Method assessment and formulation suitability. *Journal of Pharmaceutical Sciences*, **91** (3), 631–639.
9. De Bleye, C., Sacré, P.-Y., Dumont, E. *et al.* (**2014**) Development of a quantitative approach using surface-enhanced Raman chemical imaging: First step for the determination of an impurity in a pharmaceutical model. *Journal of Pharmaceutical and Biomedical Analysis*, **90**, 111–118.
10. Lewis, E.N., Schoppelrei, J., and Lee, E. (**2004**) Near-infrared chemical imaging and the PAT initiative. *Spectroscopy*, **19** (4), 26–36.
11. Elkhider, N., Chan, K.L., and Kazarian, S.G. (**2007**) Effect of moisture and pressure on tablet compaction studied with FTIR spectroscopic imaging. *Journal of Pharmaceutical Sciences*, **96** (2), 351–360.
12. Wray, P., Chan, K.L.A., Kimber, J., and Kazarian, S.G. (**2008**) Compaction of pharmaceutical tablets with different polymer matrices studied by FTIR imaging and X-ray microtomography. *Journal of Pharmaceutical Sciences*, **97** (10), 4269–4277.
13. Lakio, S., Vajna, B., Farkas, I. *et al.* (**2013**) Challenges in detecting magnesium stearate distribution in tablets. *AAPS PharmSciTech*, **14** (1), 435–444.

14. (a) Berggren, J., Frenning, G., and Alderborn, G. (**2004**) Compression behaviour and tablet-forming ability of spray-dried amorphous composite particles. *European Journal of Pharmaceutical Sciences*, **22** (2–3), 191–200;

(b) Sebhatu, T., Ahlneck, C., and Alderborn, G. (**1997**) The effect of moisture content on the compression and bond-formation properties of amorphous lactose particles. *International Journal of Pharmaceutics*, **146** (1), 101–114;

(c) Sebhatu, T. and Alderborn, G. (**1999**) Relationships between the effective interparticulate contact area and the tensile strength of tablets of amorphous and crystalline lactose of varying particle size. *European Journal of Pharmaceutical Sciences*, **8** (4), 235–242.

15. (a) Lubach, J.W., Xu, D., Segmuller, B.E., and Munson, E.J. (**2007**) Investigation of the effects of pharmaceutical processing upon solid-state NMR relaxation times and implications to solid-state formulation stability. *Journal of Pharmaceutical Sciences*, **96** (4), 777–787;

(b) Sperger, D.M. and Munson, E.J. (**2011**) Analysis of structural variability in pharmaceutical excipients using solid-state NMR spectroscopy. *AAPS PharmSciTech*, **12** (3), 821–833;

(c) Sperger, D.M., Fu, S., Block, L.H., and Munson, E.J. (**2011**) Analysis of composition, molecular weight, and water content variations in sodium alginate using solid-state NMR spectroscopy. *Journal of Pharmaceutical Sciences*, **100** (8), 3441–3452.

16. Cassidy, A.M., Gardner, C.E., Auffret, T. *et al.* (**2012**) Decoupling the effects of surface chemistry and humidity on solid-state hydrolysis of aspirin in the presence of dicalcium phosphate dihydrate. *Journal of Pharmaceutical Sciences*, **101** (4), 1496–1507.

17. Hovorka, S. and Schoneich, C. (**2001**) Oxidative degradation of pharmaceuticals: theory, mechanisms and inhibition. *Journal of Pharmaceutical Sciences*, **90** (3), 253–269.

18. Tallon, M.A., Malawer, E.G., Machnicki, N.I. *et al.* (**2008**) The effect of crosslinker structure upon the rate of hydroperoxide formation in dried, crosslinked poly(vinylpyrrolidone). *Journal of Applied Polymer Science*, **107** (5), 2776–2785.

19. Herrle, K. and Hoffmann, E. (**1973**) *Pyrrolidone-2 manufacture of insoluble and only slightly swellable poly-*n*-vinyl-pyrrolidone-2*, US Patent No. 3,759,880. Sept. 18, 1973.

20. Kothari, S., Paruchuri, S., Rao, V., and Desai, D. (**2006**) *Peroxide Scavenging Property of Croscarmellose Sodium and Its Potential to Reduce* N*-oxidation of Piperazine Ring containing Compounds*, American Association of Pharmaceutical Scientists, San Antonio.

21. Wu, Y., Dali, M., Gupta, A., and Raghavan, K. (**2009**) Understanding drug–excipient compatibility: oxidation of compound A in a solid dosage form. *Pharmaceutical Development and Technology*, **14** (5), 556–564.

22. Kibbe, A.H. and American Pharmaceutical Association, A. (**2000**) *Handbook of Pharmaceutical Excipients*, American Pharmaceutical Association.

23. Huang, G.; Wu, Y.; Dali, M., Determination of formaldehyde and glucose in pharmaceutical excipients by HPLC with pre-column derivatization. In HPLC Conference, Poster presentation: San Francisco, CA, **2007**.

24. Dubost, D.C., Kaufman, M.J., Zimmerman, J.A. *et al.* (**1996**) Characterization of a solid state reaction product from a lyophilized formulation of a cyclic heptapeptide. A novel example of an excipient-induced oxidation. *Pharmaceutical Research*, **13** (12), 1811–1814.

25. Wirth, D.D., Baertschi, S.W., Johnson, R.A. *et al.* (**1998**) Maillard reaction of lactose and fluoxetine hydrochloride, a secondary amine. *Journal of Pharmaceutical Sciences*, **87** (1), 31–39.
26. Glastrup, J. (**1996**) Degradation of polyethylene glycol. A study of the reaction mechanism in a model molecule: tetraethylene glycol. *Polymer Degradation and Stability*, **52** (3), 217–222.
27. Nassar, M.N., Nesarikar, V.N., Lozano, R. *et al.* (**2004**) Influence of formaldehyde impurity in polysorbate 80 and PEG-300 on the stability of a parenteral formulation of BMS-204352: identification and control of the degradation product. *Pharmaceutical Development and Technology*, **9** (2), 189–195.
28. Wang, G., Fiske, J.D., Jennings, S.P. *et al.* (**2008**) Identification and control of a degradation product in Avapro film-coated tablet: low dose formulation. *Pharmaceutical Development and Technology*, **13** (5), 393–399.
29. Desai, D.S., Rubitski, B.A., Bergum, J.S., and Varia, S.A. (**1994**) Effects of different types of lactose and disintegrant on dissolution stability of hydrochlorothiazide capsule formulations. *International Journal of Pharmaceutics*, **110** (3), 257–265.
30. Hoydonckx, H.E., Van Rhijn, W.M., Van Rhijn, W. *et al.* (**2007**) *Furfural and derivatives*, Ullmann's Encyclopedia of Industrial Chemistry.
31. (a) Brownley, C.A. Jr., and Lachman, L. (**1963**) Preliminary report on the comparative stability of certified colorants with lactose in aqueous solution. *Journal of Pharmaceutical Sciences*, **52**, 86–93;

    (b) Brownley, C.A. Jr., and Lachman, L. (**1964**) Browning of spray-processed lactose. *Journal of Pharmaceutical Sciences*, **53**, 452–454.
32. Blaug, S.M. and Huang, W.T. (**1972**) Interaction of dextroamphetamine sulfate with spray-dried lactose. *Journal of Pharmaceutical Sciences*, **61** (11), 1770–1775.
33. Miller, D.M., Buettner, G.R., and Aust, S.D. (**1990**) Transition metals as catalysts of "autoxidation" reactions. *Free Radical Biology and Medicine*, **8** (1), 95–108.
34. Ohyashiki, T., Kadoya, A., and Kushida, K. (**2002**) The role of Fe^{3+} on Fe^{2+}-dependent lipid peroxidation in phospholipid liposomes. *Chemical & Pharmaceutical Bulletin*, **50** (2), 203–207.
35. del Barrio, M.A., Hu, J., Zhou, P., and Cauchon, N. (**2006**) Simultaneous determination of formic acid and formaldehyde in pharmaceutical excipients using headspace GC/MS. *Journal of Pharmaceutical and Biomedical Analysis*, **41** (3), 738–743.
36. (a) Hamburger, R., Azaz, E., and Donbrow, M. (**1974**) Autoxidation of polyoxyethylenic non-ionic surfactants and of polyethylene glycols. *Pharmaceutica Acta Helvetiae*, **50** (1–2), 10–17;

    (b) Sakharov, A.M., Mazaletskaya, L.I., and Skibida, I.P. (**2001**) Catalytic oxidative deformylation of polyethylene glycols with the participation of molecular oxygen. *Kinetics and Catalysis*, **42** (5), 662–668;

    (c) Waterman, K.C., Arikpo, W.B., Fergione, M.B. *et al.* (**2008**) N-methylation and N-formylation of a secondary amine drug (varenicline) in an osmotic tablet. *Journal of Pharmaceutical Sciences*, **97** (4), 1499–1507.
37. Narang, A.S., Rao, V.M., Farrell, T. *et al.* (**2010**) *Stability implications of prolonged storage of PVA and PEG-based coating suspension* AAPS; New Orleans, LA.
38. Castello, R.A. and Mattocks, A.M. (**1962**) Discoloration of tablets containing amines and lactose. *Journal of Pharmaceutical Sciences*, **51**, 106–108.

39. (a) Hodge, J.E. (**1955**) The Amadori rearrangement. *Advances in Carbohydrate Chemistry*, **10**, 169–205; (b) Paulsen, H. and Pfulghaupt, K.W. (**1980**) Glycosamines, in *The Carbohydrates, Chemistry and Biochemistry*, Academic Press, New York, pp. 881–927.
40. Janicki, C.A. and Almond, H.R. Jr., (**1974**) Reaction of haloperidol with 5-(hydroxymethyl)-2-furfuraldehyde, an impurity in anhydrous lactose. *Journal of Pharmaceutical Sciences*, **63** (1), 41–43.
41. Killion, R.B. and Stella, V.J. (**1990**) The nucleophilicity of dextrose, sucrose, sorbitol, and mannitol with *p*-nitrophenyl esters in aqueous solution. *International Journal of Pharmaceutics*, **66** (1–3), 149–155.
42. Badawy, S.I., Williams, R.C., and Gilbert, D.L. (**1999**) Effect of different acids on solid-state stability of an ester prodrug of a IIb/IIIa glycoprotein receptor antagonist. *Pharmaceutical Development and Technology*, **4** (3), 325–331.
43. (a) Matsunaga, Y., Ohnishi, Y., Bando, N. *et al.* (**1997**) Melting granulation by addition of polyethyleneglycol for stabilization of TAT-59. *Chemical and Pharmaceutical Bulletin*, **45** (7), 1103–1107;

    (b) Patel, N.K., Patel, I.J., Cutie, A.J. *et al.* (**1988**) The effect of selected direct compression excipients on the stability of aspirin as a model hydrolyzable drug. *Drug Development and Industrial Pharmacy*, **14** (1), 77–98;

    (c) Torres, A.I. and Camacho, M.A. (**1994**) Solid state interactions of two new antineoplastic drugs (mitonafide and amonafide) and common tablet excipients in preformulation studies. *European Journal of Pharmaceutics and Biopharmaceutics*, **40** (1), 41–43.
44. George, R.C., Barbuch, R.J., Huber, E.W., and Regg, B.T. (**1994**) Investigation into the yellowing on aging of Sabril® tablet cores. *Drug Development and Industrial Pharmacy*, **20** (19), 3023–3032.
45. Chen, J.G., Markovitz, D.A., Yang, A.Y. *et al.* (**2000**) Degradation of a fluoropyridinyl drug in capsule formulation: degradant identification, proposed degradation mechanism, and formulation optimization. *Pharmaceutical Development and Technology*, **5** (4), 561–570.
46. Hartauer, K.J., Arbuthnot, G.N., Baertschi, S.W. *et al.* (**2000**) Influence of peroxide impurities in povidone and crospovidone on the stability of raloxifene hydrochloride in tablets: identification and control of an oxidative degradation product. *Pharmaceutical Development and Technology*, **5** (3), 303–310.
47. Baertschi, S.W., Brunner, H., Bunnell, C.A. *et al.* (**2008**) Isolation, identification, and synthesis of two oxidative degradation products of olanzapine (LY170053) in solid oral formulations. *Journal of Pharmaceutical Sciences*, **97** (2), 883–892.
48. Freed, A.L., Strohmeyer, H.E., Mahjour, M. *et al.* (**2008**) pH control of nucleophilic/electrophilic oxidation. *International Journal of Pharmaceutics*, **357** (1–2), 180–188.
49. Waterman, K.C., Adami, R.C., Alsante, K.M. *et al.* (**2002**) Stabilization of pharmaceuticals to oxidative degradation. *Pharmaceutical Development and Technology*, **7** (1), 1–32.
50. (a) Al-Nimry, S.S., Assaf, S.M., Jalal, I.M., and Najib, N.M. (**1997**) Adsorption of ketotifen onto some pharmaceutical excipients. *International Journal of Pharmaceutics*, **149** (1), 115–121;

    (b) Chien, Y.W., Van Nostrand, P., Hurwitz, A.R., and Shami, E.G. (**1981**) Drug–disintegrant interactions: binding of oxymorphone derivatives. *Journal of Pharmaceutical Sciences*, **70** (6), 709–710.

51. (a) Cory, W., Field, K., and Wu-Linhares, D. (**2004**) Is it the method or the process-separating the causes of low recovery. *Drug Dev Ind Pharm*, **30** (8), 891–899;

(b) Hollenbeck, R.G., Mitrevej, K.T., and Fan, A.C. (**1983**) Estimation of the extent of drug–excipient interactions involving croscarmellose sodium. *Journal of Pharmaceutical Sciences*, **72** (3), 325–327.

52. Huang, W.X., Desai, M., Tang, Q. *et al.* (**2006**) Elimination of metformin–croscarmellose sodium interaction by competition. *International Journal of Pharmaceutics*, **311** (1–2), 33–39.

53. Jackson, K., Young, D., and Pant, S. (**2000**) Drug–excipient interactions and their affect on absorption. *Pharmaceutical Science & Technology Today*, **3** (10), 336–345.

54. (a) Pesonen, T., Kanerva, H., Hirvonen, J. *et al.* (**1995**) The incompatibilities between chlorhexidine diacetate and some tablet excipients. *Drug Development and Industrial Pharmacy*, **21** (6), 747–752;

(b) Rohrs, B.R., Thamann, T.J., Gao, P. *et al.* (**1999**) Tablet dissolution affected by a moisture mediated solid-state interaction between drug and disintegrant. *Pharmaceutical Research*, **16** (12), 1850–1856.

55. Senderoff, R.I., Mahjour, M., and Radebaugh, G.W. (**1982**) Characterization of adsorption behavior by solid dosage form excipients in formulation development. *International Journal of Pharmaceutics*, **83** (1), 65–72.

56. Lessen, T. and Zhao, D.C. (**1996**) Interactions between drug substances and excipients. 1. Fluorescence and HPLC studies of triazolophthalazine derivatives from hydralazine hydrochloride and starch. *Journal of Pharmaceutical Sciences*, **85** (3), 326–329.

57. Schildcrout, S.A., Risley, D.S., and Kleemann, R.L. (**1993**) Drug–excipient interactions of seproxetine maleate hemi-hydrate: isothermal stress methods. *Drug Development and Industrial Pharmacy*, **19** (10), 1113–1130.

58. Aly, S.A.S. and Megwa, S.A. (**1987**) Drug excipient interaction, effect of adsorption of oxytetracycline hydrochloride by some tablet excipients on the physiological availability of the tablets. *STP Pharma*, **8**, 652–657.

59. (a) Gore, A.Y. and Banker, G.S. (**1979**) Surface chemistry of colloidal silica and a possible application to stabilize aspirin in solid matrixes. *Journal of Pharmaceutical Sciences*, **68** (2), 197–202;

(b) Lerk, C.F. and Bolhuis, G.K. (**1977**) Interaction of lubricants and colloidal silica during mixing with excipients. II. Its effect on wettability and dissolution velocity. *Pharmaceutica Acta Helvetiae*, **52** (3), 39–44;

(c) Lerk, C.F., Bolhuis, G.K., and Smedema, S.S. (**1977**) Interaction of lubricants and colloidal silica during mixing with excipients. I. Its effect on tabletting. *Pharmaceutica Acta Helvetiae*, **52** (3), 33–39.

60. Daniels, R., Kerstiens, B., Tischinger-Wagner, H., and Rupprecht, H. (**1986**) The stability of drug adsorbates on silica. *Drug Development and Industrial Pharmacy*, **12** (11–13), 2127–2156.

61. Sherriff, M. and Enever, R.P. (**1979**) Rheological and drug release properties of oil gels containing colloidal silicon dioxide. *Journal of Pharmaceutical Sciences*, **68** (7), 842–845.

62. Johansen, H. and Moller, N. (**1978**) Solvent deposition method for enhancement of dissolution rate: importance of drug-to-excipient ratio. *Journal of Pharmaceutical Sciences*, **67** (1), 134–136.

63. Tischinger-Wagner, H., Endres, W., Rupprecht, H., and Weingart, A. (**1987**) Oxidative degradation of linoleic acid methyl ester in suspensions of inorganic excipients. 1. Auto-oxidation in the presence of silicic acid products and aluminum oxide. *Die Pharmazie*, **42** (5), 320–324.
64. Botha, S.A. and Lötter, A.P. (**1990**) Compatibility study between naproxen and tablet excipients using differential scanning calorimetry. *Drug Development and Industrial Pharmacy*, **16** (4), 673–683.
65. (a) Rowe, R.C. and Forse, S.F. (**1983**) Pitting – A defect on film-coated tablets. *International Journal of Pharmaceutics*, **17** (2), 347–349;(b) Brittain, H.G. (**1998**) *Analytical Profiles of Drug Substances and Excipients*, vol. **25**, Academic Press.
66. Bolhuis, G.K. and Hölzer, A.W. (**1996**) Lubricant Sensitivity, in *Pharmaceutical powder compaction technology* (eds G. Alderborn and C. Nystrom), Marcel Dekker, Inc., New York, NY, pp. 517–560.
67. Mura, P., Manderioli, A., Bramanti, G. *et al.* (**1995**) Utilization of differential scanning calorimetry as a screening technique to determine the compatibility of ketoprofen with excipients. *International Journal of Pharmaceutics*, **119** (1), 71–79.
68. Miller, T.A. and York, P. (**1988**) Pharmaceutical tablet lubrication. *International Journal of Pharmaceutics*, **41** (1), 1–19.
69. Gordon, R.E., VanKoevering, C.L., and Reits, D.J. (**1984**) Utilization of differential scanning calorimetry in the compatibility screening of ibuprofen with the stearate lubricants and construction of phase diagrams. *International Journal of Pharmaceutics*, **21** (1), 99–105.
70. Pawełczyk, E. and Opielewicz, M. (**1977**) Drug decomposition kinetics. XLIX. Kinetics of auto-oxidation of drotaverine hydrochloride in the solid phase. *Acta Poloniae Pharmaceutica*, **35** (3), 311–319.
71. Osawa, Z. and Ishizuka, T. (**1973**) Catalytic action of metal salts in autoxidation and polymerization. X. The effect of various metal stearates on the thermal oxidation of 2,6,10,14-tetramethylpentadecane. *Journal of Applied Polymer Science*, **17** (9), 2897–2907.
72. Thakur, A.B., Morris, K., Grosso, J.A. *et al.* (**1993**) Mechanism and kinetics of metal ion-mediated degradation of fosinopril sodium. *Pharmaceutical Research*, **10** (6), 800–809.
73. Pragatikumar, B., Sahu, R., Murphy, K.V.R. *et al.* (**2011**) A review on mechanism, importance and methods of compatibility testing in the formulation of dosage forms. *J. Chem. Pharm. Sci.*, **4**, 141–151.
74. Hovorka, S.W., Hageman, M.J., and Schoneich, C. (**2002**) Oxidative degradation of a sulfonamide-containing 5,6-dihydro-4-hydroxy-2-pyrone in aqueous/organic cosolvent mixtures. *Pharmaceutical Research*, **19** (4), 538–545.

2

EXCIPIENTS FOR CONVENTIONAL ORAL SOLID DOSAGE FORMS

UMESH KESTUR AND DIVYAKANT DESAI

Drug Product Science and Technology, Bristol-Myers Squibb Company, New Brunswick, NJ, USA

2.1 INTRODUCTION

Excipients, which are an integral part of any formulation, can significantly impact stability, processability, and performance of a dosage form. Excipients are derived from natural, synthetic, or semisynthetic sources, and thus, any natural variability in raw materials and excipient manufacturing processes result not only in lot-to-lot variability but also variability between different vendors of the same excipient. In addition, an excipient can have multiple functions depending on the formulation and manufacturing process for a dosage form. Therefore, any inconsistency in the properties of an excipient can be reflected in the quality of a final dosage form. Ensuring consistency in an excipient involves having differentiating tests and procedures to verify their identity, purity, and quality. This has been achieved through specifications and tests in compendial monographs. However, it is well recognized that compliance to a monograph specifications alone does not guarantee that an excipient will perform its intended function. This is because compendial monographs rarely address excipient functionality even though they deal with purity, identification, and safety of excipients. Moreover, it may be a difficult task to address excipient functionality through compendial monographs since an excipient can be used in various dosage forms for various functions. Therefore, in addition to the standard tests in monographs, there has been increased interest in developing tests for excipients to confirm

Pharmaceutical Excipients: Properties, Functionality, and Applications in Research and Industry, First Edition. Edited by Otilia M.Y. Koo.

functional equivalence between lot-to-lot and various sources. However, developing functional tests for excipients requires understanding of their physical and chemical properties and how it correlates to their performance. Therefore, this chapter highlights the excipient properties that impact the performance of an excipient in addition to providing general guidelines on their usage with greater emphasis on physical properties of an excipient. The focus will be only on excipients used in traditional oral solid dosage forms while excipients for enabling formulations are highlighted in other chapters. Greater emphasis in this chapter is placed on major common components in oral solid dosage forms, namely, diluents, binders, disintegrants, and lubricants.

2.2 DILUENTS/FILLERS

Diluents or fillers are materials that are added to bulk up a dosage form when the active ingredient is not present in sufficient quantity to make a compressible tablet or fill a capsule shell. In the case of direct compression, use of the term "filler-binders" is more appropriate because they not only bulk up a dosage form but also provide binding property by imparting strength to a dosage form. In addition, diluents play an important role in imparting manufacturability by impacting properties such as powder flow, compactability, compressibility, and homogeneity by locking active pharmaceutical ingredient (API) in granules. They also impact quality and performance of the dosage form with respect to tablet friability, content uniformity, dissolution, and stability among other properties. For any oral solid dosage form, a single diluent or a combination of diluents may be used. A mixture of diluents is generally used to circumvent a shortcoming of a single diluent. For example, if lactose is the only diluent used in a formulation, the blend will have a sharp granulation endpoint. By adding some microcrystalline cellulose (MCC), the granulation endpoint is more forgiving due to its high water absorption capacity. Mixture of diluents may also be used for economical reasons where an expensive diluent may be partially replaced by an inexpensive diluent without impacting the performance of a dosage form. In general, the nature of materials used as diluents generally falls in the broad category of cellulosic materials, sugars, or inorganic salts. Some of the more commonly used diluents are highlighted as follows.

2.2.1 Types of Diluents

2.2.1.1 Microcrystalline Cellulose MCC is a purified, partially depolymerized cellulose derived from α-cellulose. MCC is a versatile excipient with application as a diluent, binder, and disintegrant in oral solid dosage forms [1–4]. It is primarily used as a diluent/binder in direct compression and wet granulation process [2, 3, 5]. As a diluent, it is used in tablet formulations in the range of 20–90% w/w. MCC undergoes plastic deformation and shows good compactability even at low compression pressures [6, 7]. However, it has poor flow characteristics. It is available in different particle sizes and densities, with larger particle size and higher bulk density helping in the flow characteristics of blends but at the cost of reduced compactability. In direct compression process, it is often used with other excipients due to its comparatively high cost. In dry granulation, MCC is often combined with a diluent such as lactose with brittle compaction property to complement the plastic deformation behavior of MCC. Wet granulation has been known to reduce the compactability of MCC due to change in structure and loss of bonding surfaces [8–10]. Being hygroscopic in nature, it is important to control the moisture content in MCC especially for moisture-sensitive drug substances. From a manufacturability perspective, having optimum moisture is helpful since low and high moisture may compromise compactability [11]. Due to the ability of MCC to undergo plastic deformation, it is sensitive to magnesium stearate, with finer size fraction being more sensitive to lubricant and mixing effects [12]. Blending colloidal silica with MCC prior to lubrication has been known to reduce magnesium stearate sensitivity of MCC due to preferential binding of colloidal silica to magnesium stearate [13, 14]. In addition, being a plastically deforming material, tabletability of MCC is also adversely impacted with increasing tablet press speeds due to time-dependent nature of plastic flow [15]. A popular method for overcoming some of the undesirable properties of MCC discussed here is through coprocessing with other excipients. Coprocessing is briefly discussed at the end of this chapter. A recent more comprehensive review on MCC has been published by Thoorens et al. [16].

2.2.1.2 Dibasic Calcium Phosphate Anhydrous and dihydrated forms of dibasic calcium phosphate (DCP) are used as fillers for oral solid dosage forms. It is more commonly used as a source of calcium in nutraceuticals than in the pharmaceutical industry. The popularity of DCP in the pharmaceutical industry is due to its excellent flow and compaction properties. Comparison between the two forms of DCP shows that anhydrous form has better intraparticular porosity, mean yield pressure, and better compressibility than the dihydrate form [17]. In addition, due to higher intraparticular porosity, disintegration of anhydrous DCP is better than the dihydrate form. However, both forms of DCP do not generate good disintegration force and needs a swelling-type disintegrant in the formulation when used [18]. Different grades of DCP are available, with coarse grade used for direct compression and milled grade for roller compaction or wet granulation. The milled grade has an alkaline pH and cannot be used with API incompatible with high pH. Being an inorganic salt, DCP can be abrasive on the tablet tooling. However, compared to MCC, it is less sensitive to magnesium stearate levels.

2.2.1.3 Lactose Lactose is one of oldest and most widely used diluents in an oral solid dosage form. It is commonly used as a diluent and is known to exist in four different forms: α-lactose monohydrate, anhydrous α-lactose, anhydrous β-lactose, and amorphous lactose [19, 20]. Even though pure forms are present, some of the commercially available lactose may be a mixture of multiple forms. α-Lactose monohydrate is prepared by crystallization of a supersaturated solution of lactose below 93.5 °C. α-Lactose monohydrate is primarily used in wet granulation process. Even though α-lactose monohydrate contains 5% water of crystallization, the water is unavailable for reaction even if used with moisture-sensitive materials. Lactose reacts with primary amines to undergo browning or Maillard-type reaction. Agglomerated or granulated form of α-lactose monohydrate has been used to prepare direct compression grades of α-lactose monohydrate. Lactoses, in general, are brittle materials and has greater fragmentation propensity than that of MCC.

Anhydrous lactose is a directly compressible excipient made by roller drying of lactose solution followed by milling and sieving [21]. They exist in two isomeric forms: α and β, with the commercially available anhydrous lactose high in β content. The β form, in general, is more soluble, while the α shows poor disintegration properties [19]. Since the material has no water of hydration, they can be used to formulate with moisture-sensitive materials. However, on exposure to high humidity, they can pick up moisture and convert into the monohydrate. Once compacted, it does not lose its compactability as much compared to other types of lactose. Being a reducing sugar, it can undergo reaction with primary and secondary amines.

Spray-dried lactose is prepared by spray-drying a slurry of α-lactose monohydrate crystals in lactose solution. Spray-dried lactose consists of a mixture of α-lactose monohydrate and amorphous lactose with α-lactose monohydrate being the primary component. Crystalline component of the spray-dried lactose along with the spherical nature of the particles impart good flow to a formulation. Amorphous part of spray-dried lactose helps with compression. Even though the material is relatively nonhygroscopic, at higher relative humidity (more than 50% RH) the material can pick up moisture, resulting in crystallization of the amorphous part, which can negatively impact the compaction properties. Reworking potential for spray-dried lactose is poor with the material losing its compactability once it undergoes a process such as roller compaction. Similar to other lactoses, it will interact with amines to undergo the Maillard reaction.

2.2.1.4 Mannitol It is a widely used diluent in both pharmaceutical and food industries. It is nonhygroscopic and a good choice of diluent for moisture-sensitive drugs. Mannitol also gives up moisture when dried after wet granulation. Due to its negative heat of solution, it provides a cooling sensation and as a result of its good mouth feel, it is a preferred diluent in chewable tablets. It is present in different polymorphic forms and they have different compression characteristics [22]. Crystalline grades are generally used for wet granulation with the spray-dried grades being used for direct compression. Without spray drying, mannitol has poor flow and binding properties to be used as direct compression excipient. Higher lubricant levels are needed for mannitol-containing granulations than granulations made with other diluents.

2.2.2 Diluent/Filler Performance

Physicochemical compatibility between the diluents and the active ingredient will be an overriding factor for the choice of a diluent. However, there are many properties of a diluent that can directly impact the performance of an oral solid dosage form. Based on the USP/NF, there are several functionality-related characteristics that have been deemed to be important including crystallinity, polymorphic form,

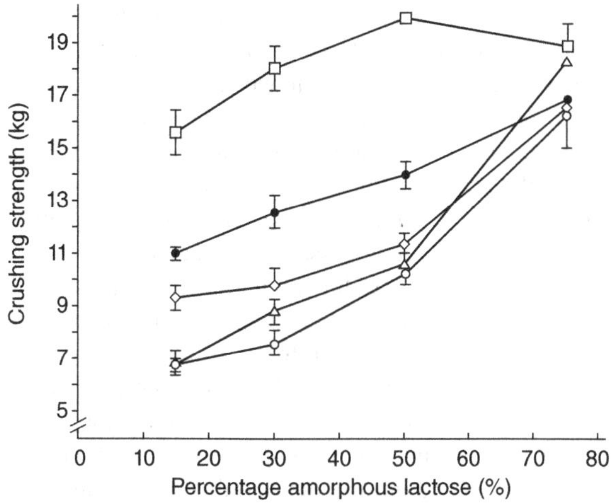

Figure 2.1 Crushing strength of tablets of spray-dried lactose samples with varying particle size, compressed at a compaction pressure of 75 MPa and containing 6% water content. From top to bottom (□) 1–8 μm, (•) 8–16 μm, (◇) 16–24 μm, (Δ) 24–32 μm and (○) 32–45 μm. Reproduced from Ref. [26] with permission from Elsevier.

particle size and size distribution, moisture content, particle shape, density (true bulk and tap), specific surface area, degree of polymerization (DP), flow properties, solubility, and compaction properties among others. Some of these properties are interdependent and impact both manufacturability and performance of a dosage form. For example, changes in crystallinity of a diluent can impact properties such as moisture absorption, compactability, disintegration, dissolution, and potential stability for a dosage form [23–25]. In the case of lactose, Vromans et al. showed that increasing the amorphous content of lactose prepared by spray drying increased both the crushing strength (Figure 2.1) and disintegration time of tablets without disintegrants [20, 26]. In the presence of a disintegrant, for tablets containing greater than 50% amorphous lactose, the disintegration time was dependent on the type of disintegrant and in the order crospovidone > sodium starch glycolate > croscarmellose sodium. Changes in crushing strength was attributed to behavioral difference of amorphous and crystalline content under compression pressure, with amorphous material being more plastic in nature and the crystalline material being more brittle. Suzuki and Nakagami studied the effect of MCC crystallinity on compactability and dissolution of acetaminophen tablets [25]. The authors reported that the compression energy was lowered with decrease in crystallinity. However, the impact on dissolution was determined by the level of crystallinity. A decrease in dissolution rate was reported with a decrease in degree of crystallinity from 65.5% to 37%. While for MCC samples containing less than 26% crystallinity, the dissolution rate was faster than the standard MCC with 65.5% degree of crystallinity. In addition to diluents showing differences in crystallinity, it is possible that a diluent can have different polymorphic forms. Therefore, characterization of the polymorphic form and proportion of each form in a diluent is important. In the case of lactoses, anhydrous β form has higher initial water solubility, dissolution rate, and higher crushing strength than the anhydrous α form [19]. This becomes especially important when excipients from vendors are made of a mixture of forms. For example, Bolhuis and Zuurman studied two commercially available agglomerated lactoses (Tablettose from Meggle and Pharmatose® DCL 15 from DMV) and concluded that better compactability of Pharmatose was attributed to the material containing 15% β-lactose versus Tablettose containing only 3% [27].

Another aspect that may impact processability and quality of a final dosage form is particle size and size distribution of a diluent. Particle size and size distribution have been shown to impact several properties including moisture content, flow, blending, compactability, compressibility, tablet hardness, friability, tablet ejection, and dissolution [25, 26, 28–30]. However, it is to be noted that even if particle size and size distribution are different between different lots or vendors, whether they have an impact on product quality may be formulation dependent. In a study, Gamble et al. studied batch-to batch and intravendor variabilities of anhydrous lactose from three vendors (DMV-Fonterra, Kerry Bioscience, and Friesland Food Domo) and their subsequent impact on functionality [31]. Characterization of the material from the different vendors indicated that there were differences between the solid state properties such as specific surface area, particle size, and dispersive surface energy between vendors. Additional testing also showed that electrostatic charging (triboelectrification approach) and chloroform content were found to be different between the vendors. However, the study demonstrated that even though there were statistically significant differences in these properties, the differences did not translate to any significant impact on powder flow or compression/compaction characteristics. Another property that can impact functionality of a dosage form is particle shape. For mannitol, granulated powder that had a less transparent and more irregular shape was shown to have higher hardness for the same compaction pressure than the native crystal powders with a smoother surface [22]. In the case of MCC, tablet tensile strength showed an increase with an increase in the length to width ratio though other factors such as specific surface area and bulk density may be impacted with change in morphology [32]. Compactability is also impacted by bulk density with lower bulk density improving compactability [33, 34].

Differences in raw materials used and changes to the manufacturing process also play an important role in determining the functionality of an excipient [30, 35, 36]. In the case of MCC, lignin content that depends on the source of wood and also the manufacturing process was shown to impact the dissolution efficiency of tablets [36]. In the case of soluble starch, introduction of an acetone solvent wash step improved the tensile strength and tabletability of compacts compared to untreated starch [37]. This was attributed to change in surface energetic properties (measured by inverse gas chromatography) of the excipient as a result of washing with a solvent. Thus, it is important to assess the range of acceptable parameters when changes are made to the manufacturing site or process. In the case of polymeric diluents, the impact of DP (degree of polymerization) on performance of diluents is not clear. Shlieout et al. showed that for MCC, the types with higher DP had greater water absorption and compressibility than the types with lower DP [38]. However, in another study,

Dybowski showed that DP did not impact MCC characteristics while the differences were related to origin of raw materials and method of preparation [39]. In general, since diluents are used at a range of concentrations, the level at which the diluents are being used and properties and interactions of the other excipients in the formulation will determine how they impact product quality attributes.

2.3 BINDERS

Binders are materials that are added to impart cohesiveness to the drug and excipient combination during the formulation of an oral solid dosage form. Cohesiveness not only assists in the formation and flow of granules during the manufacturing process but also helps in maintaining the integrity of tablets upon compression from these granules. The nature of materials used as binders may be sugars such as sorbitol, sucrose, and glucose; or natural polymeric materials such as starch, pregelatinized starch, gelatin, acacia; or semi and synthetic polymers such as povidone (PVP), hydroxypropyl cellulose (HPC), methylcellulose (MC), ethylcellulose (EC), and hydroxypropyl methylcellulose (HPMC) among others. Several factors influence the choice and amount of a binder used in a formulation including physiochemical compatibility of the binder with the drug and other excipients, processability, cohesiveness, and the impact of the binder on functional properties of a dosage form including friability, disintegration, and dissolution. Among the range of polymers available today, synthetic polymers are preferred over natural sources to minimize lot-to-lot variability and also potential microbial contamination from natural materials. In this section, different types of binders, their usage, and performance-related properties are discussed.

2.3.1 Types of Binders

2.3.1.1 Natural Polymers *Starch*: Starch, a carbohydrate made up of linear amylose and branched amylopectin, had been a popular choice for a binder historically. Being insoluble in cold water, starch is prepared in the form of paste using hot water. Aqueous starch paste is used in the concentration range of 5–10% w/w. However, the use of starch as a binder has been less preferred in recent years. Use of starch as a binder can lead to softer tablets with higher friability. Moreover, high viscosity of starch paste not only makes the distribution of binder difficult but also leads to uneven distribution in granules in addition to time-consuming process of preparing starch paste itself. Due to these reasons, it has been largely replaced by pregelatinized starch, which can be added as a dry powder as well.

Pregelatinized starch: It is a starch that has been chemically or mechanically modified to partially or completely rupture all of the starch granules. In contrast to starch, fully pregelatinized starch is soluble in cold water. Pregelatinized starch is used in the concentration range of 5–10% w/w for wet granulation purposes and up to 20 wt% as a binder for direct compression. In addition to being used as a paste, partial pregelatinization of starch also results in a freely flowable powder form, which can be used as a directly compressible material [40, 41]. However, functional differences may be seen between partially and fully pregelatinized starch from different sources [42]. Symecko and Rhodes studied the effect of compaction force on the dissolution of acetaminophen using three fully gelatinized starches (National 1551, Lycatab PGS, and Pregeflo M) and a partially pregelatinized starch, Starch 1500. The study showed that National 1551 and Lycatab PGS showed a similar behavior. In the case of Starch 1500, the percent dissolved at 30 minutes was lower than that of National 1551 and Lycatab PGS. In contrast, Pregeflo M source not only showed a delay in dissolution performance, but these delays also increased with compaction force, which was not seen with other starch sources. Even though there can be differences in performances with different sources, compared to native starch, pregelatinized starch has a greater ability to prevent lamination in tablets because of their lower brittle fracture index [43]. Stearic acid is the preferred lubricant for pregelatinized starch because magnesium stearate at concentrations of 0.25% w/w can soften the tablets made with pregelatinized starch and potentially impact tablet strength and dissolution [44]. Pregelatinized starch undergoes plastic deformation during tableting and is therefore more sensitive to magnesium stearate concentration.

2.3.1.2 Synthetic Polymers *Povidone:* Polyvinylpyrrolidone (PVP), a polymer of 1-vinyl pyrrolidone, is one of the most widely used binders. PVP is available in a range of average molecular weights, which gives them different viscosities in solution. Though different molecular weights are available, low to medium viscosity grades are used more often than high viscosity grades because of the potential for dissolution slowdown with high viscosity grades and also due to difficulty in handling high-viscosity solutions. As a binder, it is used in the concentration range of 0.5–5% w/w in a formulation. A binder solution of PVP may be made using either water or a hydroalcoholic solution. If the PVP concentration in water is high (e.g., 50% w/w in water), the binder solution is heated to 50–60 °C and heated jacketed vessel and tubing systems are used to ensure delivery of the viscous binder solution. PVP may also be added as a dry powder, with water added during granulation to activate the binder. However, higher amounts of dry powder PVP may be necessary to achieve the same level of binder functionality compared to PVP solution. One of the major drawbacks with the use of PVP is its hygroscopicity. Under high humidity storage conditions, PVP-containing formulations can take up water and eventually have an impact on tablet hardness, disintegration, and dissolution [45, 46]. In addition, solubilization effect of PVP can negatively impact the stability of dosage forms as reported for hydrolytic degradation of hydrochlorothiazide (HCTZ) tablets [47].

Methylcellulose (MC): MC is substituted cellulose where 27–32% of the hydroxyl groups are substituted by methyl ether. It is available in a variety of molecular weights resulting in solutions of different viscosities. As a binder, low and medium viscosity grades are used in the concentration range of 1–5% w/w. They may be used in the form of dry powder or as a binder solution. MC is practically insoluble in hot water and forms viscous colloidal dispersion in cold water. Therefore, to prepare a solution, MC is first mixed with a portion of water at 70–80 °C. To get a clear solution, cold water is added to the slurry to bring the temperature below 20 °C. One of the advantages of using MC is that it does not cause hardening of tablets on storage.

Hydroxypropyl cellulose (HPC): HPC is hydroxypropylated substituted ether of cellulose, which is used as a binder typically in the concentration range of 2–6% w/w. HPC is used in wet and dry granulation and in direct compression. Particle size of HPC can influence whether the binder is added in wet or dry form. For example, in the case of acetaminophen wet granulated tablets, fine particle grade of HPC (Klucel-EXF) added in the dry mix produced similar hardness versus compaction force profile compared to Klucel-EF distributed as solution [48]. Aqueous solutions of HPC can undergo acid hydrolysis at low pH or base-catalyzed oxidation at high pH with best stability in the range of pH 6–8 [49]. Improved binding properties are observed with higher hydroxypropyl group substitution.

For HPC, cloud point test has been reported to be a functional test to distinguish performance of the binder from different vendors [50]. Desai et al. studied the impact of two different sources of HPC (Klucel-EF and HPC-L) on dissolution performance of HCTZ tablets [50]. The authors reported that tablets made with Klucel-EF exhibited faster dissolution compared to HPC-L as binder. The differences were attributed to cloud point differences for Klucel-EF and HPC-L in water (Table 2.1). Klucel-EF has a cloud point temperature closer to dissolution medium temperature resulting in less viscous layer of binder on HCTZ tablets.

Ethylcellulose (EC): EC is an ethyl ether of cellulose. Similar to other polymeric materials, it is available in different viscosity grades. Low viscosity grades are used as binder while EC has other applications as a hydrophobic coating material and modified release polymer among others [51, 52]. As a tablet binder, it is used in the dry form or distributed from a solvent in the concentration range of 2–15% w/w. Due to its very low solubility in water, alcoholic solutions are used to disperse EC. Due to its low chemical reactivity in water, it is useful to formulate EC as an excipient for moisture-sensitive drugs. However, its hydrophobic nature can cause delayed

TABLE 2.1 Properties of Various Klucel-EF and HPC-L Lots Sourced from Hercules and Nippon Soda

HPC Type (Source)	Lot Number	Hydroxy–Propoxy Group (%)	Molecular Substitution (MS)	Cloud Point in Water (°C)
		Certificate of Analysis		
Klucel-EF	1108	74.4	3.8	39
(Hercules)	8508	73.4	3.7	39
	9150	72.8	3.6	39
	9878	74.4	3.8	39
	8870	71.9	3.5	39
	9945	74.4	3.8	39
	8210	73.9	3.7	40
HPC-L	NBC-0121	64.3	2.8	48
(Nippon Soda)	NJL-1621	66.6	3.0	48

Adapted from Ref. [50] with permission from Elsevier.

penetration of water into the tablets with a potential to slow down dissolution. This slow dissolution could be an issue for an immediate release tablet but desirable for a sustained release formulation. EC is available in a micronized form. Fine particle grade has been shown to have better binding capacity both as a dry binder and as an aqueous binder compared to coarse EC [53–55]. Desai et al. studied the performance of fine particle ethylcellulose (FPEC) as a dry binder to that of coarse particle EC on the mechanical properties and performance of acetaminophen tablets. Compared to tablets containing the coarse grade EC that showed significant friability, tablets containing 10–15% w/w FPEC produced harder tablets with low friability. Increasing the concentration of FPEC beyond 15% w/w resulted in flow issues during compaction in addition to tablets with extensive capping. The authors hypothesized that 10–15% w/w of FPEC is effective in improving the mechanical properties of acetaminophen tablets compared to coarse grade of EC. Capping seen with higher concentrations of FPEC was a result of release of entrapped air during compression process. Aqueous dispersions of EC have been suggested as an alternative to improve flow issues with the fine particle grade [54].

Hydroxypropyl methylcellulose (HPMC): HPMC or hypromellose is cellulose hydroxypropyl methyl ether. It is available in different grades that differ in viscosity and substitution. As a binder, it is used in the concentration range of 2–5% w/w in wet or dry granulation. Water or hydroalcoholic solvents are used to distribute or activate the binder. The binder efficacy is similar to MC. At higher concentrations, HPMC is routinely used as a drug release controlling polymer for modified release dosage forms.

Polyethylene glycol (PEG): The application of polyethylene glycol (PEG) as a binder is limited compared to its use for preparation of solid dispersions, hot melt granulation, and also as plasticizer in other oral solid dosage forms. They are available in different molecular weight grades with the lower-molecular-weight grades being liquids and higher molecular weights being semisolids and solids. Olsson et al. studied the effect of different molecular weights of PEG on tensile strength of sodium chloride and sodium bicarbonate tablets [56]. The addition of PEG increased the tensile strength of sodium chloride and sodium bicarbonate tablets. However, for sodium bicarbonate tablets, increasing the molecular weight of PEG improved the tensile strength while in the case of sodium chloride lower molecular weight improved the tensile strength more than did the higher molecular weight. These results were attributed to volume reduction mechanism of sodium chloride in the presence of PEG and the impact on the tensile strength of PEG on sodium bicarbonate, respectively. PEG imparts plasticity to granules, but its use at high concentrations can adversely impact tablet disintegration.

2.3.1.3 Sugars Sugars such as glucose, sucrose, and sorbitol may be used as binders. Sucrose syrup is used at the concentration range of 50–67% w/w for wet granulation purposes. Sucrose at 2–20% w/w is used for dry granulation. Water or hydroalcoholic solvent is used as the granulating solvent. They are also used in combination with other binders such as starch. Liquid glucose at the concentration range of 5–10% w/w is used during wet granulation. In general, sugar binders produce hard and brittle tablets especially at higher concentrations of the binder. Over time, there can be increase in hardness leading to slower disintegration. Reducing sugars have a tendency to react with amines [57]. Due to its pleasant taste, these binders are used in chewable tablets and for bitter-tasting drugs.

2.3.2 Binder Performance

USP/NF identifies several functionality-related characteristics of a binder that influence the granulation process, granule properties, and eventually manufacturability and performance of a dosage form and are summarized in Table 2.2. There are several other factors related to processing and other components used in the formulation that influence the effectiveness of a binder including mode of delivery, solvent used, binder concentration, solubility of excipients, mechanical properties of the drug and other excipients, and moisture among others. Binders are added by dispersing a solution or suspension containing a binder or in the dry form where the binder is activated by the addition of a solvent into the dry mix. In the commonly used wet massing method, the binder solution is delivered either through the use of tubes or is sprayed from nozzles. Delivery of binder through nozzles results in distribution of binder over a larger surface area. This is especially important for a high-solubility drug where delivery through a dripping method instead of a nozzle can result in localized overwetting. Delivery through tubes may be the preferred route when handling high-viscous solutions. Heated jacketed vessels and tubings may be used to ensure smooth delivery of a binder especially when using a high-viscous solution. When a binder is used in the dry form, its activation and effectiveness depends on solubility and wettability of the excipients in the mixture. Typically, higher concentrations of a binder are necessary in the dry form to achieve a similar level of effectiveness compared to a binder distributed through a solution. Water, ethanol, or hydroalcoholic solvents are used for distribution or activation of a binder. However, differences in solvent can impact the properties of a tablet. For example, theophylline microcrystalline system showed differences in manufacturability and performance depending on the use of aqueous or hydroalcoholic solvent system [58]. Harder and stronger

TABLE 2.2 Functionality-Related Characteristics of a Binder That Could Impact the Manufacturing and Performance of an Oral Solid Dosage Form

Functionality-Related Characteristics of a Binder
Particle size and distribution
Molecular weight and distribution
Solubility
Viscosity
Surface tension and wettability

pellets were obtained with increasing water content, but ethanol-based granulations showed better compactability and faster dissolution. Similar results were obtained with PVP–lactose-based granulations [59]. Even though organic solvents can be used in wet granulation, their use has been limited because of the potential economical, regulatory, environmental, and safety implications. Another less common way of binder activation involves melt granulation. This method involves the use of molten liquid as a binder. Low melting materials such as waxes and PEGs are added to the system followed by heating and mixing simultaneously. As the material melts, they act as a granulating agent and mixing helps in the distribution of the binder. The melting point of the materials used is in the range of 50–100 °C. This method has a disadvantage that the numbers of binders that meet the temperature requirements are limited, and moreover heat labile drugs cannot be processed using this technique. A more detailed discussion on melt granulation can be found elsewhere [60].

In addition to wet massing methods, a binder can be distributed through spray drying or foam generation in wet granulation method, or via roller compaction in dry granulation method. Binders that are surface active can be foamed and added to a wet granulation process. It has been shown that the mode of binder distribution can impact the strength of granules and/or tablets differently [59, 61, 62]. In the example of acetaminophen–gelatin system, it was shown that for a given compaction pressure, binder delivered through spray-dried system resulted in harder tablets compared to other methods of binder distribution (wet massing or roller compaction). This result was attributed to greater plastic deformation of spray-dried granules resulting in greater binder–binder contact [61]. In contrast, for lactose–PVP system, wet massing method produced tablets of higher tensile strength compared to granules made by spray-drying granules [59]. A factor that can impact the choice of a binder distribution method is the intrinsic mechanical property of a drug [62]. Cantor et al. studied the impact of intrinsic drug mechanical properties and binder delivery method (conventional wet massing vs foam granulation) on mechanical properties of high drug loading compacts using HPC as the binder. Metformin, aspirin, and acetaminophen were chosen as the model drug compounds representing viscoelastic, brittle, and ductile materials, respectively, for wet granulation. Based on the strength of the compacts, foam granulation enhanced the plasticity of brittle drugs such as acetaminophen while wet massing method appeared to favor viscoelastic drugs such as metformin. Aspirin showed both plastic and brittle characteristics.

Binder concentration also impacts the effectiveness of a binder [63, 64]. Increasing the binder concentration generally improves the mechanical strength of tablets. There is increased plastic deformation and available bonding area with increasing binder concentration [65, 66]. However, the extent of change in mechanical strength and the concentration of binder needed vary with the type of binder used. In addition to improved mechanical strength, the tablet manufacturing conditions may change with increasing binder concentration. Joneja and others studied the impact of several binders including HPC, MCC, PVP, and starch on the performance of acetaminophen tablets prepared by wet granulation [67]. Increasing HPC concentration not only improved the strength of tablets but also improved the tablet manufacturing conditions. HPC showed lowest ejection force during tableting operation with the ejection force reducing with increasing binder concentration. Moreover, HPC-based tablets were less sensitive to increasing tableting speed. Higher binder concentrations can also help in reducing the dusting tendency of tablets during a coating operation for a given shape of a tablet. Even though binder concentration can have a positive impact on mechanical strength and tableting manufacturing conditions, it can negatively impact disintegration time and dissolution. Therefore, a balance between the binder concentration and optimum tablet performance is desirable. Another property that has a similar impact on tablet strength as that of binder concentration is the molecular weight of binder [56, 68, 69]. In general, increasing the molecular weight improves the strength of tablets. However, in general, low viscosity grades are preferred because of the potential to cause dissolution slowdown with higher viscosity grades.

Particle size of binder can impact the strength of tablets especially in a dry granulation process [70, 71]. This has been attributed to increasing number of contacts with decreasing particle size thus improving the strength. Polarity between a binder and its substrate also impacts the performance of a formulation [72, 73]. Horisawa et al. studied the impact of binder physiochemical properties on granule friability and strength of different hydrophilic and hydrophobic powders prepared by wet granulation method [72]. The authors reported that surface polarities as calculated from contact angle measurement indicated that binders and powders with similar polarities produced stronger granules. In another experiment, Rowe studied the effect of spreading coefficients of different binders on granule and tablet strength of paracetamol [73]. The study indicated that higher spreading coefficient resulted in stronger tablets. Tablet properties are also influenced by behavior of a binder under compaction pressure. Mattsson and Nystrom studied binder properties affecting the compactability of tablets made from sodium bicarbonate/binder mixtures. The authors found that binders with greater deformability produced strong tablets compared to binders with low deformability. Another factor that can influence the functionality of a binder is moisture. Healey et al. studied the impact of moisture on the tensile strength of binder films made of PVP, gelatin, acacia, methylhydroxyethyl cellulose, and starch (Figure 2.2) [74]. The authors showed gelatin films had the highest values of tensile strength and Young's modulus but was brittle in nature. PVP showed lowest strength and low Young's modulus when exposed to increasing amounts of moisture. In addition to impacting the tensile strength of binder, the effect of moisture uptake by binder can negatively impact dissolution. Fitzpatrick et al. studied the impact of accelerated stability conditions on the dissolution performance of their wet granulated model compound with PVP and HPC as binders [46]. At 40 °C/75% RH after 1 year of storage, PVP-containing tablets showed significant slowdown in dissolution, while HPC-containing tablet was not impacted. The slowdown was attributed to change in physical state of the PVP resulting in densification and reduction in porosity.

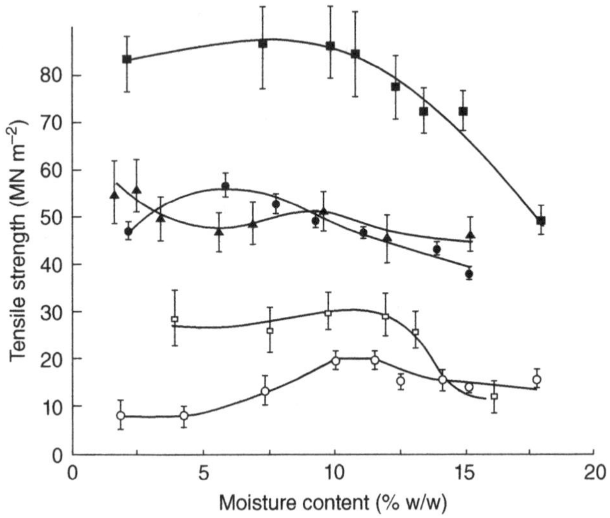

Figure 2.2 The effect of moisture content on the tensile strength of binder films. (■) Gelatin; (▲) methylhydroxyethylcellulose; (•) starch; (□) acacia; (○) PVP. The vertical error bar shows limits of error of the means at $P = 0.95$. Reproduced from Ref. [74] with permission from Wiley.

2.4 DISINTEGRANTS

Disintegrants are materials that are added to tablet and hard gelatin capsule formulations to aid in the breakup of a tablet or granules after oral administration. Disintegrant rapidly takes up water and breaks apart the formulation into smaller fragments upon contact with an aqueous environment thus making the drug available for dissolution over a larger surface area. The mechanism of disintegrant action may include wicking, swelling, or structure recovery. It is likely that the disintegrant exhibits more than one type of mechanism. A disintegrant may be added intragranularly prior to wet granulation or extragranularly prior to lubrication or included in both ways. In direct compression of tablet, disintegrants are added prior to lubrication to facilitate the breakup of the tablet formulation.

2.4.1 Types of Disintegrants

2.4.1.1 Starch Starch is one of the oldest disintegrants used in a solid dosage form. As a disintegrant, it is used at the concentration range of 3–25% w/w in a formulation. Typically, 10–15% w/w is needed to get good disintegrant action. At high concentrations, they can negatively impact the tablet hardness. In addition, starches used at high concentrations can cause issues of flow and compactability. Moreover, intragranular starch is not good as a disintegrant [75]. Therefore, they have been largely replaced by modified starches and other superdisintegrants that can take up water faster even at low concentrations. Pregelatinized starch at 5–10% w/w concentration is used as disintegrant. They mainly act through swelling mechanism and are more effective extragranularly.

2.4.1.2 ***Sodium Starch Glycolate (SSG)*** Sodium starch glycolate (SSG) is the sodium salt of carboxymethyl ether of starch or of a cross-linked carboxymethyl ether of starch. Carboxymethylation (degree of substitution) induces hydrophilicity by disrupting hydrogen bonding and allowing water access to the molecule. Cross-linking reduces the water-soluble fraction and gel formation that also reduces viscosity in water. The disintegrant efficiency has been inversely correlated to its levels of cross-linking [76]. Since carboxymethylation and cross-linking have opposing effects on water solubility, water access, and viscosity, a balance between the two is important to ensure optimal performance. The mechanism of its disintegrant action has been attributed to its high rate of water uptake and rapid swelling property [77]. In addition to degree of substitution and cross-linking, purity of SSG also affects its disintegrant efficiency [76, 78]. Sodium chloride, sodium glycolate, and sodium citrate or acetate are the by-products of the synthesis of SSG [49]. USP32-NF27 describes two types of SSG based on differences in pH, sodium and sodium chloride content. SSG is generally used at a concentration of 2–8% w/w in wet granulated and direct compression systems [79, 80]. Particle size, porosity, viscosity, and settling volume of SSG can vary depending on the vendor source [81]. The rate and extent of liquid uptake is pH dependent. Mechanical properties of SSG also can vary depending on the vendor source [82]. This is important only when large quantities are used in a formulation. Lower amounts may not have impact on the tablet hardness. Exposure to high humidity can impact dissolution; however, removal of moisture can increase hardness in some cases because of its binding action [83].

2.4.1.3 ***Crospovidone*** It is cross-linked homopolymer of 1-vinyl-2-pyrrolidone. As a disintegrant, it is used at a concentration of 2–5% w/w. It is water-insoluble material but has a high wicking tendency, causing it to take up water [84]. In addition, structure recovery can disrupt tablet integrity by generating more pores. It swells without gelling and this property is advantageous for developing orally disintegrating tablets and in cases where gelling can delay dissolution. Being nonionic in nature, the disintegrant action is independent of pH of the media and is a potential choice of disintegrant for cationic drugs [85]. Coarse particles of cross-linked PVP have better disintegrant efficiency [86]. In the case of insoluble fillers, crospovidone with higher porosity showed faster disintegration than disintegrant with lower porosity [87]. Disintegration force decreased with decrease in porosity and particle size. For tablets containing insoluble system, liquid uptake, settling volume, and disintegration force are critical for dissolution.

2.4.1.4 ***Croscarmellose Sodium (CCS)*** Croscarmellose sodium (CCS) is a cross-linked polymer of carboxymethylcellulose sodium. The carboxymethyl groups that substitute the hydroxyl groups determine the degree of substitution. Sodium chloride and sodium glycolate are by-products of the cross-linking reaction. It is used at a concentration of 0.5–5% w/w as a disintegrant both in wet granulated and direct compression systems [80, 88]. It is insoluble in water but hydrophilic and rapidly takes up water and swells. Carboxymethyl substitution increases the swelling ability of CCS, with basic substituents in CCS having a greater tendency to swell than acidic substituents [89]. The swelling ability of CCS is also adversely impacted at lower pH due to conversion of carboxymethyl sodium to its less hydrophilic free form. Particle size of CCS can impact disintegrant action with larger particles of CCS having a greater tendency to swell. Similarly, morphologies that favor greater moisture absorption or have more sites for moisture uptake enhances its wicking ability. Molecular weight can impact water uptake capacity, with higher molecular weight CCS having slightly higher water uptake capacity than lower molecular weight [90]. Source of cellulose (wood pulp vs cotton linters) also can result in differences in physical properties. CCS derived from wood pulp has lower molecular weight, increased water solubility, slightly decreased pH, decreased water capacity, and swelling rate compared to CCS from cotton linters [90].

2.4.2 Disintegrant Performance

Incorporation of disintegrant in wet granulation may be done intragranularly, extragranularly, or in a combination of both. However, there is no clear consensus on placement of a disintegrant to maximize its efficiency. For example, in the case of wet granulated tablets with lactose as filler, Van Kamp et al. reported that the mode of incorporation of SSG and crospovidone had no effect on disintegration time [91]. For paracetamol tablets prepared by fluid bed wet granulation, Khattab et al. showed that the order of efficiency of disintegrant (crospovidone, CCS, and SSG) incorporation followed the order combined mode > extragranular > intragranular [92]. Gordon et al. studied the mode of incorporation of disintegrants on tablets containing active pharmaceutical materials with three different solubilities. Irrespective of the solubility of the drug used, the order of effectiveness was extragranular > both > intragranular [80]. Therefore, it is clear that the method of placement of disintegrant alone cannot determine the efficiency of a disintegrant. One of the factors that can impact the efficiency of a disintegrant is particle size [76, 86, 93]. Rudnic et al. studied the effect of different particle sizes of crospovidone on direct compressed tablets of acetylsalicylic acid and found that increase in the mean particle size of crospovidone enhanced disintegration [86]. Similar result was reported for the efficiency of different particle sizes of starch as a disintegrant [93]. This was attributed to efficient formation of hydrophilic networks by larger size particles [94]. The efficiency of disintegrant can be impacted by processes such as dry and wet granulation [95]. Zhao and Augsburger studied the impact of granulation process on superdisintegrant performance on lactose–dicalcium phosphate formulation. The authors found that dry granulation of superdisintegrants (crospovidone, SSG, and CCS) resulted in increased disintegration time especially for crospovidone and SSG with lesser impact of CCS. However, a larger impact on the disintegration time was seen for all three disintegrants that were wet granulated. The decreased efficiency of disintegrants was attributed to lower and less uniform density of disintegrants in the granulated material. Moreover, disintegration efficiency can decrease if the granules break apart into larger particles, which can result in slower dissolution.

Several authors have studied the effect of compression force on the efficiency of a disintegrant [18, 83, 88, 96, 97]. Khan and Rooke studied the effect of compression force on dissolution efficiency of different disintegrants [18]. For tablets containing dicalcium phosphate dihydrate when crospovidone and calcium carboxymethylcellulose were used as disintegrants, disintegration time decreased with increasing compression force. In the case of casein formaldehyde system, disintegration time showed minima at 1000–2000 kg cm^{-2} compression force and then increased with increasing compression force. In the case of furosemide/microcrystalline tablets containing CCS as disintegrant, an increase in compression force increased disintegration time [88]. These results indicate that the effect of compression force depends on the mechanism of disintegrant action and their deformation behavior. This is because compression force impacts porosity with higher compression force leading to lower porosity. Very low porosity can reduce the water uptake capacity of tablets especially for disintegrants that has wicking as the main mechanism of action. In some cases, at low compression and high porosity even though water can be taken up quickly, high porosity can prevent generation of high disintegrating force necessary for disintegrant action.

The solubility of excipients used in the formulation can also impact the efficiency of a disintegrant. Disintegrants are more effective in the presence of insoluble excipients [98]. In the case of soluble excipients, tablets dissolve rather than disintegrate in the presence of water. Tablet excipients dissolving can lead to generation of higher viscosity of the liquid and also generation of more porous matrix thus reducing the efficiency of a swelling-type disintegrant [98, 99]. The ability of soluble excipients to compete for available water has also been speculated as a reason for lower disintegrant efficiency in the presence of soluble excipients [79]. The effect of concentration of disintegrant on disintegration time has been studied by several authors [88, 100–102]. Setty et al. studied the effect of functionality of crospovidone, CCS, and SSG on the development of fast dispersible aceclofenac tablets [100]. Increasing the concentration of CCS decreased the disintegration time while increasing the concentration of crospovidone had no effect on disintegration time. Increasing the concentration of SSG increased the disintegration time and was attributed to formation of viscous gel layer forming a barrier to penetration of water.

Another factor that can impact the efficiency of disintegrant is pH. For example, swelling capacity of disintegrants is impacted by pH (Figure 2.3) [80, 103, 104]. This is especially true for CCS and SSG than for crospovidone [103]. This has been attributed to lower liquid holding capacity of unionized form of cellulosic and modified starches. Greater impact can be seen with soluble filler than when insoluble fillers are used. Formulating tablets with extremely alkaline or acidic excipients could potentially lead to dissolution issues with these ionic excipients. Bindra et al. studied the impact of alkalinity of excipient (sodium carbonate) on tablets formulated with CCS [105]. There was a significant slowdown in dissolution upon storage, with the

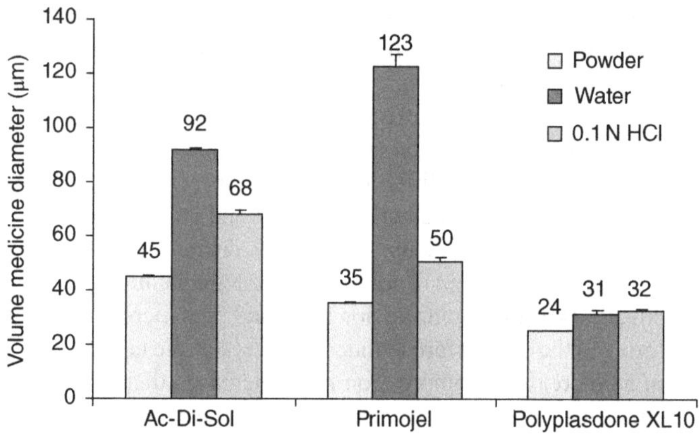

Figure 2.3 Volume median diameter of different disintegrants in different media. Taken from Ref. [103] with permission from Springer.

decrease being proportional to the increase in alkalinity. This was attributed to partial or complete hydrolysis of the ester cross-links in CCS, leading to by-products of increasing hydrogel characteristics thus leading to a viscous barrier on tablets to delay dissolution.

Lubricants can also impact the efficiency of a disintegrant. Coating of particles by a lubricant can interfere with wetting of tablets. The extent of impact of a lubricant also depends on swellability of the disintegrant. Slightly swelling disintegrants are more impacted than strongly swelling disintegrants [106, 107]. Rojas et al. studied the effect of different disintegrants on spironolactone tablets release. The results showed that spironolactone tablets containing crospovidone and CCS were impacted by magnesium stearate while tablets containing SSG were not impacted by magnesium stearate [108]. Another factor that can impact the efficiency of a disintegrant is moisture [18, 80, 83, 109]. Disintegrant by its natural tendency to absorb water can lead to swelling of tablets, resulting in softer and more friable tablets. In cases where disintegrants also have binding ability, there can be an increase in hardness after exposure to moisture [77]. Aging also reduces the dissolution efficiency due to loss of absorption and swelling ability from moisture uptake (Figure 2.4).

Disintegrants with a higher water affinity also tend to show a greater slowdown in disintegration. This impact is greater when tablets contain soluble excipients. Coating operation can also cause changes in tablets containing disintegrants [110]. Coating can cause swelling of tablets, surface erosion, and pitting. It has been postulated that problems during coating with tablets containing disintegrants are due to ineffective distribution of disintegrants [111]. There are functionality-related characteristics recommended in USP/NF for disintegrants and they include purity, LOD, pH, settling volume, and water-soluble components among others. Additional tests to understand the disintegrant efficiency include water uptake rate and capacity, molecular weight, and swelling index. Table 2.3 summarizes the functionality-related characteristics of disintegrants and their significance.

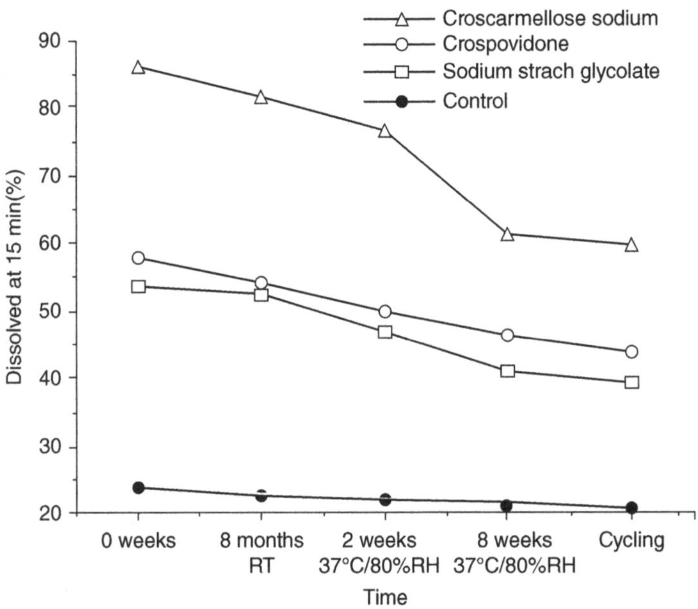

Figure 2.4 The impact of tablet aging on the effectiveness of disintegrants on tablet dissolution. Reproduced from Ref. [109] with permission from Elsevier.

TABLE 2.3 Functionality-Related Characteristics of Disintegrants and Their Potential Impact in an Oral Solid Dosage Form

Functionality-Related Characteristics	Significance
↑ Degree of cross-linking	↓ Water solubility
↑ Degree of substitution	↑ Hydrophilicity, ↑ water access
↑ Ionization	↑ Water holding capacity
Morphology	Longer morphology/more sites: ↑ water absorption capacity
Molecular weight	Impacts water uptake capacity
Particle size and distribution	↑Size and coarseness: potential for increased swelling

2.5 LUBRICANTS

Lubricants are the essential part of any tablet, capsule, or powder formulation. Their main function is to reduce the friction between the metal parts of machines such as tablet, capsule, or powder filler and the blend. Many theories proposed their mode of action [112]. The theory that has greater experimental evidence is that lubricants get themselves deposited on the metallic surfaces that come in direct contact with the powder blend during tableting or capsule filling operation and reduce the friction between the metallic parts and the blend. Type of materials used as lubricants include fatty acids, metallic salts of fatty acids, fatty acid esters, and inorganic materials [113]. Stearic acid, which is a fatty acid and magnesium stearate (metallic salt of fatty acid), are the most commonly used lubricants and are discussed in more detail.

2.5.1 Types of Lubricants

2.5.1.1 Stearic Acid Stearic acid is a straight-chain saturated monobasic acid with 18 carbon chain. It is manufactured from hydrolysis of animal fat or from hydrogenation of cottonseed or vegetable oil [49]. Commercially available stearic acid is present as a mixture of stearic acid with palmitic and myristic acid. Depending on the ratio of stearic to palmitic acid, it can vary from macrocrystalline to microcrystalline structure [114]. It is present in different polymorphic forms (forms A, B, and C), with form C being most stable [115].

2.5.1.2 Magnesium Stearate Most of the commercially available magnesium stearates are crystalline. This includes anhydrate, dihydrate, and trihydrate forms. Anhydrous form is known to convert hydrates in the presence of high amount of moisture. However, commercially available magnesium stearates are generally a mixture of forms. Amorphous magnesium stearate is also commercially available. The amorphous magnesium stearate absorbs high amount of moisture and shows different sorption characteristics [116]. Depending on the vendor source, magnesium stearates may differ in their morphology and also degree of crystallinity (Figure 2.5). Certain morphologies may be more prone to particle breakage during unit operations such as blending, resulting in increased lubricant coverage on granules. In addition to crystallinity differences, magnesium stearate can show batch-to-batch and vendor-to-vendor variability in particle size, surface area, bulk strength, and fatty

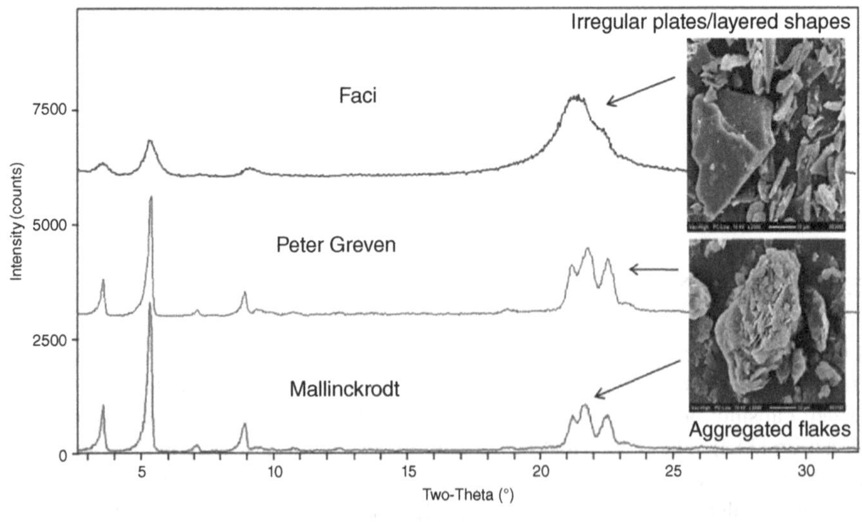

Figure 2.5 PXRD and SEM images of magnesium stearate from three different vendors.

TABLE 2.4 Functionality-Related Characteristics of Magnesium Stearate That Could Impact the Manufacturing and Performance of an Oral Solid Dosage Form

Functionality-Related Characteristics of MgSt
Particle size and distribution
Hydration state
Moisture content
Specific surface area
Crystallinity
Crystal form
Morphology
Composition
Bulk yield strength

acid composition [117–121]. These differences in physical properties may result in differences in lubrication efficiency and also have different compression profiles leading to difference in hardness and ultimately to tablet friability. Lower strength of tablets and higher friability issues may be minimized by optimizing the moisture content and increasing the binder concentration to improve the tablet hardness [119]. Magnesium stearate manufacturing methods are proprietary information; however, some information has been published [122]. Functionality-related characteristics of magnesium stearate are summarized in Table 2.4.

2.5.2 Lubricant Performance

Compared to other excipients in tablet or capsule formulations, lubricants are used at low levels. For example, magnesium stearate is generally used in the range of 0.1–1% w/w and stearic acid because of lower surface area used at 1–3% w/w range. They are screened through a 30-mesh screen to break up any lumps. Many times they are mixed with a small amount of the blend before distributing to the entire blend. Pharmaceutical scientists have struggled whether to add lubricants on weight basis or surface area basis. There are some practical difficulties in using surface area measurement in manufacturing settings. Moreover, the surface area of lubricants is bound to change as they are further mixed with the rest of excipients in the tablet press or capsule press hopper and feed frame of the tablet press. Although lubricants are used mainly to lubricate blends, they can play an unintended role in tablets and capsule formulations. For example, magnesium stearate and stearic acid are almost insoluble in water, their coating of powder blend can cause decrease in dissolution due to "water-proofing" effects. The water-proofing effect is dependent on the solubility of the API. The maximum effect is seen with the API with comparatively low solubility [123].

The process parameters in the blending operation, more specifically blending time, speed, and scale, can exaggerate the deleterious effects of lubricants [124–126]. Overmixing can prevent bonding of powder blend during tablet compression or slug formation during capsulation, resulting in softer tablets or softer slugs. This

TABLE 2.5 Dissolution of 40 mg Potency Capsules Hand Filled with Granules Containing 1% w/w Magnesium Stearate at Various Time Points During Capsule Filling Showing Impact of Overmixing

Drug and Sampling Time of Granules	Mean % of Drug Dissolved (% RSD)					
	5 min	10 min	20 min	30 min	45 min	60 min
Hydrochlorothiazide						
Before capsule filling run	11.8 (34.2)	35.9 (29.1)	79.0 (14.5)	93.4 (3.5)	95.6 (3.5)	96.4 (2.2)
End of 30 min filling run	3.6 (22.5)	7.9 (13.3)	13.8 (12.4)	20.8 (20.4)	30.3 (31.6)	35.9 (28.9)
SQ32756						
Before capsule filling run	15.8 (12.5)	45.7 (12.9)	86.1 (15.9)	96.8 (0.2)	97.7 (0.4)	97.7 (0.5)
End of 30 min filling run	4.9 (20.1)	13.8 (16.6)	26.5 (18.6)	40.4 (21.9)	55.2 (18.3)	65.5 (13.4)
Aztreonam						
Before capsule filling run	46.0 (31.0)	94.5 (10.6)	103.7 (1.0)	103.9 (1.0)	104.2 (1.3)	104.4 (0.8)
End of 30 min filling run	17.5 (6.7)	31.0 (3.7)	58.2 (14.2)	73.8 (15.2)	92.9 (12.0)	103.2 (3.5)

Order of API solubility is aztreonam > SQ32756 > hydrochlorothiazide.
Reproduced from Ref. [123] with permission from Elsevier.

effect is more prominent with materials that deform rather than materials that undergo brittle fracture. Materials that are brittle will create newer surfaces due to fragmentation. These newer surfaces will bond better since they are not covered by lubricants. Materials such as colloidal silica have been used to reduce the sensitivity of formulations to magnesium stearate [112, 127, 128]. Overmixing can also lead to dissolution variability within a batch as the portion of the blend that is overlubricated with magnesium stearate tend to dissolve slowly compared to the blend that is not overmixed [123]. Moreover, sensitivity to overmixing is less for API with high water solubility or in the formulation containing high-swelling superdisintegrant. The overmixed blend has been shown to exhibit lower dissolution in capsules (Table 2.5) but dissolves rapidly upon tableting as the compression of the blend creates fresh surfaces that are not coated with magnesium stearate [123].

Lubricants, especially stearic acid, have also been reported to cause dissolution slowdown on storage for capsule formulations due to their low melting point [129]. To circumvent the overmixing of lubricants with the blend, many tablet press manufacturers are designing the tablet press in such a way that the lubricant can be sprayed directly on the tablet punches and/or die cavities, just prior to compression. In addition to blending-related issues, the concentration of magnesium stearate can also impact the quality of coated tablets. Due to its hydrophobic nature, magnesium stearate can reduce the adhesion of coating film to the tablet substrate resulting in logo bridging [130–133]. Magnesium stearate and stearic acid are generally not known to cause any chemical stability issue because of their poor solubility and lack of reactivity. Magnesium stearate, however, was implicated in magnesium ion-mediated degradation of fosinopril sodium [134]. In addition, magnesium stearate with some impurities such as magnesium oxide can create alkaline microenvironmental pH, causing degradation of drugs such as quinapril hydrochloride [135]. In spite of its chemical inertness, magnesium stearate was implicated in the phase transformation of a BMS proprietary drug substance leading to slowdown in tablet dissolution [136].

As described above, lubricants can adversely impact the dissolution of tablet and capsule formulations and tablet hardness. Therefore, their target amount and the range needs to be carefully determined. Under the quality by design (QbD) paradigm, the expectation is to assess the robustness of the formulation by demonstrating that critical quality attributes (CQA) of a product such as dissolution and tablet hardness are not impacted by vendor-to-vendor or lot-to-lot variability. These studies can be done at the pilot scale and if applicable verifying the key findings at the commercial scale. Scale-up process is tricky for formulations in general, but much so for lubricants since the blender size, blender rpm, and blender fill volume can have significant impact on properties of the blend and ultimately to that of the dosage form. Some mathematical models have been developed to provide guidance on how to adjust the blender fill volume and blender revolutions at various scales [137].

2.6 COATING-RELATED EXCIPIENTS

Coating is applied to tablets or capsules for various purposes. A nonfunctional coat may be applied for various reasons including for esthetic purposes, brand identification, and create a surface for printing among others. Coatings can also be functional in nature and applied to alter the product performance. This includes coatings applied to control the release profile of the drug, to prevent photolytic degradation, or to mask the taste [52, 138, 139]. Historically, sugar coating has been used to mask the taste of medicines and to improve patient compliance. However, in the last few years, sugar coating has been replaced by film coating due to its ease of application. Within the film coating, due to the safety hazards associated with the organic solvent coating and also advances in polymer chemistry, aqueous film coating has gained strong foothold in pharmaceutical industry. When coating is applied to any solid dosage form, two competing forces are in operation. One is cohesive forces between the molecules of the film-forming material and the other is adhesive forces, the forces between the surface of the solid dosage form and the molecules of the film-forming material [140]. Nature of these forces is primarily determined by the polymer used in the coating formulation. Other additives will contribute to modify the properties of the polymer. A typical coating formulation contains a polymer, plasticizer, adjuvant such as an opacifier, a detackifier, and a colorant.

2.6.1 Coating Polymers

Polymer material constitutes 40–60% of the coating formulation. Chemistry and molecular weight of the polymers used determine important film-related properties including water solubility, mechanical properties, viscosity of coating solution, and water vapor and oxygen permeability. If the film-forming polymer is water soluble, the coating is used for immediate release. Based on chemistry of the polymer, the materials most commonly used in coating systems are either vinyl-based polymers or cellulosic polymers. Vinyl polymers include polyvinyl alcohol (PVA), PVP, copolymer of PVA, and polyvinyl acetate. Cellulosic polymers commonly used include HPMC, HPC, and hydroxyethyl cellulose (HEC). Water-insoluble polymers such as ECare used for the controlled release coating.

HPMC is the most commonly used cellulosic polymer for film coating purposes. HPMC has a high glass transition temperature and needs a plasticizer to make the film more flexible. HPMC has high viscosity in water and therefore using high solids percentage (greater than 15% w/w) in water is not recommended. HPMC films, in general, have high tensile strength but poor adhesion properties [141–143]. They show greater moisture and oxygen permeability compared to PVA [144]. HPMC blend with other polymers such as HPC can improve the adhesion and moisture barrier properties. Additives such as lactose or polydextrose also help to reduce viscosity and improve adhesion at the cost of reduced strength. In contrast, PVA-based coatings are tacky. Therefore, instead of plasticizers, antitacking agents are added to the coating formulations. PVA-based materials have good adhesion but poor tensile strength compared to HPMC. PVA-based coating solutions are less viscous compared to HPMC coatings for the same solids percentage in water and thus PVA-based coatings allow for a higher solids percentage and shorter coating time. PVA also has good oxygen and moisture barrier properties. PVA has self-polishing property and hence at the end of coating if the tablets are allowed to rub against each other by slowly rotating the pan, it will result in tablets with shiny polished appearance.

Another important application of coating is to protect the drug from the acidic environment to release later in the alkaline environment of intestine. The type of coating used for this purpose is called enteric coating. The polymers used for enteric coating normally have carboxylic groups. Depending on the type of acid and number of carboxylic groups, different release profiles can be obtained. These acid and acid ester functional groups are unionized at low pH and hence insoluble. As they pass along the gastrointestinal tract and encounter higher pH, these groups become ionized, making the polymer soluble. Commonly used polymers for this purpose are cellulose based, polyvinyl derivatives, and polymethacrylates [145]. The cellulose-based polymers are cellulose acetate phthalate (CAP), cellulose acetate trimellitate (CAT), cellulose acetate succinate (CAS), hydroxypropylmethylcellulose phthalate (HPMCP), and hydroxypropylmethylcellulose acetate succinate (HPMCAS) and polyvinyl derivatives such as polyvinyl acetate phthalate (PVAP) [145]. Polymethacrylates are mixtures of cationic and anionic polymers of dimethylaminomethacrylates, methacrylic acid, and methacrylic acid esters in various ratios [146]. Depending on the type of polymer used, films of different solubility characteristics can be produced to provide pH-dependent or pH-independent drug release profiles. Many times different polymers are mixed to obtain desired release profiles.

2.6.2 Plasticizers

Most polymers used for film coating are brittle and require plasticizers or adjuvants to improve their handling and processing [147]. Plasticizers are present in films to decrease the glass transition temperature to make them softer and more flexible [147]. They also improve the strength and toughness of films. Commonly used plasticizers belong to one of the general chemical classes: (i) Acetate and phthalate esters such as glyceryl triacetate (triacetin), triethyl citrate, and diethyl phthalate; (ii) polyhydric alcohols such as glycerol, propylene glycol, and PEG and their different molecular weight; (iii) glycerides such as acetylated monoglycerides; and (iv) oils such as castor and mineral oil. Amount of plasticizer and effectiveness of plasticizer–polymer interactions determine its effectiveness. Their concentrations vary typically in the range of 10–25%. Affinity of the plasticizer to water is also important in determining the moisture permeability of polymer films. For example, triacetin is hydrophobic and therefore not only reduces the water vapor transmission rate but also reduces the amount of water absorbed by HPMC films, thus providing better protection against moisture compared to PEG [148].

2.6.3 Other Additives

Insoluble excipients are added in the coating formulation to prevent or reduce aggregation of solids during the coating process [149]. For example, talc is added as anti-adherent. Other insoluble additives added to reduce the tackiness of the film includes glycerol monostearate (GMS), magnesium stearate, silicon dioxide, and kaolin [149]. Titanium dioxide is a common opacifier used in the coating formulation. It can help in protecting light-sensitive molecules against photolytic degradation [150]. Surfactants have also been added to coating formulations. Surfactants may improve wettability, provide blend homogeneity, and improve the spreading of polymeric films on tablet substrate [140, 151]. Coating formulations containing high concentrations of insoluble solids or waxes also provide better protection against moisture.

2.7 COLORANTS

Colorants are included in dosage forms for many reasons such as to enhance esthetic appeal, differentiations with similar size dosage forms, create a brand image, and, to some extent, to improve compliance with pediatric patients. Their concentrations in formulations vary from 0.01% to 2.0% depending on the color intensity desired. Early colors in pharmaceutical industry were derived from the natural sources such as turmeric, caramel, and indigo. However, later on, they were replaced by synthetic dyes or lakes of dyes because of their high coloring power, low cost, and availability in different colors. Majority of the colorants are certified Food Drug and Cosmetic (FD&C) or Drug and Cosmetics (D&C) colorants [152]. They can be synthetic dyes that impart colors when they are dissolved [153]. They contain 80–93% pure colorant material and their tinctorial (coloring) strength directly proportional to the dye content [153]. From a chemistry perspective, the certified FD&C colors fall into five classes: azo, triphenylmethane, indigoid, xanthene, and pyrazolone [154]. The knowledge about their chemical structures can help to predict not only their physical properties but also their chemical interactions with other formulation components.

The colorants allowed in food may not be allowed in pharmaceuticals and therefore it is advisable to check their status based on Codes of Federal Regulations and other similar forum in Europe and Japan. Lakes are derived by precipitating colorants in the presence of inert carriers such as alumina and they contain 10–30% dyes. Lakes have become very popular in coating material because of the color reproducibility they can impart [152]. Unlike dyes, lakes are insoluble and therefore their particle size is very important for their coloring strength [153]. Since dyes can be delisted for safety reasons by regulatory authorities, multinational pharmaceuticals have been leaning toward using inorganic colorants such as red and yellow ferric oxides and the colors derived using their various proportions. They are accepted worldwide and easy to handle during the commercial-scale manufacturing. One important factor to consider for ferric oxides is their maximum daily allowable intake, which is 5 mg of elemental iron per day in the United States.

Colorants are added in tablet core or coating. Some of the colorants are difficult to remove from equipment after the manufacturing is completed. Such colorants should be added toward the end of the manufacturing cycle to minimize cleaning burden. Colorants are also added in capsule shells to provide differentiation, but some of them can interact with gelatin [155]. For example, all FD&C dyes reacted with a type A gelatin but not with a type B gelatin. The major impact of the reaction was on the delayed disintegration of the gelatin capsule shell. In liquid formulations, they are used in combination with suitable flavors to enhance patients' acceptance and compliance. In addition to enhancing the esthetic appeal of the dosage forms, they can also enhance the stability of the dosage forms. It was shown that incorporation of 0.2% w/w yellow iron oxide in tablet cores improved the light stability of sorivudine and nifedipine [139]. It was also shown that sorivudine tablets with 0.2% w/w yellow, red, or black iron oxide in tablet cores had better light stability than those coated with 11% w/w Opadry® white (Figure 2.6). Moreover, a combination of 0.05% w/w red and 0.04% w/w yellow iron oxides in tablet cores had better light stability than those core tablets containing either 0.2% yellow or red iron oxides alone. Synthetic iron oxides are strong absorbers of radiation wavelength below 400 nm. The photostabilization effects of iron oxides were attributed to this property. A combination of red and yellow iron oxides was found to be superior in photostabilization of modified release omapatrilat tablets during *in vitro* dissolution compared to FD&C Blue No. 2 dye (indigo carmine) and FD&C Blue No. 2 lake [156].

2.8 pH MODIFIERS

pH modifiers are excipients with basic or acidic property and are added into the formulation to create the desired pH in the vicinity of an active ingredient either to

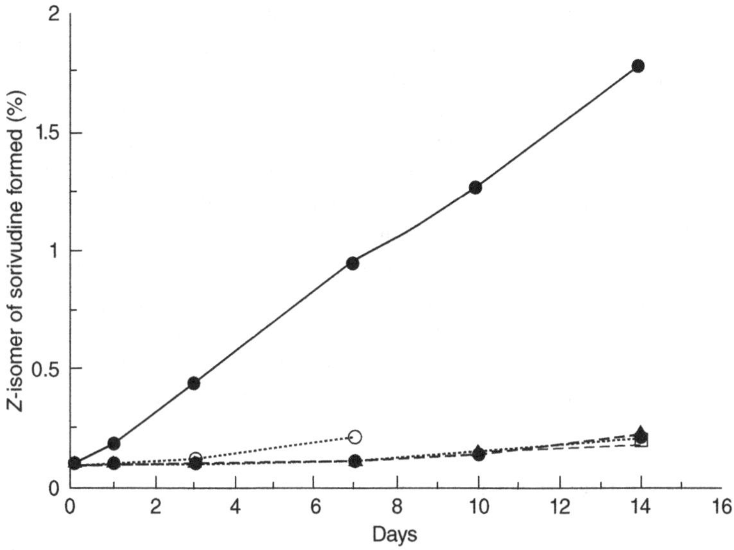

Figure 2.6 Increase in the concentration of *Z*-isomer in uncoated 10 mg potency sorivudine tablets containing no colorant (•); tablets without iron oxide but with 11% w/w coating of Opadry® white (o); 0.2% w/w yellow iron oxide (▲) or (□) 0.2% w/w red iron oxide after 14-day exposure to fluorescent room light (110-ft candle light). Reproduced from Ref. [139] with permission from Elsevier.

stabilize the molecule [157] or to influence its solubility and dissolution profile [158]. They are normally used in the range of 0.5–5% w/w. The commonly used excipients to increase acidity are citric acid, ascorbic acid, and tartaric acid. To increase the alkaline environment, sodium citrate, magnesium oxide, calcium carbonate, and sodium bicarbonate are commonly used excipients. The pH in the vicinity is also called microenvironmental pH, which is determined by suspending 10% or 20% w/w the formulation blend in water (slurry) and measuring the pH of the resulting suspension. Other sophisticated techniques are also available, but for all practical purposes the slurry method is adequate [157]. The uniform distribution of pH modifier is challenging given their low amount in the formulation. It was shown that the wet granulation was superior to dry granulated process to incorporate disodium citrate in improving the stability an ester prodrug [159]. Stability of levothyroxine tablets, a drug with very small doses, was significantly enhanced by using pH modifiers such as sodium carbonate, sodium bicarbonate, or magnesium oxide even though the formulation already had a basic diluent DCP [160]. The pH modifiers ascorbic acid and magnesium oxide were shown to be effective in stabilizing quinapril hydrochloride when used judiciously with different pH grades of silicates (Neusilin) [161]. The challenge of distributing a pH modifier was circumvented by using citric acid pellets as starting cores and then applying functional coats to control the release of a weakly basic drug. Thus, by maintaining acidic pH in cores, solubility of the weakly basic drug was maintained, which provided more controlled release of the drug [162]. Retardation of diffusion of weakly basic drug noscapine in alkaline pH medium was avoided from the coated tablets by the addition of organic acids such as succinic, adipic, tartaric, or citric acid in tablet cores and maintaining microenvironmental pH acidic. The acidic pH dissolved the drug and facilitated its release [163]. It was also shown that buffered formulation can be used to improve the dissolution rate of acidic drugs in acidic medium and decreasing the rate in the alkaline medium [164]. For example, the release of acidic drug indomethacin from ethyl cellulose microcapsules was better controlled by using internal buffer, dibasic sodium phosphate to improve its solubility and release [165]. Achlorhydria, where the stomach pH is not acidic enough, can retard the bioavailability of basic drugs such as dipyrimadole. Acidic microenvironmental pH modifiers were successfully used to improve the bioavailability of dipyrimadole in dogs under achlorhydric conditions [166].

2.9 ANTICAKING AGENTS

Anticaking agents are also known as antiagglomeration agents. They are used to prevent lump formulation in granulation blend or in API. Agglomeration is an issue with respect to flow, particle size, and in general processability for highly water-soluble compounds. In the presence of small amount of moisture, the API gets dissolved and the dissolved API acts as a binder and forms lumps in API itself or within the blend. The anticaking agents because of their high surface area cover the API particles and prevent caking. Even in day-to-day use, people encounter lumpy sugar or salt. Commonly used anticaking agents are silicon dioxide, PEGs, metallic stearates such as magnesium stearate, and stearic acid. They are used in concentrations ranging from 0.5% to 5% w/w. Higher concentrations of some anticaking agents can reduce the compressibility of the blend especially with metallic stearates and stearic acid [167]. The anticaking agents should not react chemically with the API or other excipients of the formulation [168].

2.10 ANTIOXIDANTS

Many excipients such as PEG, crospovidone, and povidone are known to contain trace levels of peroxides. Peroxides are very reactive and can form degradants following their reactions with an API. Peroxides in the form of organperoxides and hydroperoxides act as initiators of free radical oxidation reactions [169]. The European Pharmacopoeia does not allow for more than 400 ppm of peroxides in povidone. Even though such a limit does not exist for crospovidone in USP/NF, many vendors provide "peroxide-free" crospovidone [122]. While crospovidone and povidone could be a source of residual peroxide, excipients such as CCS has been reported to reduce an oxidative degradant in a tablet formulation [122]. Treatment with silicates reduces peroxide levels in excipients [170]. The most common approach to reduce oxidative degradants is to include antioxidants in formulations. Most common antioxidants are butylated hydroxy anisole (BHA) and butylated hydroxy toluene (BHT). They are water insoluble and used at concentrations as low as 0.01% w/w. On the other hand, ascorbic acid, propyl gallate, and sodium sulfite are water soluble and used at a higher concentration. Water-soluble antioxidants are found to be more effective than water-insoluble ones in reducing peroxide concentration in povidone [170].

2.11 COPROCESSED EXCIPIENTS

Previous sections in this chapter have focused on excipients for specific purposes (a disintegrant, binder filler, etc.). In general, excipients are used to improve stability, manufacturability, and performance of a dosage form. However, as highlighted in the previous sections, there are no ideal excipients that exhibit all these properties. For example, MCC shows loss in compactability with wet granulation, certain lactoses have poor reworking potential and so on. In addition, several factors including increasing popularity of direct compression process, reduced manufacturing steps, and improvements in tablet and capsule machinery place greater burden on excipients to ensure a robust manufacturing process [171]. Therefore, there are clearly opportunities to develop excipients with improved functionality. There are three routes by which new high functional excipients can be developed [172]. One route is to make new grades of the same excipients or specialized grades of excipients. The limitations to this approach are inherent properties of the excipients determine the functionality and making a new grade of the same excipients might result in only limited improvements in functionality. Another approach is to introduce new chemical entities as excipients. Even though new chemical entities may result in improved functionality, there is greater regulatory burden due to the necessity to prove safety profiles and address toxicity issues. This makes it a very expensive and less attractive process. Another more practical approach to produce high functional excipients is by combining two or more excipients. A simple physical mixture of two or more excipients also falls under this category but in this case the excipients are just blends with physical separation at particle level. A direct result of this is that synergistic effects in functionality may not be obtained. On the other hand, a more convenient and economical way of developing new high functionality excipients is through coprocessing.

Coprocessed excipients involve tailored combinations of two or more excipients with no physical separation at the particulate level [173]. The properties of the resulting excipients are a synergistic combination of the desirable functionalities (in most cases complementary) of the participating excipients without significant chemical change, that is, the physical form and functionality may change without a chemical change or new chemical bonds. The ratio of the components will depend on the final desirable characteristics. The process used for the manufacture of these excipients are not novel and are routinely used methods such as granulation, spray drying, melt extrusion, and cocrystallization. Improvements in functionality may be due to changes in properties such as particle size, size distribution, shape, and porosity, leading to improved flow, compressibility, hardness, disintegration, and so on among others [171]. An example of such an improvement is seen with one of the widely used coprocessed excipient silicified MCC. It is a coprocessed material obtained from codrying of MCC and a glidant colloidal silicon dioxide. Colloidal silicon dioxide is present at 2% level. Compared to MCC alone, it shows better compactability after wet granulation, better flow, and low lubricant sensitivity [174, 175]. It has been shown that improved functionality of silicified MCC is related to intrinsic properties rather than any chemical interaction [176]. Table 2.6 show a list of available coprocessed excipients and their claimed benefits.

The advantages of coprocessed materials are well recognized [177]. It is obvious that coprocessing provides a synergistic effect that is not achievable through simple physical blends. Since each excipient has its disadvantages or shortcomings in terms of its properties, choosing excipients with complementary properties results in utilizing the best property of excipients while masking their disadvantages. For example, to improve compaction performance and compressibility, a combination of a brittle and a plastic diluent has been a popular choice. In addition, having a multifunctional excipient may not only reduce the number of excipients but also reduce a processing step such as blending. Moreover, it reduces the inventory of excipients that needs to be maintained. Lot-to-lot variability of individual components is now reduced since fewer excipients are used in general. Since coprocessed materials are made with the purpose of having higher functionalities, it should inherently improve the quality of the product. Analytical burden is also reduced since fewer tests are necessary due to lower number of excipients [173]. Coprocessing has its disadvantage as well. Ratio of the excipients is fixed in the coprocessed material thus reducing the flexibility for the formulator.

Even though coprocessed material has been available since the 1980s, the biggest hurdle to their greater usage in dosage forms has been its noninclusion in monographs. For a new chemical entity as excipient, additional safety and toxicological information is necessary for regulatory approval and its broader usage. However, for a coprocessed material, absence of chemical change and use of monograph excipients to make the coprocessed material reduce the need for new toxicological and safety studies compared to a totally new chemical entity. There are additional analytical considerations associated with coprocessed excipients compared to a traditional approved excipient. Analytical tests to identify the parent components and establish the absence of chemical change and also confirmation of a synergistic effect for a coprocessed material are important [177]. An acceptable range of compositions of parent excipients that does not impact the functionality of a dosage form should be established along with the understanding of the impact of the variability of parent excipients on coprocessed material variability. It is clear that there are challenges and opportunities for the development and acceptability of a coprocessed excipient. The International Pharmaceutical Excipient Council (IPEC) Federation, a global organization comprising regional associations organized to promote quality in pharmaceutical excipients, has been working on guidelines to to deal with technical, safety, and regulatory concerns related to the development and commercialization of coprocessed excipients. Excipient manufacturers can use these guidelines to anticipate technical, safety, and regulatory challenges for the development of high functionality excipients. In addition to improving the functionality of excipients by

TABLE 2.6 Details of Marketed Coprocessed Excipients and Their Claimed Benefits

Trade Name	Supplier	Composition	Main Application	Added Benefit
Avicel®CE-15	FMC biopolymer	85% MCC, guar gum	Chewable tablets	Better mouth feel, less grittiness, minimal chalkiness, improved palatability
StarCap 1500®	Colorcon®	85–95% Corn starch, pregelatinized starch	Capsule fill, DC[a]	Improved tablet disintegration, pH-independent dissolution
StarLac®	Meggle/Roquette Pharma	85% α-Lactose, white maize starch	DC	Improved flow, improved tablet hardness, disintegration
MicroceLac® 100	Meggle	75% α-Lactose monohydrate, MCC	DC, roller compaction, capsule fill	Better compaction, improved flow
Cellactose® 80	Meggle	75% α-Lactose monohydrate and cellulose powder	DC, roller compaction, capsule fill	Compactability, improved flow, increased adherence, decreased segregation
Ludipress®	BASF	93% Lactose, 3.5% Kollidon® 30, Kollodon CL	DC	Improved flow, low hygroscopicity better binding
Ludipress LCE	BASF	96.5% Lactose, 3.5% Kollodin® 30	DC	Improved flow, binding
Ludiflash®	BASF	90% Mannitol, 5% crospovidone, PVA	ODT[b]	Tablet hardness, rapid disintegration for ODT, mouth feel
Prosolv®	JRS Pharma	98% MCC, 2% colloidal silicon dioxide	DC	Improved flow, less sensitive to wet granulation, better tablet hardness

(*continued*)

TABLE 2.6 (*Continued*)

Trade Name	Supplier	Composition	Main Application	Added Benefit
Pharmatose DCL	DFE Pharma	95% β-Lactose, lactitol	DC	Better compressibility, low lubricant sensitivity
Xylitab 200		98% Xylitol, NaCMC	DC	DC
Celocal®	FMC Biopolymer	30% MCC, calcium sulfate	DC	Improved compressibility, hardness, low lubricant sensitivity
Pharmaburst®	SPI Pharma™	Mannitol, sorbitol	ODT	Superior organoleptics, rapid disintegration, highly compactable, soluble matrix
Retalac®	Meggle	Lactose, HPMC	Modified release-DC, roller compaction, spheronization, extrusion	DC of modified release, pH-independent release, superior processability
PVAP-T	Colorcon	90% PVAP, Titanium dioxide	Enteric coating	Less processing steps for coating suspension, less tacky, preplasticized powder blend enteric coating composition
Emdex	JRS Pharma	92% Dextrose, 4% maltose, 4% maltodextrin	DC	Superior flow, compaction, lower impact on blood sugar
PanExcea™ MHC 300G	Avantor™	MCC, HPMC, crospovidone	ODT, chewable tablet	Improved flow, compressibility for ODT
Di-Pac®	Domino specialty ingredients	97% Sucrose, maltodextrin	DC	High flow, low hygroscopicity, low reactivity

[a]DC, direct compression.
[b]ODT, orally disintegrating tablet.

coprocessing them, efforts have been made to improve API property by coprocessing it with an excipient. For example, metformin hydrochloride, which is used as frontline treatment for type 2 diabetes, requires high doses. This challenge is further aggravated by the fact that metformin hydrochloride has poor compactability. This challenge was successfully overcome by coprocessing metformin with HPMC, a release controlling polymer [178]. Moreover, this approach reduced the amount of HPMC needed to control the release of metformin hydrochloride. Coprocessing of gatifloxacin with stearic acid and/or palmitic acid was carried out to mask the taste of gatifloxacin for a pediatric formulation [179]. Coprocessing approach was also used to combine two antiviral drugs, nevirapine and stavudine, by spray drying method to overcome the content uniformity issue [180]. The coprocessed drugs were then combined with another antiviral drug lamivudine to develop a triple combination tablet.

An API-excipient coprocessing approach faces two big hurdles. First, the API manufacturing plant also needs to be approved as a drug product manufacturing plant with relevant GMP measures in place. The reason for this requirement is that regulatory agencies consider API-excipient coprocessed material as a drug product intermediate. Second hurdle is that once the API-excipient coprocessing step is complete, the shelf-life clock starts regardless when the coprocessed API-excipient is converted into a final product. Even though this is the current line of thinking for regulatory agencies, it will likely evolve further with increasing number of products containing API-excipient coprocessed material.

2.12 FUTURE DIRECTIONS

As one look toward the future horizon for excipients, two important issues remain to be addressed. The first issue is how to address the functionality aspects of excipients and the second one is how to develop a product-independent pathway for regulatory approvals of new excipients.

Functionality is the most difficult aspect to address, both from the aspects of excipient vendors and pharmaceutical compendia. The main reason is that any given excipient can be used in different dosage forms for different purposes. For example, mannitol can be used in a tablet as well as in an injectable formulation. In addition, even if an excipient is used for the same purpose, the extent of its impact may differ based on the formulation. For example, water-proofing effect of magnesium stearate may affect the dissolution of a drug with low solubility because of overmixing, but may not impact the dissolution of a drug with high solubility. Based on the aforementioned examples, it is clear that functionality of an excipient is always linked to the property of the drug used in a dosage form. Therefore, in the QbD approach, excipient functionality can be established by using multiple lots and multiple source of an excipient during product development. Then, there are excipients that are used for purposes such as wetting agent, pH modifier, or antioxidant. In these specific cases, their functionality is already established, so the focus is more on their uniform distribution in the dosage form. Even though excipients by definition are inert, there are some excipients that are known to alter bioavailability of a drug. Therefore, their amount and content uniformity in a dosage form are as important as that of drug itself.

The second most important issue facing excipient vendors and formulators is that there is no independent pathway to get the approval of a new excipient prior to its use in a formulation. Thus, an approval of an excipient is always tied with the approval of a drug product. Therefore, a formulator has to take a risk by using an unapproved excipient in the formulation and be prepared to generate long-term toxicity data to support the excipient use in the drug product. Pharmaceutical companies are reluctant to take such risks. Therefore, very few new excipients have been introduced into the market in the last few years. Excipients that have been introduced and accepted in recent years including sulfobutylether β-cyclodextrin (Captisol®), hydroxystearic acid PEG ester (Solutol® HS15), and polyvinyl caprolactam – polyvinyl acetate – PEG graft copolymer (Soluplus®) were developed to address solubility or stability issue of a specific drug. Their use in a product was warranted since the approved available excipients could not address the specific issue.

In the coming years, it is hoped that more attention will be given to develop and better understand functionality tests for excipients used in oral solid dosage forms. Development of such functionality tests and their wide disseminations will help pharmaceutical scientists in developing robust formulations. It is also hoped that regulatory agencies, excipient vendors, and pharmaceutical companies will collaborate to develop a product-independent pathway for the approval of a new excipient. With the new chemical entities becoming very challenging to develop, excipients are expected to play an even greater role in enabling their development.

REFERENCES

1. Reier, G.E. and Shangraw, R.F. (**1966**) Microcrystalline cellulose in tableting. *Journal of Pharmaceutical Sciences*, **55** (5), 510–514.
2. Lerk, C., Bolhuis, G., and De Boer, A. (**1979**) Effect of microcrystalline cellulose on liquid penetration in and disintegration of directly compressed tablets. *Journal of Pharmaceutical Sciences*, **68** (2), 205–211.
3. Jivraj, M., Martini, L.G., and Thomson, C.M. (**2000**) An overview of the different excipients useful for the direct compression of tablets. *Pharmaceutical Science & Technology Today*, **3** (2), 58–63.
4. Gissinger, D. and Stamm, A. (**1980**) A comparative evaluation of the properties of some tablet disintegrants. *Drug Development and Industrial Pharmacy*, **6** (5), 511–536.
5. Bolhuis, G.K. and Chowhan, Z.T. (**1996**) Materials for direct compaction. *Drugs and the Pharmaceutical Sciences*, **71**, 419–500.
6. David, S. and Augsburger, L. (**1977**) Plastic flow during compression of directly compressible fillers and its effect on tablet strength. *Journal of Pharmaceutical Sciences*, **66** (2), 155–159.
7. Hiestand, E., Wells, J., Peot, C., and Ochs, J. (**1977**) Physical processes of tableting. *Journal of Pharmaceutical Sciences*, **66** (4), 510–519.

8. Staniforth, J.N., Baichwal, A.R., Hart, J.P., and Heng, P.W. (**1988**) Effect of addition of water on the rheological and mechanical properties of microcrystalline celluloses. *International Journal of Pharmaceutics*, **41** (3), 231–236.

9. Westermarck, S., Juppo, A.M., Kervinen, L., and Yliruusi, J. (**1999**) Microcrystalline cellulose and its microstructure in pharmaceutical processing. *European Journal of Pharmaceutics and Biopharmaceutics*, **48** (3), 199–206.

10. Buckton, G., Yonemochi, E., Yoon, W.L., and Moffat, A.C. (**1999**) Water sorption and near IR spectroscopy to study the differences between microcrystalline cellulose and silicified microcrystalline cellulose before and after wet granulation. *International Journal of Pharmaceutics*, **181** (1), 41–47.

11. Sun, C.C. (**2008**) Mechanism of moisture induced variations in true density and compaction properties of microcrystalline cellulose. *International Journal of Pharmaceutics*, **346** (1), 93–101.

12. Van der Watt, J. (**1987**) The effect of the particle size of microcrystalline cellulose on tablet properties in mixtures with magnesium stearate. *International Journal of Pharmaceutics*, **36** (1), 51–54.

13. Pingali, K., Mendez, R., Lewis, D. *et al.* (**2011**) Mixing order of glidant and lubricant – Influence on powder and tablet properties. *International Journal of Pharmaceutics*, **409** (1–2), 269–277.

14. Rowe, R. (**1988**) Interactions in the ternary powder system microcrystalline cellulose, magnesium stearate and colloidal silica—a solubility parameter approach. *International Journal of Pharmaceutics*, **45** (3), 259–261.

15. Roberts, R.J. and Rowe, R.C. (**1985**) The effect of punch velocity on the compaction of a variety of materials. *Journal of Pharmacy and Pharmacology*, **37** (6), 377–384.

16. Thoorens, G., Krier, F., Leclercq, B. *et al.* (**2014**) Microcrystalline cellulose, a direct compression binder in a quality by design environment – A review. *International Journal of Pharmaceutics*, **473** (1–2), 64–72.

17. Doldán, C., Souto, C., Concheiro, A. *et al.* (**1995**) Dicalcium phosphate dihydrate and anhydrous dicalcium phosphate for direct compression: A comparative study. *International Journal of Pharmaceutics*, **124** (1), 69–74.

18. Khan, K. and Rooke, D. (**1976**) Effect of disintegrant type upon the relationship between compressional pressure and dissolution efficiency. *Journal of Pharmacy and Pharmacology*, **28** (8), 633–636.

19. Van Kamp, H.V., Bolhuis, G.K., Kussendrager, K.D., and Lerk, C.F. (**1986**) Studies on tableting properties of lactose. IV. Dissolution and disintegration properties of different types of crystalline lactose. *International Journal of Pharmaceutics*, **28** (2–3), 229–238.

20. Vromans, H., Bolhuis, G.K., Lerk, C.F., and Kussendrager, K.D. (**1987**) Studies on tableting properties of lactose. VIII. The effect of variations in primary particle size, percentage of amorphous lactose and addition of a disintegrant on the disintegration of spray-dried lactose tablets. *International Journal of Pharmaceutics*, **39** (3), 201–206.

21. Shangraw, R., Wallace, J.W., and Bowers, F. (**1981**) Morphology and functionality in tablet excipients for direct compression: Part I. *Pharmaceutical Technology*, **5** (9), 69–78.

22. Debord, B., Lefebvre, C., Guyot-Hermann, A. *et al.* (**1987**) Study of different crystalline forms of mannitol: Comparative behaviour under compression. *Drug Development and Industrial Pharmacy*, **13** (9–11), 1533–1546.

23. Mihranyan, A., Llagostera, A.P., Karmhag, R. *et al.* (**2004**) Moisture sorption by cellulose powders of varying crystallinity. *International Journal of Pharmaceutics*, **269** (2), 433–442.
24. Kothari, S.H., Kumar, V., and Banker, G.S. (**2002**) Comparative evaluations of powder and mechanical properties of low crystallinity celluloses, microcrystalline celluloses, and powdered celluloses. *International Journal of Pharmaceutics*, **232** (1), 69–80.
25. Suzuki, T. and Nakagami, H. (**1999**) Effect of crystallinity of microcrystalline cellulose on the compactability and dissolution of tablets. *European Journal of Pharmaceutics and Biopharmaceutics*, **47** (3), 225–230.
26. Vromans, H., Bolhuis, G., Lerk, C. *et al.* (**1987**) Studies on tableting properties of lactose. VII. The effect of variations in primary particle size and percentage of amorphous lactose in spray dried lactose products. *International Journal of Pharmaceutics*, **35** (1), 29–37.
27. Bolhuis, G. and Zuurman, K. (**1995**) Tableting properties of experimental and commercially available lactose granulations for direct compression. *Drug Development and Industrial Pharmacy*, **21** (18), 2057–2071.
28. Kushner, J., Langdon, B.A., Hiller, J.I., and Carlson, G.T. (**2011**) Examining the impact of excipient material property variation on drug product quality attributes: A quality-by-design study for a roller compacted, immediate release tablet. *Journal of Pharmaceutical Sciences*, **100** (6), 2222–2239.
29. Hlinak, A.J., Kuriyan, K., Morris, K.R. *et al.* (**2006**) Understanding critical material properties for solid dosage form design. *Journal of Pharmaceutical Innovation*, **1** (1), 12–17.
30. Landin, M., Martinez-Pacheco, R., Gomez-Amoza, J. *et al.* (**1993**) Effect of country of origin on the properties of microcrystalline cellulose. *International Journal of Pharmaceutics*, **91** (2), 123–131.
31. Gamble, J. F., Chiu, W.-S., Gray, V. *et al.* (**2010**) Investigation into batch-to-batch and intra-vendor variability of lactose anhydrous. Paper presented at the AAPS, New Orleans.
32. Obae, K., Iijima, H., and Imada, K. (**1999**) Morphological effect of microcrystalline cellulose particles on tablet tensile strength. *International Journal of Pharmaceutics*, **182** (2), 155–164.
33. Zuurman, K., Riepma, K.A., Bolhuis, G.K. *et al.* (**1994**) The relationship between bulk density and compactibility of lactose granulations. *International Journal of Pharmaceutics*, **102** (1–3), 1–9.
34. Hwang, R.-C. and Peck, G.R. (**2001**) A systematic evaluation of the compression and tablet characteristics of various types of microcrystalline cellulose. *Pharmaceutical Technology*, **25** (3), 112–132.
35. Landin, M., Martinez-Pacheco, R., Gomez-Amoza, J. *et al.* (**1993**) Effect of batch variation and source of pulp on the properties of microcrystalline cellulose. *International Journal of Pharmaceutics*, **91** (2), 133–141.
36. Landin, M., Martínez-Pacheco, R., Gómez-Amoza, J. *et al.* (**1993**) Influence of microcrystalline cellulose source and batch variation on the tabletting behaviour and stability of prednisone formulations. *International Journal of Pharmaceutics*, **91** (2), 143–149.
37. Chamarthy, S.P., Pinal, R., and Carvajal, M.T. (**2009**) Elucidating raw material variability – Importance of surface properties and functionality in pharmaceutical powders. *AAPS PharmSciTech*, **10** (3), 780–788.
38. Shlieout, G., Arnold, K., and Müller, G. (**2002**) Powder and mechanical properties of microcrystalline cellulose with different degrees of polymerization. *AAPS PharmSciTech*, **3** (2), 45–54.

39. Dybowski, U. (**1997**) Does polymerisation degree matter? *Manufacturing Chemist*, **68** (12), 19–21.

40. Sakr, A., Elsabbagh, H., and Emara, K. (**1974**) Sta-Rx 1500 starch: A new vehicle for the direct compression of tablets. *Arch. Pharm. Chem.(Sci)*, **2**, 14–24.

41. Bolhuis, G., Lerk, C., Moes, J., and Mulder, C.W.A. (**1979**) Comparative evaluation of excipients for direct compression. *Pharmacy World & Science*, **1** (1), 1223–1233.

42. Symecko, C. and Rhodes, C. (**1997**) The effect of compaction force and type of pregelatinized starch on the dissolution of acetaminophen. *Drug Development and Industrial Pharmacy*, **23** (3), 229–238.

43. Alebiowu, G. and Itiola, O.A. (**2002**) Compressional characteristics of native and pregelatinized forms of sorghum, plantain, and corn starches and the mechanical properties of their tablets. *Drug Development and Industrial Pharmacy*, **28** (6), 663–672.

44. Colorcon. (**1997**). Starch 1500. Technical Literature.

45. Chowhan, Z. (**1980**) Role of binders in moisture-induced hardness increase in compressed tablets and its effect on in vitro disintegration and dissolution. *Journal of Pharmaceutical Sciences*, **69** (1), 1–4.

46. Fitzpatrick, S., McCabe, J.F., Petts, C.R., and Booth, S.W. (**2002**) Effect of moisture on polyvinylpyrrolidone in accelerated stability testing. *International Journal of Pharmaceutics*, **246** (1), 143–151.

47. Desai, D., Rubitski, B., Varia, S., and Jain, N. (**1996**) Povidone-and poloxamer-mediated degradation of hydrochlorothiazide in an antihypertensive combination tablet product. *International Journal of Pharmaceutics*, **142** (1), 61–66.

48. Guo, J.-H., Skinner, G.W., Harcum, W.W., and Barnum, P.E. (**1998**) Pharmaceutical applications of naturally occurring water-soluble polymers. *Pharmaceutical Science & Technology Today*, **1** (6), 254–261.

49. Rowe, R.C., Sheskey, P.J., and Quinn, M.E. (eds) (**2009**) *Handbook of pharmaceutical excipients*, Sixth edn, Pharmaceutical Press and American Pharmacists Association, Grayslake, IL and Washington DC.

50. Desai, D., Rinaldi, F., Kothari, S. *et al.* (**2006**) Effect of hydroxypropyl cellulose (HPC) on dissolution rate of hydrochlorothiazide tablets. *International Journal of Pharmaceutics*, **308** (1), 40–45.

51. Tiwari, S.B., Murthy, T.K., Raveendra Pai, M. *et al.* (**2003**) Controlled release formulation of tramadol hydrochloride using hydrophilic and hydrophobic matrix system. *AAPS PharmSciTech*, **4** (3), 18–23.

52. Porter, S.C. (**1989**) Controlled-release film coatings based on ethylcellulose. *Drug Development and Industrial Pharmacy*, **15** (10), 1495–1521.

53. Desai, R.P., Neau, S.H., Pather, S.I., and Johnston, T.P. (**2001**) Fine-particle ethylcellulose as a tablet binder in direct compression, immediate-release tablets. *Drug Development and Industrial Pharmacy*, **27** (7), 633–641.

54. Pollock, D., and Balwinski, K. (**1997**) Aqueous granulation of an ethylcellulose containing inert matrix formulation. Paper presented at the Annual Association of Pharmaceutical Scientist Annual Meeting, Boston, MA.

55. Agrawal, A.M., Neau, S.H., and Bonate, P.L. (**2003**) Wet granulation fine particle ethylcellulose tablets: Effect of production variables and mathematical modeling of drug release. *The AAPS Journal*, **5** (2), 48–60.

56. Olsson, H., Mattsson, S., and Nyström, C. (**1998**) Evaluation of the effect of addition of polyethylene glycols of differing molecular weights on the mechanical strength of sodium chloride and sodium bicarbonate tablets. *International Journal of Pharmaceutics*, **171** (1), 31–44.
57. Desai, D., Rao, V., Guo, H. *et al.* (**2007**) Stability of low concentrations of guanine-based antivirals in sucrose or maltitol solutions. *International Journal of Pharmaceutics*, **342** (1), 87–94.
58. Millili, G.P. and Schwartz, J.B. (**1990**) The strength of microcrystalline cellulose pellets: The effect of granulating with water/ethanol mixtures. *Drug Development and Industrial Pharmacy*, **16** (8), 1411–1426.
59. Wikberg, M. and Alderborn, G. (**1993**) Compression characteristics of granulated materials. VII. The effect of intragranular binder distribution on the compactibility of some lactose granulations. *Pharmaceutical Research*, **10** (1), 88–94.
60. Parikh, D.M. (ed.) (**2005**) *Handbook of Pharmaceutical Granulation Technology*, 2nd edn, Taylor and Francis group, LLC, Boca Raton, FL.
61. Rue, P., Seager, H., Burt, I., Ryder, J., Murray, S., and Beal, N. (**1979**) The effect of granule structure on the mechanical properties of tablets. *Journal of Pharmacy and Pharmacology*, **31** (S1), 73P. doi: 10.1111/j.2042-7158.1979.tb11621.x.
62. Cantor, S.L., Kothari, S., and Koo, O.M.Y. (**2009**) Evaluation of the physical and mechanical properties of high drug load formulations: Wet granulation vs. novel foam granulation. *Powder Technology*, **195** (1), 15–24.
63. Becker, D., Rigassi, T., and Bauer-Brandl, A. (**1997**) Effectiveness of binders in wet granulation: A comparison using model formulations of different tabletability. *Drug Development and Industrial Pharmacy*, **23** (8), 791–808.
64. Upadrashta, S.M., Katikaneni, P.R., and Nuessle, N.O. (**1992**) Chitosan as a tablet binder. *Drug Development and Industrial Pharmacy*, **18** (15), 1701–1708.
65. Itiola, O. and Pilpel, N. (**1991**) Formulation effects on the mechanical properties of metronidazole tablets. *Journal of Pharmacy and Pharmacology*, **43** (3), 145–147.
66. Boutell, S., Newton, J., Bloor, J., and Hayes, G. (**2002**) The influence of liquid binder on the liquid mobility and preparation of spherical granules by the process of extrusion/spheronization. *International Journal of Pharmaceutics*, **238** (1), 61–76.
67. Joneja, S., Harcum, W., Skinner, G. *et al.* (**1999**) Investigating the fundamental effects of binders on pharmaceutical tablet performance. *Drug Development and Industrial Pharmacy*, **25** (10), 1129–1135.
68. Baykara, T. and Acarturk, F. (**1989**) The effect of the binding agents on the friability and compressibility of granules. *Drug Development and Industrial Pharmacy*, **15** (9), 1341–1351.
69. Danjo, K., Kozaki, K., Sunada, H., and Otsuka, A. (**1994**) Influence of the molecular weight of binding agents on the physical properties of granules and tablets. *Chemical and Pharmaceutical Bulletin*, **42** (10), 2121–2125.
70. Nyström, C., Mazur, J., and Sjögren, J. (**1982**) Studies on direct compression of tablets. II. The influence of the particle size of a dry binder on the mechanical strength of tablets. *International Journal of Pharmaceutics*, **10** (3), 209–218.
71. Nyström, C. and Glazer, M. (**1985**) Studies on direct compression of tablets. XIII. The effect of some dry binders on the tablet strength of compounds with different fragmentation propensity. *International Journal of Pharmaceutics*, **23** (3), 255–263.

72. Horisawa, E., Komura, A., Danjo, K., and Otsuka, A. (**1995**) Effect of binder characteristics on the strength of agglomerates prepared by the wet method. II. *Chemical and Pharmaceutical Bulletin*, **43** (3), 488–492.

73. Rowe, R. (**1990**) Correlation between predicted binder spreading coefficients and measured granule and tablet properties in the granulation of paracetamol. *International Journal of Pharmaceutics*, **58** (3), 209–213.

74. Healey, J., Rubinstein, M., and Walters, V. (**1974**) The mechanical properties of some binders used in tableting. *Journal of Pharmacy and Pharmacology*, **26** (S1), 41P–46P.

75. Chowhan, Z. and Yang, I.C. (**1983**) Effect of intergranular versus intragranular cornstarch on tablet friability and in vitro dissolution. *Journal of Pharmaceutical Sciences*, **72** (9), 983–988.

76. Rudnic, E., Kanig, J., and Rhodes, C. (**1985**) Effect of molecular structure variation on the disintegrant action of sodium starch glycolate. *Journal of Pharmaceutical Sciences*, **74** (6), 647–650.

77. Khan, K. and Rhodes, C. (**1975**) Water sorption properties of tablet disintegrants. *Journal of Pharmaceutical Sciences*, **64** (3), 447–451.

78. Bolhuis, G.K., Van Kamp, H., Lerk, C. *et al.* (**1984**) Effect of variation of degree of substitution, crosslinking and purity on the disintegration efficiency of sodium starch glycolate. *Acta Pharmaceutica Technologica*, **30** (1), 24–32.

79. Gordon, M., and Chowhan, Z. (**1987**) Effect of tablet solubility and hygroscopicity on disintegrant efficiency in direct compression tablets in terms of dissolution. *Journal of Pharmaceutical Sciences*, **76** (12), 907–909.

80. Gordon, M.S., Rudraraju, V.S., Dani, K., and Chowhan, Z.T. (**1993**) Effect of the mode of super disintegrant incorporation on dissolution in wet granulated tablets. *Journal of Pharmaceutical Sciences*, **82** (2), 220–226.

81. Shah, U. and Augsburger, L. (**2002**) Multiple sources of sodium starch glycolate, NF: Evaluation of functional equivalence and development of standard performance tests. *Pharmaceutical Development and Technology*, **7** (3), 345–359.

82. Edge, S., Steele, D.F., Staniforth, J.N. *et al.* (**2002**) Powder compaction properties of sodium starch glycolate disintegrants. *Drug Development and Industrial Pharmacy*, **28** (8), 989–999.

83. Sheen, P.-C. and Kim, S.-I. (**1989**) Comparative study of disintegrating agents in tiaramide hydrochloride tablets. *Drug Development and Industrial Pharmacy*, **15** (3), 401–414. doi: 10.3109/03639048909040220.

84. Kornblum, S.S. and Stoopak, S.B. (**1973**) A new tablet disintegrating agent: Crosslinked polyvinylpyrrolidone. *Journal of Pharmaceutical Sciences*, **62** (1), 43–49.

85. Balasubramaniam, J., Bindu, K., Rao, V., Ray, D., Haldar, R., and Brzeczko, A. (**2008**) Effect of superdisintegrants on dissolution of cationic drugs. *Dissolution Technologies*, **15**(2), 18–25.

86. Rudnic, E., Lausier, J., Chilamkurti, R., and Rhodes, C. (**1980**) Studies of the utility of cross linked polyvinlpolypyrrolidine as a ablet disintegrant. *Drug Development and Industrial Pharmacy*, **6** (3), 291–309.

87. Shah, U. and Augsburger, L. (**2001**) Evaluation of the functional equivalence of crospovidone NF from different sources. I. Physical characterization. *Pharmaceutical Development and Technology*, **6** (1), 39–51.

88. Marais, A.F., Song, M., and De Villiers, M.M. (**2005**) Effect of compression force, humidity and disintegrant concentration on the disintegration and dissolution of directly compressed furosemide tablets using croscarmellose sodium as disintegrant. *Tropical Journal of Pharmaceutical Research*, **2** (1), 125–135.
89. Zhao, N. and Augsburger, L. (**2006**) The influence of product brand-to-brand variability on superdisintegrant performance a case study with croscarmellose sodium. *Pharmaceutical Development and Technology*, **11** (2), 179–185.
90. Impact of cellulose source on properties and disintegration performance of croscarmellose sodium. FMC BioPolymer Technical Bulletin, Accessed online Aug 13 2015 at: http://www.fmcbiopolymer.com/Portals/Pharm/Content/Docs/SDW-802%20Bulletin%2010-13-11.pdf
91. Van Kamp, H., Bolhuis, G., and Lerk, C. (**1983**) Improvement by super disintegrants of the properties of tablets containing lactose, prepared by wet granulation. *Pharmacy World & Science*, **5** (4), 165–171.
92. Khattab, I., Menon, A., and Sakr, A. (**1993**) Effect of mode of incorporation of disintegrants on the characteristics of fluidbed wet granulated tablets. *Journal of Pharmacy and Pharmacology*, **45** (8), 687–691.
93. Smallenbroek, A., Bolhuis, G., and Lerk, C. (**1981**) The effect of particle size of disintegrants on the disintegration of tablets. *Pharmacy World & Science*, **3** (1), 1048–1051.
94. Swarbrick, J. (ed.) (**2007**) *Encyclopedia of Pharmaceutical Technology*, 3rd edn, vol. **6**, Informa Healthcare USA, New York.
95. Zhao, N. and Augsburger, L.L. (**2006**) The influence of granulation on super disintegrant performance. *Pharmaceutical Development and Technology*, **11** (1), 47–53.
96. Hill, P.M. (**1976**) Effect of compression force and corn starch on tablet disintegration time. *Journal of Pharmaceutical Sciences*, **65** (11), 1694–1697. doi: 10.1002/jps.2600651134.
97. Lowenthal, W. (**1972**) Disintegration of tablets. *Journal of Pharmaceutical Sciences*, **61** (11), 1695–1711.
98. Johnson, J.R., Wang, L.H., Gordon, M.S., and Chowhan, Z.T. (**1991**) Effect of formulation solubility and hygroscopicity on disintegrant efficiency in tablets prepared by wet granulation, in terms of dissolution. *Journal of Pharmaceutical Sciences*, **80** (5), 469–471.
99. Chebli, C. and Cartilier, L. (**1998**) Cross-linked cellulose as a tablet excipient: A binding/disintegrating agent. *International Journal of Pharmaceutics*, **171** (1), 101–110.
100. Setty, C. M., Prasad, D., Gupta, V., and Sa, B. (**2008**) Development of fast dispersible aceclofenac tablets: Effect of functionality of superdisintegrants. *Indian Journal of Pharmaceutical Sciences*, **70** (2), 180–185.
101. Bolhuis, G., Zuurman, K., and Te Wierik, G. (**1997**) Improvement of dissolution of poorly soluble drugs by solid deposition on a super disintegrant. II. The choice of super disintegrants and effect of granulation. *European Journal of Pharmaceutical Sciences*, **5** (2), 63–69.
102. Patel, S., Kaushal, A.M., and Bansal, A.K. (**2011**) The effect of starch paste and sodium starch glycolate on the compaction behavior of wet granulated acetaminophen formulations. *Journal of Excipients and Food Chemicals*, **2** (3), 64–72.
103. Zhao, N. and Augsburger, L.L. (**2005**) The influence of swelling capacity of superdisintegrants in different pH media on the dissolution of hydrochlorothiazide from directly compressed tablets. *AAPS PharmSciTech*, **6** (1), 120–126.

104. Chen, C.-R., Lin, Y.-H., Cho, S.-L. *et al.* (**1997**) Investigation of the dissolution difference between acidic and neutral media of acetaminophen tablets containing a super disintegrant and a soluble excipient. *Chemical and Pharmaceutical Bulletin*, **45** (3), 509–512.

105. Bindra, D., Hsieh, D., Pandey, P., and Barbour, N. (**2014**) Incompatability of croscarmellose with alkaline excipients in a tablet formulation. *Pharmaceutical Development and Technology*, **19** (3), 285–289.

106. Bolhuis, G., Van Kamp, H., Lerk, C., and Sessink, F. (**1982**) On the mechanism of action of modern disintegrants. *Acta Pharmaceutica Technologies*, **28**, 111–114.

107. Bolhuis, G.K., Smallenbroek, A.J., and Lerk, C.F. (**1981**) Interaction of tablet disintegrants and magnesium stearate during mixing. I. Effect on tablet disintegration. *Journal of Pharmaceutical Sciences*, **70** (12), 1328–1330. doi: 10.1002/jps.2600701210.

108. Rojas, J., Guisao, S., and Ruge, V. (**2012**) Functional assessment of four types of disintegrants and their effect on the spironolactone release properties. *AAPS PharmSciTech*, **13** (4), 1054–1062.

109. Gordon, M.S., Rudraraju, V.S., Rhie, J.K., and Chowhan, Z.T. (**1993**) The effect of aging on the dissolution of wet granulated tablets containing super disintegrants. *International Journal of Pharmaceutics*, **97** (1), 119–131.

110. Pourkavoos, N. and Peck, G.E. (**1993**) The effect of swelling characteristics of superdisintegrants on the aqueous coating solution penetration into the tablet matrix during the film coating process. *Pharmaceutical Research*, **10** (9), 1363–1371.

111. Hadfield, P., Bayhan, N., Porter, S., and Drefko, W. (**2008**) An investigation into the effects of increasing the super-disintegrant content in a tablet formulation on the ease of film coating using surface-roughness analysis to evaluate surface effects. Paper presented at the American Association of Pharmaceutical Scientists, Atlanta.

112. Wang, J., Wen, H., and Desai, D. (**2010**) Lubrications in tablet formulations. *European Journal of Pharmaceutics and Biopharmaceutics*, **75**, 1–15.

113. Miller, T. and York, P. (**1988**) Pharmaceutical tablet lubrication. *International Journal of Pharmaceutics*, **41** (1), 1–19.

114. Li, J. and Wu, Y. (**2014**) Lubricants in pharmaceutical solid dosage forms. *Lubricants*, **2** (1), 21–43.

115. Garti, N., Wellner, E., and Sarig, S. (**1980**) Stearic acid polymorphs in correlation with crystallization conditions and solvents. *Kristall und Technik*, **15** (11), 1303–1310.

116. Swaminathan, V. and Kildsig, D.O. (**2001**) An examination of the moisture sorption characteristics of commercial magnesium stearate. *AAPS PharmSciTech*, **2** (4), 1–7.

117. Dansereau, R. and Peck, G.E. (**1987**) The effect of the variability in the physical and chemical properties of magnesium stearate on the properties of compressed tablets. *Drug Development and Industrial Pharmacy*, **13** (6), 975–999.

118. Barra, J. and Somma, R. (**1996**) Influence of the physicochemical variability of magnesium stearate on its lubricant properties: Possible solutions. *Drug Development and Industrial Pharmacy*, **22** (11), 1105–1120.

119. Kestur, U., Wong, B.L.Z., Tang, D., Sharif, S.Guo, H., Desai, D. (**2012**) Effect of povidone concentration on magnesium stearate sensitivity to magnesium stearate. Paper presented at the AAPS annual meeting, Chicago.

120. Rao, K.P., Chawla, G., Kaushal, A.M., and Bansal, A.K. (**2005**) Impact of solid-state properties on lubrication efficacy of magnesium stearate. *Pharmaceutical Development and Technology*, **10** (3), 423–437.

121. Ertel, K. and Carstensen, J. (**1988**) Chemical, physical, and lubricant properties of magnesium stearate. *Journal of Pharmaceutical Sciences*, **77** (7), 625–629.

122. Narang, A.S., Desai, D., and Badawy, S. (**2012**) Impact of excipient interactions on solid dosage form stability. *Pharmaceutical Research*, **29** (10), 2660–2683.

123. Desai, D.S., Rubitski, B.A., Varia, S.A., and Newman, A.W. (**1993**) Physical interactions of magnesium stearate with starch-derived disintegrants and their effects on capsule and tablet dissolution. *International Journal of Pharmaceutics*, **91** (2–3), 217–226.

124. Bolhuis, C., Lerk, C., and Broersma, P. (**1980**) Mixing action and evaluation of tablet lubricants in direct compression. *Drug Development and Industrial Pharmacy*, **6** (1), 15–33.

125. Bossert, J. and Stains, A. (**1980**) Effect of mixing on the lubrication of crystalline lactose by magnesium stearate. *Drug Development and Industrial Pharmacy*, **6** (6), 573–589.

126. Mitrevej, K.T. and Augsburger, L. (**1982**) Adhesion of tablets in a rotary tablet press. II. Effects of blending time, running time, and lubricant concentration. *Drug Development and Industrial Pharmacy*, **8** (2), 237–282.

127. Lerk, C., Bolhuis, G., and Smedema, S. (**1977**) Interaction of lubricants and colloidal silica during mixing with excipients. I. Its effect on tabletting. *Pharmaceutica Acta Helvetiae*, **52** (3), 33–39.

128. Johansson, M.E. and Nicklasson, M. (**1987**) Influence of mixing time, particle size, and colloidal silica on the surface coverage and lubrication of magnesium stearate, in *Pharmaceutical Technology: Tableting Technology*, Halstead Press, New York, pp. 43–50.

129. Desai, D., Kothari, S., Huang, M. (**2008**) Solid-state interaction of stearic acid with povidone and its effect on dissolution stability of capsules. *International Journal of Pharmaceutics*, **354**, 77–81.

130. Rowe, R. (**2007**) Film/tablet adhesion, film thickness, internal stresses and bridging of the intagliations – A unified model with practical implications. *European Journal of Pharmaceutical Sciences*, **30** (3), 236–239.

131. Rowe, R. (**1977**) The adhesion of film coatings to tablet surfaces-the effect of some direct compression excipients and lubricants. *Journal of Pharmacy and Pharmacology*, **29** (1), 723–726.

132. Rowe, R. (**1983**) The coating of tablet surfaces by lubricants as determined by a film/tablet adhesion measurement. *Acta Pharmaceutica Suecica*, **20** (1), 77–80.

133. Pandey, P., Bindra, D.S., Gour, S. *et al.* (**2014**) Excipient–process interactions and their impact on tablet compaction and film coating. *Journal of Pharmaceutical Sciences*, **103** (11), 3666–3674.

134. Thakur, A.B., Morris, K., Grosso, J.A. *et al.* (**1993**) Mechanism and kinetics of metal ion-mediated degradation of fosinopril sodium. *Pharmaceutical Research*, **10** (6), 800–809.

135. Poznan, S.B. (**2005**) The influence of pharmaceutical excipients on quinapril hydrochloride stability. *Acta Poloniae Pharmaceutica*, **62**, 189–193.

136. Wang, J., Davidovich, M., Desai, D. *et al.* (**2010**) Solid-state interactions of a drug substance and excipients and their impact on tablet dissolution: A thermal–mechanical facilitated process induced transformation or PTI. *Journal of Pharmaceutical Sciences*, **99** (9), 3849–3862.

137. Kushner, J. and Morre, F. (**2010**) Scale-up model describing the impact of lubrication on tablet tensile strength. *International Journal of Pharmaceutics*, **399**, 19–30.

138. Roche, E. J. (**1991**) Taste masking and sustained release coatings for pharmaceuticals. US patent 5,075,114 A.

139. Desai, D., Abdelnasser, M., Rubitski, B., and Varia, S. (**1994**) Photostabilization of uncoated tablets of sorivudine and nifedipine by incorporation of synthetic iron oxides. *International Journal of Pharmaceutics*, **103** (1), 69–76.

140. Banker, G.S. (**1966**) Film coating theory and practice. *Journal of Pharmaceutical Sciences*, **55** (1), 81–89.

141. Cunningham, C. (**2007**) Reducing coated tablet defects from laboratory through production scale: Performance of hypromellose or polyvinyl alcohol-based aqueous film coating systems. Paper presented at the Annual Association of Pharmaceutical Scientists, San Diego.

142. Felton, L.A., O'Donnell, P.B., and McGinity, J.W. (**2008**) Mechanical properties of polymeric films prepared from aqueous dispersions, in *Aqueous Polymeric Coatings for Pharmaceutical Dosage Forms*, 3rd edn (eds L.A. Felton and J.W. McGinity), Informa Healthcare, pp. 105–128.

143. Obara, S. and Kokubo, H. (**2008**) Application of HPMC and HPMCAS to aqueous film coating of pharmaceutical dosage forms, in *Aqueous Polymeric Coatings for Pharmaceutical Dosage Forms*, 3rd edn (eds L.A. Felton and J.W. McGinity), Informa Healthcare, Boca Raton, FL, pp. 279–322.

144. Rajabi-Siahboomi, A.R. and Farrell, T.P. (**2008**) The applications of formulated systems for the aqueous film coating of pharmaceutical oral solid dosage forms, in *Aqueous Polymeric Coating for Pharmaceutical Dosage Forms*, 3rd edn (eds L.A. Felton and J.W. McGinity), Informa Healthcare, Boca Raton, FL Vol. 176, pp. 323–344.

145. Thoma, C., Betchtold K. (**2000**) *Enteric Coated Hard Gelatin Capsules*, in *Capsugel Library*, Capsugel pp. 1–17, Accessed online Aug 13 2015 at: http://owndoc.com/pdf/enteric-coated-hard-gelatin-capsules.pdf.

146. Shukla, A.J. (ed.) (**1994**) *Handbook of Pharmaceutical Excipients*, 2nd edn, The Pharmaceutical Press, London.

147. Jenquin, M.R., Sarabia, R.E., Liebowitz, S.M., and McGinity, J.W. (**1992**) Relationship of film properties to drug release from monolithic films containing adjuvants. *Journal of Pharmaceutical Sciences*, **81** (10), 983–989.

148. Johson, K., Hathaway, R., Leung, P., and Franz, R. (**1991**) Effect of triacetin and polyethylene glycol 400 on some physical properties of hydroxypropyl methylcellulose free films. *International Journal of Pharmaceutics*, **73**, 197–208.

149. Felton, L.A. and McGinity, J.W. (**2002**) Influence of insoluble excipients on film coating systems. *Drug Development and Industrial Pharmacy*, **28** (3), 225–243.

150. Bechard, S.R., Quraishi, O., and Kwong, E. (**1992**) Film coating: Effect of titanium dioxide concentration and film thickness on the photostability of nifedipine. *International Journal of Pharmaceutics*, **87**, 133–139.

151. Felton, L.A., Austin-Forbes, T., and Moore, T.A. (**2000**) Influence of surfactants in aqueous-based polymeric dispersions on the thermomechanical and adhesive properties of acrylic films. *Drug Development and Industrial Pharmacy*, **26** (2), 205–210.

152. Seitz, J.A., Mehta, S.P., and Yeager, J.L. (eds) (**1986**) *Tablet Coating*, 3rd edn, Lea & Febiger, Philadelphia.

153. Allam, K.V. and Kumar, G.P. (**2011**) The cosmetics for pharmaceutical dosage forms. *International Journal of Pharmacy and Pharmaceutical Sciences*, **3**, 12–21.

154. Swartz, C.J. and Cooper, J. (**1962**) Colorants for pharmaceuticals. *Journal of Pharmaceutical Sciences*, **51** (2), 89–99.

155. Cooper, J.W. Jr., Ansel, H.C., and Cadwallader, D.E. (**1973**) Liquid and solid solution interactions of primary certified colorants with pharmaceutical gelatins. *Journal of Pharmaceutical Sciences*, **62** (7), 1156–1164.

156. Lobo, M., Patel, J., Kamins, G. *et al.* (**2007**) Interactions of omapatrilat with FD&C Blue No. 2 lake during dissolution of modified release tablets. *International Journal of Pharmaceutics*, **339**, 168–174.

157. Badawy, S.I.F. and Hussain, M.A. (**2007**) Minireview: Microenvironmental pH modulation in solid dosage forms. *Journal of Pharmaceutical Sciences*, **96** (5), 948–959.

158. Hawley, M. and Morozowich, W. (**2010**) Modifying the diffusion layer of soluble salts of poorly soluble basic drugs to improve dissolution performance. *Molecular Pharmaceutics*, **7** (5), 1441–1449.

159. Badawy, S.I.F., Williams, R.C., and Gilbert, D.L. (**1999**) Chemical stability of an ester prodrug of a glycoprotein IIb/IIIa receptor antagonist in solid dosage forms. *Journal of Pharmaceutical Sciences*, **88** (4), 428–433.

160. Patel, H., Stalcup, A., Dansereau, R., and Sakr, A. (**2003**) The effect of excipients on the stability of levothyroxine sodium penta hydrate tablets. *International Journal of Pharmaceutics*, **264**, 35–43.

161. Hailu, S. and Bogner, R. (**2009**) Effects of pH grade of silicates on chemical stability of coground amorphous quinapril hydrochloride and its stabilization using pH-modifiers. *Journal of Pharmaceutical Sciences*, **98** (9), 3358–3372.

162. Ploen, J., Andersch, J., Heschel, M., and Leopold, C. (**2009**) Citric acid as a pH-modifying additive in an extended release pellets formulation containing a weakly basic drug. *Drug Development and Industrial Pharmacy*, **35** (10), 1210–1218.

163. Thoma, K. and Zimmer, T. (**1990**) Retardation of weakly basic drugs with diffusion tablets. *International Journal of Pharmaceutics*, **58**, 197–202.

164. Doherty, C. and York, P. (**1989**) Microenvironmental pH control of drug dissolution. *International Journal of Pharmaceutics*, **50**, 223–232.

165. Tirkkonnen, S., Urtti, A., and Paronen, P. (**1995**) Buffer controlled release of indomethacin from ethylcellulose microcapsules. *International Journal of Pharmaceutics*, **124**, 219–229.

166. Taniguchi, C., Inoue, R., Kaeabata, Y. *et al.* (**2012**) Novel formulations of dipyrimadole with microenvironmental pH-modifiers for improved dissolution and bioavailability under hypochlorhydria. *International Journal of Pharmaceutics*, **434**, 148–154.

167. El-Say, K.M., Refaey, T.A., Samy, A.M., and Badawy, A.A. (**2010**) Comparative study among different techniques to improve the physical and technical properties prevailing to compression of poorly flowing and highly cohesive drug. *International Journal of Pharmaceutical Sciences Review and Research*, **4**, 67–76.

168. Lipasek, R.A., Taylor, L.S., and Mauer, L.J. (**2011**) Effects of anticaking agents and relative humidity on the physical and chemical stability of powdered vitamin C. *Journal of Food Science*, **76** (7), C1062–C1074.

169. Wu, Y., Levons, J., Narang, A.S. *et al.* (**2011**) Reactive impurities in excipients: Profiling, identification and mitigation of drug–excipient incompatibility. *AAPS PharmSciTech*, **12** (4), 1248–1263.

170. Narang, A.S., Rao, V.M., and Desai, D.S. (**2012**) Effect of antioxidants and silicates on peroxides in povidone. *Journal of Pharmaceutical Sciences*, **101**, 127–139.

171. Nachaegari, S.K. and Bansal, A.K. (**2004**) Coprocessed excipients for solid dosage forms. *Pharmaceutical Technology*, **28** (1), 52–65.

172. Moreton, R.C. (**1996**) Tablet excipients to the year 2001: A look into the crystal ball. *Drug Development and Industrial Pharmacy*, **22** (1), 11–23.

173. Reimerdes, D. (**1993**) The near future of tablet excipients. *Manufacturing Chemist*, **64** (7), 14–15.

174. Khalaf, A., Tobyn, M., and Staniforth, J. (**1997**) Measurement of the flow properties of silicified microcrystalline cellulose. Paper presented at the Proceedings of the AAPS conference

175. Staniforth, J. and Chatrath, M. (**1996**) Towards a new class of high functionality tablet binders. I: Quasi-hornification of microcrystalline cellulose and loss of functionality. *Pharmaceutical Research*, **13** (9), S208.

176. Tobyn, M.J., McCarthy, G.P., Staniforth, J.N., and Edge, S. (**1998**) Physico-chemical comparison between microcrystalline cellulose and silicified microcrystalline cellulose. *International Journal of Pharmaceutics*, **169** (2), 183–194. doi: 10.1016/S0378-5173(98)00127-6

177. IPEC-Americas Co-processed Excipients Workshop. (Apr 29, **2013**). Retrieved Oct 2014 http://ipecamericas.org/system/files/coprocessed-excipient-workshop.pdf

178. Erdemir, D., Chang, S. Y., Desai, D., and Kiang, S. (**2013**, WO2013130785A3).

179. Raghavan, K.S., Ranadive, S.A., Bembenek, K.S. *et al.* (**2003**) US6589955 B2US6589955 B2, Google patents.

180. Mohammed, G., Puri, V., and Bansal, A.K. (**2008**) Coprocessing of nevirapine and stavudine by spray drying. *Pharmaceutical Development and Technology*, **13** (4), 299–310.

3

EXCIPIENTS AND THEIR FUNCTIONALITY FOR ENABLING TECHNOLOGIES IN ORAL DOSAGE FORMS

KUNNAL SHARMA AND SHREYA THAKKAR

Department of Pharmaceutical Technology (Formulations), National Institute of Pharmaceutical Education and Research (NIPER), Punjab, India

SAKSHI KHURANA AND ARVIND KUMAR BANSAL

Department of Pharmaceutics, National Institute of Pharmaceutical Education and Research (NIPER), Punjab, India

3.1 PHARMACEUTICAL EXCIPIENTS

3.1.1 Introduction

The term excipient is derived from Latin word excipiens, which has its origin from the verb excipiere, which means "to mix". United States Pharmacopeia-National Formulary (USP-NF) defines pharmaceutical excipients as the substances other than the active ingredients that are intentionally included in a drug delivery system and are appropriately evaluated for safety [1]. Excipients play an important role in processing, stability, and performance of drug products. Over the years, excipients have undergone a paradigm shift from being "inert ingredients" to "functional ingredients" in a dosage form. Excipients are being increasingly used to enhance bioavailability and overall effectiveness of drug products. Understanding this metamorphosis in the role of excipients, the traditional quality specification of excipients were realized to be

Pharmaceutical Excipients: Properties, Functionality, and Applications in Research and Industry, First Edition. Edited by Otilia M.Y. Koo.

insufficient. Pharmacopoeial monographs for excipients have traditionally focused on minimum standards of identity, purity, and quality. But in these monographs, there is little emphasis on "critical material attributes" (CMAs) necessary for intended functionality of an excipient. Mere compliance to pharmacopoeial specifications provides no guarantee of functionality of the excipient. The situation is further complicated by the fact that same excipient may perform different function in different dosage forms and manufacturing processes. For example, microcrystalline cellulose may act as a diluent, disintegrant, or compaction aid. This has led to the emergence of the concept of "functionality-related characteristics" (FRCs) and "functionality-related testing" (FRT) for ensuring performance.

3.1.2 Functionality of Excipients

Functionality is defined as the property of an excipient that helps in achieving objectives of quality, manufacturability, and performance of a dosage form [2]. It includes chemical and physical properties of the excipient. Functionality depends on many factors such as

- Chemical and physical properties of excipients
- Amount of by-products or additives present
- Interaction of excipients with other constituents of formulation
- Stress applied during the processing of product.

Excipient functionality is specific for a drug product and a manufacturing process. Hence, functionality has to be evaluated in relation to the formulation and manufacturing process. Adequate knowledge of functionality helps in effective application of process analytical technology (PAT) for achieving objectives of quality by design (QbD) [3].

3.1.3 FRCs and FRTs

FRCs are controllable chemical or physical characteristics of excipients that affect their functionality [4]. It was long realized that traditional specifications of excipients were not sufficient to control their functionality in dosage forms. These specifications focused on ensuring identification, purity, and quality of excipients. European Pharmacopoeia (EP), in 2007, introduced FRCs in the monograph of excipients, as a nonmandatory portion [4]. Parallel to EP, United States Pharmacopeia (USP) also adopted the concept of FRCs by including a general chapter on "excipient performance" [5]. This has brought harmonization among major pharmacopoeias vis-à-vis concept of FRCs. However, diversity in pharmaceutical excipients, their multiple functions, multiplicity of manufacturing process, and different approaches adopted by pharmacopoeias shall keep posing challenges to complete harmonization of requirements of FRCs.

FRCs are included to aid manufacturers of pharmaceutical products in establishing specification based on a specific application. Excipient manufacturers have also contributed to the concept of FRCs by including FRT in their specifications. However, the path is riddled with difficulties, as the same excipient can be used by different drug product manufacturers in different ways. This makes adoption of "universal" FRCs extremely difficult for an excipient. FRCs can include parameters related to chemical and physical properties. Some of the nonlimiting examples include chemical properties such as chemical composition (for chemically heterogeneous excipients), molecular weight, molecular weight distribution, degree of polymerization, and degree of substitution. Similarly, physical properties such as particle size, particle size distribution, particle shape, crystallinity/amorphous content, and compressibility may contribute to functionality [3]. FRCs should be supported with appropriate FRTs to obtain measurable parameters. Some examples of FRCs along with corresponding FRTs are included in Table 3.1.

3.1.4 FRCs and Pharmacopoeial Harmonization

International Pharmaceutical Excipient Council (IPEC) is a global organization that brings together producers, distributors and users of pharmaceutical excipients. It promotes better quality and optimal use of pharmaceutical excipients for better pharmaceutical manufacturing of a drug product. IPEC promotes best use of excipients in dosage forms so that patient treatment can be improved without affecting the efficacy, safety, and stability of active ingredients. IPEC also ensures that dosage form deliver the promised benefit to patients [6].

IPEC has been promoting inclusion of FRCs and FRT in the mandatory section of monographs of excipients. It supports pharmacopoeia to include a general chapter on systematic process for evaluation of critical properties of excipients and test methods related to these properties that are important for particular formulation [7]. However,

TABLE 3.1 Examples of FRCs and Their Testing Methods [3]

S.No.	Functionality-Related Characteristics	Functionality-Related Testing
1	Molecular mass and mass distribution	Size exclusion chromatography (SEC)
2	Degree of substitution	C^{13}NMR, FTIR spectroscopy
3	Particle size	Laser light diffraction
4	Particle size distribution/powder fineness	Sieve analysis
5	Specific surface area	Gas adsorption method
6	Water sorption	Gravimetric analysis
7	Wettability	Gravimetric analysis
8	Viscosity	Viscometers
9	Crystallinity	XRD, Solution Calorimetry
10	True density	Pycnometers
11	Flowability	Angle of repose, Hausner ratio
12	Compressibility	Compressibility index, Hausner ratio

EP and USP have adopted diverse approaches to address the issue of FRCs.

EP has included FRCs as nonmandatory portion of the excipient monographs. EP while dealing with FRCs mentions that "*the following tests are not mandatory requirements but in view of their known importance for achieving consistency in manufacture, quality and performance of medicinal products, it is recommended that suppliers should verify these characteristics and provide information on the results and analytical method applied to users. The methods indicated below have been found suitable however, other methods may be used.*"

The concept of FRCs can be further exemplified by taking an example of EP monograph of anhydrous lactose. Anhydrous lactose finds main applications as a filler/diluent in powder/compressed solid dosage forms. Those characteristics that have been suggested to be relevant for this type of application are particle size distribution (by laser diffraction or sieve analysis) and bulk/tapped density (by determining the Hausner ratio) [3].

In contrast to EP, USP does not include nonmandatory sections in excipient monographs. Rather, it has adopted a slightly different approach of including a general chapter on "excipient performance." This general chapter of USP provides an overview of the key functional categories of excipients, tests for assessing excipient performance, and test procedures that may not be presented in compendial monographs. Functional categories for most common dosage forms such as tablets and capsules, oral liquids, semisolids, topicals and suppositories, parenterals and aerosols have been included. This provides greater specificity for each functional category. Each functional category, apart from general description, provides the mechanisms of activity of the excipients, common physical and chemical properties of these excipients. For each functional category, a list of pharmacopoeial general chapters are also provided that are helpful in the development of specific tests, procedures, and acceptance criteria, which in turn help to ensure that the FRCs (referred as CMAs in USP) are adequately monitored and controlled. Details of physical and chemical properties for each functional category are provided in these general chapters. For example, under dosage form category – "tablets and capsules" various categories such as diluent, binder, disintegrant, lubricant, glidant, and/or anticaking agent and coloring agent are included [5].

3.1.5 Excipients Used in Novel Drug Delivery Systems

Novel drug delivery systems (NDDS) are sophisticated products that modulate the delivery of drugs, in terms of solubility, dissolution kinetics, release kinetics, spatial control over release, and drug targeting. Excipients are essential ingredients of these systems and perform critical role in their functioning. Excipients play a role of enabler of processability and performance in all types of dosage forms, but their role becomes even more critical in the case of NDDS. The expectations from excipients in NDDS are much higher than in conventional dosage forms. For example, polymeric excipient(s) and their FRCs such as molecular weight, viscosity, hydration, and thermal gelation are vital for their function of retarding the release of drug in modified release (MR) systems. Similarly, physical stability of amorphous form in amorphous solid dispersions (ASDs) can only be achieved due to functionality of the excipient used as "stabilizer."

The same excipient can perform different functions in different NDDS drug products. Excipient functionality can only be understood in context to a specific drug product and manufacturing process. It is thus imperative to understand the mechanisms involved in excipient functionality. This is a prerequisite for identification of meaningful FRCs and their testing. Hydroxypropyl methyl cellulose (HPMC) can act as release controlling matrix in MR products, "stabilizer" of amorphous API in ASDs, and surface stabilizer in nanocrystal-based products. The expected functionality of HPMC in these three drug products is quite different. The chemical and physical parameters contributing to these functionalities, that is, FRCs, would obviously be different. Glass transition temperature would be an FRC for ASDs as it governs anti-plasticization effect, which is critical for the physical stability of the ASDs. In contrast, "hydrophobicity" is an FRC for its use as a stabilizer in nanocrystals, as it governs the adsorption of polymer on the drug surface.

Another critical topic affecting functioning of NDDS is the phenomenon of "excipient variability." Excipient variability is a well-reported phenomenon in the case of pharmaceutical excipients. This can be contributed by variability in source, manufacturing process, or lot-to-lot variability. This variability in the FRCs of excipients, especially those performing critical function in NDDS, can have a magnified impact on performance of drug product. For example, an increase in viscosity, nominal phthalyl content, and molecular weight of HP-55S (grade of hypromellose phthalate) resulted in higher film strength and higher resistance to simulated gastric fluid in an enteric-coated formulation [8]. It thus becomes extremely important to understand the FRCs of excipients and mechanism(s) involved in their functionality. This, in turn, would allow adopting robust control strategies. It is important to understand the contribution of excipient FRC to "critical process parameters" and "critical quality attributes (CQAs)" of the delivery system. Achievement of objectives of QbD is dependent on the CQAs. The design space is defined as the multidimensional combination and interaction of material attributes and process parameters that enable achievement of desired quality. The design space is dependent on the CQAs, which in turn are affected by CPPs and CMAs. The acceptable ranges of individual CMAs of functional polymers can alter the design space, thus affecting the pharmaceutical development. Thus, overall product quality is closely hinged to CMAs of not only the API but also that of the excipients.

Bioavailability, that is, rate and extent of drug absorption in the systemic circulation is an indicator of *in vivo* performance of the drug. Solubility and permeability are two key factors that govern oral bioavailability. A significant percentage (around 70%) of new chemical entities suffer from poor aqueous solubility and pose challenge for development of optimum drug delivery system. Poor aqueous solubility of these drugs leads to erratic absorption and consequently poor oral bioavailability [9–11].

Biopharmaceutical classification system (BCS) provides a framework for classification of drugs based on their solubility and permeability. BCS class II and IV drugs have solubility-limited oral bioavailability [9, 12]. Various pharmaceutical approaches have been used to improve apparent solubility and/or dissolution rate of poorly water-soluble drugs, and these include salt formation, particle size reduction, complexation, emulsions, cosolvents, hydrotropy, nanoparticles, and solid dispersions (SDs) [10–15].

This chapter discusses the topic of NDDS from the perspective of excipients. The chapter handles various delivery systems from the perspective of excipients used, desired functionality of the excipients used, mechanistic understanding of functionality, chemical/physical parameters affecting functionality (FRCs), and suggested FRT for these FRCs.

3.2 SOLID DISPERSIONS

SDs can be classified into ASDs and crystalline solid dispersions (CSDs) based upon the properties and physical state of the drug and carrier, which can be amorphous or crystalline respectively. SDs can also be classified into four generations on the basis of development stages and compositions [11, 12, 15]. Figure 3.1 captures various classes of SDs.

First-generation SDs consist of drug particles dispersed in crystalline matrix either in amorphous or in crystalline form. Dispersed drug can form eutectic or monotectic mixture with carrier. This generation of SDs enhances solubility and dissolution rate because of the better wettability and particle size reduction. First-generation SDs have an disadvantage of lower solubility as compared to the other generations owing to the higher thermodynamic stability of system [11, 12, 15, 16].

To overcome disadvantages of first-generation SDs, second-generation SDs were developed as ASDs. They have faster dissolution rate and/or solubility owing to higher free energy of amorphous systems compared to crystalline systems [12, 15, 16]. In these systems, drug particles can be present in molecular form, amorphous particles, or small crystalline particles in amorphous carrier system. Based on miscibility

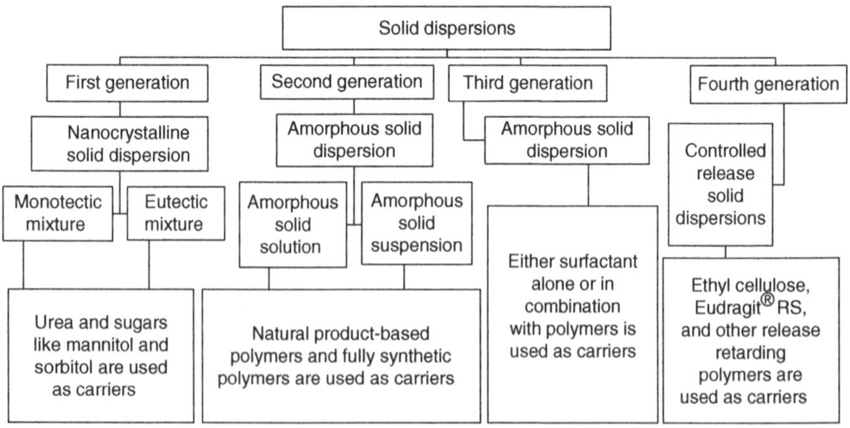

Figure 3.1 Classification of solid dispersions.

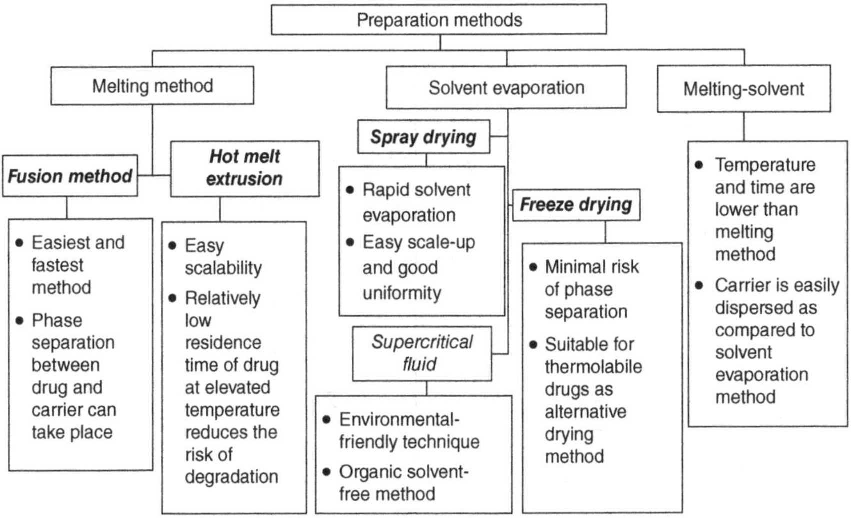

Figure 3.2 Methods of preparation for SDs.

of drug particles with carriers, these systems are further classified into amorphous solid solution and amorphous solid suspension [12, 15].

Second-generation SDs can enhance dissolution rate and/or solubility but the supersaturation of drug in carrier can lead to drug precipitation thus compromising solubility and bioavailability advantage. This limitation can be overcome by third-generation SDs by including surfactants/emulsifiers as a part of carrier system. Surfactants/emulsifiers can also be used alone for generation of SDs or as a whole in these SDs [11, 12, 15].

Fourth generation of SDs is still ill-defined but includes controlled release solid dispersions (CRSDs). These systems are beneficial for drugs that are poorly water soluble and at the same time have short biological half-life. These systems provide extended release of drug with enhanced solubility and/or dissolution rate. Rate controlling polymers are used as carriers here [12].

3.2.1 Methods of Preparation

There are three methods for the preparation of SDs: solvent evaporation, melting, and melting solvent method. Of these three methods, melting and solvent evaporation methods are most widely used (Figure 3.2) [9, 12].

3.2.2 Excipients Used in SDs

A broad range of carriers is used for the SDs as excipients in four generations of SDs. Table 3.2 captures various types of carriers used in different generations of SDs.

TABLE 3.2 Excipients Used in SDs [17]

S.No.	Excipients		Function	Usual Concentration Used (%)	Pharmacopoeial Status	Preferred Processing Method
Crystalline carriers – first-generation SDs						
1	Sugars	Sorbitol, mannitol, sucrose, xylitol, inulin, galactose, trehalose	Provide better wettability of drug particles and faster release	20–70	USP-NF, EP	Not used frequently these days
2	Others	Urea		10–30		Appropriate for solvent evaporation owing to solubility in both organic and aqueous solvents
Polymers as amorphous carriers – second-generation SDs						
3	Synthetic polymers	PEG, PVP, polymethacrylates	Provide better wettability, codissolution of drug and molecular dispersion of drug	5–20	USP-NF, EP	PEG due to its lower melting point is suitable for melting method and others are utilized in solvent evaporation method
		Eudragit® E, Carbopol 940, crospovidone, and polyvinylacetate phthalate		5–20	[a]	
4	Natural product-based polymers	Methyl cellulose (MC), HPC, hypromellose or HPMC, hypromellose acetate succinate (HPMCAS), hypromellose phthalate (HPMCP), chitosan, and some sugar derivatives		5–30	USP-NF, EP	Mainly utilized in solvent evaporation method

Surfactants + Polymers[b]– third-generation SDs					
5	Poloxamer 407	Prevent precipitation triggered by supersaturation	5–10	USP-NF, EP	Solvent evaporation method or melting method
	Gelucire 44/14		2–5	[a]	
	Compritol 888 ATO			[a]	
	Inutec SP1			[a]	
Release retarding polymers – fourth-generation SDs					
6	HPC	Retard release of drugs having poor water solubility and short half-life	5–35	USP-NF, EP	Solvent evaporation method
	EC		3–20	USP-NF, EP	
	Eudragit® RS, RL		5–20	[a]	
	PEO		5–40	USP-NF, EP	

[a]Not official in USP-NF or EP.
[b]Mentioned under second-generation SDs.

3.2.2.1 First-Generation SDs Urea and sugars such as sorbitol, mannitol, lactose, sucrose, xylitol, galactose, and inulin are used as carriers in first-generation SDs. Melting point, solubility in various solvents, and ability to form eutectic mixture are important parameters for these carriers. Urea was the first carrier used to prepare eutectic mixture and is preferred over sugars because of its good solubility in both the aqueous and organic solvents, while sugars have poor solubility in most of the organic solvents [10–12, 15]. Sugars generally have high melting point, which is not suitable while formulating SDs through hot melt method [10, 12]. Sorbitol and xylitol are moderately hygroscopic in nature [17].

3.2.2.2 Second-Generation SDs Carriers used in second-generation SDs are generally amorphous polymers, belonging to two types, that is, synthetic polymers and polymers based on natural products [10, 12]. Synthetic polymers include polyethylene glycol (PEG) [10–12], povidone (PVP) [10–12], and polymethacrylates [10, 12], while polymers based on natural products include hydroxypropylcellulose (HPC), hydroxypropylmethyl cellulose phthalate (HPMCP), ethyl cellulose (EC) and sugar derivative such as trehalose [10, 12]. Among all these PEGs, PVP and HPMC are the most widely used polymers for preparation of SDs. PVP is very hygroscopic and picks up moisture even in low humidity environments. Liquid PEGs are also hygroscopic which with increase in average molecular weight. PEG 4000 and above are nonhygroscopic in nature [17].

ASDs: These occupy a significant place in the armory of formulation development scientists as they allow enhancement of apparent solubility as well as dissolution kinetics. ASDs currently enjoy a prominent place for improvement of solubility of poorly water-soluble drugs. ASDs are defined as SDs in which the amorphous drug is molecularly dispersed in an excipient matrix [12]. Polymers are commonly used as the excipients for development of ASDs. In addition, surfactants or plasticizers may be included in the formulation to provide synergistic enhancement in solubility. These systems are characterized by miscibility of drug in the polymeric matrix. ASDs have been utilized for commercialization of numerous products such as Sporanox® (itraconazole), Prograf® (Tacrolimus), Rezulin® (pioglitazone and metformin), and Kaletra® (lopinavir and ritonavir) [14, 16].

Challenges in Development of ASDs Amorphous form of a material is the high energy state compared to crystalline form. This confers superior solubility profile to amorphous form. However, higher free energy also drives amorphous state toward lower energy crystalline form by way of recrystallization, during storage and dissolution. This conversion is associated with loss of solubility advantage from amorphous form. Efforts are required to "stabilize" the amorphous form in formulation so that higher solubility of the amorphous form can be exploited [16].

Many excipients such as polymers, surfactants, lipids, carbohydrates, proteins, sugars, amino acids, and organic acids have been used for "stabilization" of ASDs [16]. However, greatest success has been achieved with polymers for "stabilization" of amorphous form in ASDs. The most important functionality of the polymeric excipient in ASDs is the "stabilization" of the amorphous form. Excipients also need to meet criteria of safety, processability using hot melt extrusion or spray drying, nonhygroscopicity, dispersibility in dissolution medium, ensuring release of drug and inhibition of postdissolution recrystallization of drug.

Polymers achieve "stabilization" of the amorphous form via multiple mechanisms such as antiplasticization (increase of T_g value), reduced molecular mobility, and intermolecular interactions between drug–polymer such as hydrogen bonding. Studies carried out by Kakumanu *et al.* [18] showed effect of various polymers (PVP, HPMC) on the structural relaxation of amorphous celecoxib. Differential scanning calorimetry (DSC) analysis of generated batches indicated change in enthalpy relaxation rate owing to the antiplasticization effect of polymers and reduced molecular mobility of drug. Miscibility of drug with polymer is prerequisite for development of a molecular level drug–polymer solid dispersion. Most commonly, processes such as "hot melt extrusion" and "spray drying" are used to generate ASDs with polymers.

3.2.2.3 Third-Generation SDs Third-generation SDs utilize surfactants/emulsifiers along with polymers as drug carrier system. Polymers used are same as described in the preceding section. Surfactants used as carriers include Poloxamer® 407, Gelucire® 44/14 (lauryl macrogolglycerides), Compritol® 888 ATO (glyceryl behenate) and Inutec® SP1 (inulin lauryl carbonate) [10, 12]. Studies conducted by Ali *et al.* [19] showed that SDs of ibuprofen and ketoprofen with different ratios of poloxamer 407 and 188 when analyzed by Fourier transform infrared spectroscopy (FTIR) showed hydrogen bond formation between drug and carrier, and improved dissolution of both drugs.

3.2.2.4 Fourth-Generation SDs Release retarding polymers are used as carriers in CRSDs that include HPC, EC, Eudragit®, and polyethylene oxide (PEO) [12].

3.2.3 FRCs of Excipients Used in ASDs

SDs may be of various types but ASDs are the most commonly used and mainly employ polymers as carriers. Most important functionality of polymers used in ASDs includes "stabilization" of amorphous state, processability, nontacky nature, and nonhygroscopicity. Some of the important FRCs of these carriers are molecular weight, melting point or T_g, viscosity, solubility, hygroscopicity and interfacial properties such as wettability. Table 3.3 captures the chemical and physical parameters that act as FRCs of various classes of excipients used in SDs. Corresponding FRTs for these FRCs are also captured in the table.

3.3 LIPID-BASED SYSTEMS

3.3.1 Introduction

Lipid-based formulations (LBFs) include a broad range of formulations ranging from simple oily solution of drug to complex compositions consisting of mixtures of oils

TABLE 3.3 FRCs and FRTs of Excipients Used in SDs

S.No.	Excipient	Chemical/Physical Property	Significance	FRTs
1	Polymers	Molecular weight, viscosity, chain length, degree of substitution	Molecular weight/chain length of polymer affects "stabilization potential," processability, and release behavior	SEC, viscometer, chemical analysis as given in respective monographs in EP (8th edition)
		T_g (for amorphous polymers)	Polymers having high T_g are suitable as they offer greater antiplasticization and greater reduction in molecular mobility	DSC
		Melting point (for crystalline polymers)	Melting point affects miscibility of drug with carrier and its dispersibility in SDs	DSC
		Hydration capacity (especially critical for polymers that swell in aqueous medium)	Affects release of API during dissolution stage	Chemical analysis as given in respective monographs in EP (8th edition)
2	Surfactants	HLB value, surface free energy	Affects recrystallization tendency of drug and release kinetics	HLB calculated or experimentally determined using titrimetry, surface free energy by tensiometry

Source: France [3].

(triglycerides or mixed glycerides), cosolvents, surfactants, and cosurfactants (water soluble and/or water insoluble). LBFs are commonly presented as liquids but can also be converted into solid dosage form by adsorbing liquid formulation onto carrier excipients [20, 21]. LBFs offer improved bioavailability for BCS class II and class IV drugs. Drug molecules having high log *P*, low melting point, and low dose are most suitable candidates for LBFs [20].

3.3.2 Classification System for LBFs

A lipid formulation classification system (LFCS) was introduced by Pouton in 2000 that provide insights into various LBFs [22]. Additional type of formulation (Type IV) was added to this original classification in 2006 [20, 21]. LFCS provides a useful tool for selection of suitable type of LBFs for a particular drug molecule [20, 21, 23]. According to this classification system, there are four types of LBFs. Figure 3.3 captures various classes of LBFs. Most of the marketed formulations are Type III systems and they can be further subdivided into Type III A and Type III B based on the ratio of oil- and water-soluble fractions. Type III A formulations have a higher proportion of oils while Type III B have a higher proportion of water-soluble fraction [20–23].

Type I systems are blends of oils/lipids (triglycerides or mixed glycerides) and they themselves have little or no solubility in aqueous media. These systems require rapid digestion by pancreatic lipase/colipase and are almost completely absorbed after digestion due to the formation of mixed micelles. Owing to the absence of surfactants, these systems solely depend upon digestion that influences formation

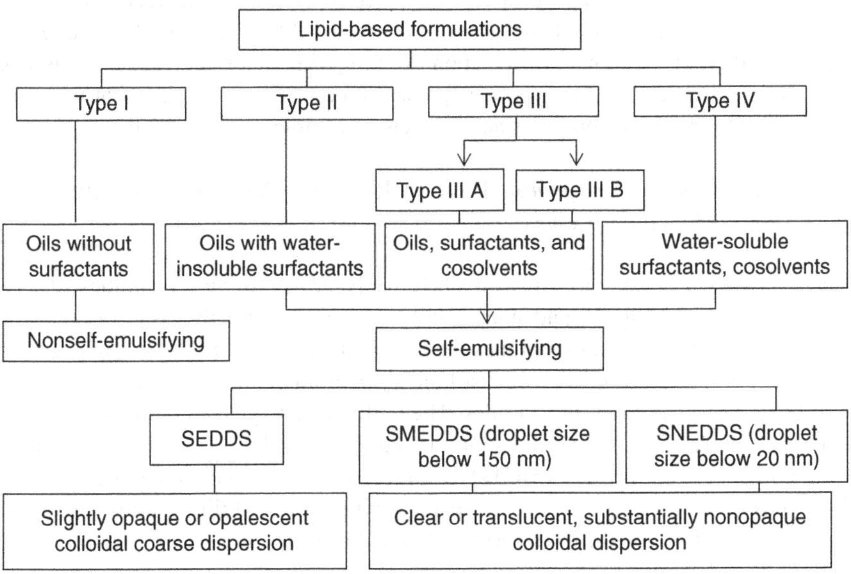

Figure 3.3 Classification of lipid-based formulations. Source: Tarate [24]. Reproduced with permission of Bentham Science.

of colloidal dispersion. Major disadvantage of these systems is that they are more suitable for highly lipophilic drugs because of the low solvent capacity of excipients used [20–23].

Type II systems are known as self-emulsifying drug delivery systems (SEDDS) formed without water-soluble components. SEDDS are isotropic blends of oils, lipophilic surfactants (HLB < 12), and cosurfactants. Concentration of surfactants is crucial in this case as self-emulsification is generally obtained at concentrations of about 25% w/w. Further increase in the surfactant concentration to 50–60% w/w impedes emulsification due to formation of viscous crystalline gel at oil/water interface. These systems form a turbid oil/water dispersion having a globule size of 0.25–2 μm [20–23].

Type III formulations are known as self-microemulsifying drug delivery systems (SMEDDS) or SEDDS with water-soluble components and are comprised of oils, surfactants, cosurfactants, and cosolvents. These systems are further divided into Type III A and Type III B as described earlier. These systems upon dilution with water form nonopaque dispersions with the droplet size below 150 nm [20–23].

Type IV formulations were added into the classification system of LFCS in 2006. These consist of water-soluble surfactants and cosolvents and are devoid of oils; thus, these are suitable for drugs that are hydrophobic but are not lipophilic. This type of system has a good solvent capacity for many drugs [20–23]

3.3.3 Excipients Used in LBFs

A wide range of excipients such as vegetable oils, surfactants (water soluble and water insoluble), cosolvents, and cosurfactants are used in LBFs. Various factors affecting selection of excipients include drug solubility, miscibility with formulation components, stability, dispersion behavior upon dilution with water and regulatory issues such as safety. The following sections present a compilation of excipients used in LBFs [25, 26]. Table 3.4 provides an overview of excipients used in LBFs.

3.3.3.1 Lipid-Based Excipients These include vegetable oils, mixed glycerides (polar oils), and their derivatives.

Vegetable Oils: These encompass majorly triglycerides, minute quantities of free fatty acids, phospholipids, and nonsaponifiable matter. Effective concentration of ester group is critical as it determines their solvent capacity for drugs. Triglycerides are classified as short-chain (<5 carbons), medium-chain (6–12 carbons), and long-chain (>12 carbons) triglycerides. They offer advantages of complete digestion and absorption after the administration and have wide regulatory acceptance. Medium-chain triglycerides (MCTs) are preferred over long-chain triglycerides (LCTs) as they have higher solvent capacity and greater stability toward oxidation. An example of MCT is the triglycerides obtained from coconut oil and palm seed oil [20, 23, 25–27].

TABLE 3.4 Excipients Used in LBFs

S.No.	Excipients		Function	Usual Concentration Used (%)	Pharmacopoeial Status
Vegetable oils (triglycerides)					
1	Long-chain triglycerides (LCT)	Corn oil	Drug carrying vehicle	15–40	USP-NF
		Peanut oil			USP-NF
		Olive oil			USP-NF, EP
		Sesame oil			USP-NF, EP
2	Medium-chain triglycerides (MCT)	Derivatives of coconut oil	Drug carrying vehicle with stability toward oxidation		[a]
		Derivatives of palm seed oil			[a]
Surfactants – water insoluble (HLB < 12)					
3	Polysorbate 85		Emulsifier (by adsorption on oil water interphase)	5–10	[a]
	Labrafil® M1944CS				USP-NF, EP
	Labrafil® M2125CS				USP-NF, EP
	Lauroglycol™ 90				USP-NF
Surfactants – water soluble (HLB > 12)					
4	Polysorbate 20		Emulsifier (capable of micelle formation on contact with water)	5–10	USP-NF, EP
	Labrasol®				USP-NF, EP
	Gelucire 44/14				USP-NF, EP
	Cremophor® EL				USP-NF

(continued)

TABLE 3.4 (*Continued*)

S.No.	Excipients		Function	Usual Concentration Used (%)	Pharmacopoeial Status
Cosolvents					
5	Ethanol		Increases solvent capacity for drug	5–15	[a]
	PEG				USP-NF, EP
	Glycerol				USP-NF, EP
Other excipients					
6	Antioxidants	α-Tocopherol	Prevent oxidative rancidity of oils and fats	0.001–0.05	USP-NF
		Butylated hydroxyl toluene (BHT)		0.02	USP-NF
		Butylated hydroxyl anisole (BHA)		Not more than 0.02	USP-NF

[a]Not official in USP-NF or EP.
Source: Rowe [17].

Mixed Glycerides and Derivatives: These are also (also called polar oils) are produced by partial hydrolysis of triglycerides and are a mixture of variable quantities of monoglycerides, diglycerides, and triglycerides. Hydrogenated vegetable oils are obtained by catalytic hydrogenation of unsaturated bonds. Partial glycerides can be made by the process of glycerolysis. Chemical composition and physical characteristics of triglyceride derivatives are function of the starting material (triglyceride) and degree of hydrolysis. Sorbitantrioleate (Span 85) a lipophilic sorbitan fatty acid ester with a HLB value of 1.8 is a good example of polar oil [20, 23, 26–28].

3.3.3.2 Surfactants

Water-Soluble Surfactants These represent the class of surfactants having HLB ≥ 12 and are most commonly used in Type III or Type IV formulations. When used above their critical micelle concentration, they are capable of forming micellar solutions. These surfactants are obtained by the reaction of hydrolyzed vegetable oils with polyethylene oxides (PEG) in the presence of alkaline catalyst or by the reaction of alcohols with ethylene oxide to produce alkyl ether ethoxylates (e.g., cetostearyl alcohol ethoxylate "cetomacrogol") [20]. Fatty acid chains in water-soluble surfactants can either be saturated or unsaturated. Cremophor RH40 is a good example of a product with saturated alkyl chains, obtained from hydrogenation of castor oil [20, 23].

Water-Insoluble Surfactants These surfactants have intermediate HLB values (8–12) and are incapable of forming micellar solution, due to the insufficient hydrophilicity. However, they easily form opaque dispersions upon application of force. Typical examples of water-insoluble surfactants are polyoxyethylene – 20 (Tween 20®), and polyoxyethylene – 20 glyceryltrioleate (Tagot – TO®) [20, 23].

3.3.3.3 Cosolvents Function of cosolvents in LBFs is to enhance the solvent capacity for drugs and to facilitate dispersion process. Commonly used cosolvents are ethanol, glycerol, PEGs, and propylene glycols. Their use can be affected by immiscibility of some cosolvents with oils and precipitation of solubilized drug on dilution [25–27].

3.3.3.4 Other Excipients Lipid-soluble antioxidants such as α-tocopherol, β-carotene, and butylated hydroxyl toluene (BHT) are used to protect LBFs against oxidation [20, 26, 27].

3.3.4 FRCs of Lipidic Excipients

Type III LBFs, that is, SEDDS are the most commonly developed LBF formulations. A typical formula of Type III LBF consists of vehicle, surfactant, cosurfactant, cosolvent, and miscellaneous (mostly stabilizer) excipients. Each excipient contributes a specific functionality to the overall functioning of LBFs. Some of the important critical functionality criteria of LBFs are drug carrying capacity, viscosity,

TABLE 3.5 FRCs and FRTs for Excipients Used in LBFs

S.No.	Excipient	Chemical/Physical Property	Significance	FRTs
1	Vegetable oils and their derivatives	Composition of fatty acid, effective concentration of ester group, melting point	Affects solvent capacity for drug	Chemical analysis and DSC
2	Surfactants (water insoluble)	HLB value, molecular weight, nonionizable versus ionizable surfactant	Higher molecular weight surfactants exhibit poor adsorption on oil–water interphase	HLB calculated or experimentally determined using titrimetry
3	Surfactants (water insoluble)	HLB value, surface free energy	Surfactants having same free surface energy as drug provide good dispersibility	HLB calculated or experimentally determined using titrimetry, tensiometry
4	Cosolvents	Viscosity	Drug carrying capacity and dispersion kinetics	Viscometer

Source: France [3].

ease of dispersibility, kinetics of drug release, postdispersion precipitation of the drug, postdispersion globule size, and zeta potential of the globule. The overall performance of these criteria is interplay of qualitative and quantitative composition of the LBF. An optimized formulation is required to achieve a fine balance between all these criteria. Table 3.5 captures the chemical and physical parameters that act as FRCs of various classes of excipients used in LBFs. Corresponding FRTs for these FRCs are also captured in the table.

3.4 NANOCRYSTALS

3.4.1 Introduction

Size reduction of a drug is a nonspecific formulation strategy that can be applied to any drug for enhancement of dissolution kinetics. In the case of many hydrophobic drugs, micronization is not able to sufficiently increase the dissolution velocity, and reduction of particle size to nanometer size might be required. "Drug nanocrystals" refers to crystals of size ranging from 10 to 1000 nm and composed of mainly a crystalline drug stabilized with excipients. Nanocrystals and ASDs are the most preferred solubility enhancement techniques, and many products have been marketed using them [13, 29, 30].

3.4.2 Advantages of Nanocrystals as Drug Delivery System

Various approaches are used to enhance apparent solubility and/or dissolution rate of poorly water-soluble drugs that have their own advantages and limitations. It is very challenging to develop a dosage form for drugs that are poorly water soluble and at the same time have a poor solubility in organic solvents. Hence, drugs having high log *P*, high melting point and high dose are suitable candidates for nanocrystal approach. Nanocrystals provide improvement in oral bioavailability by increasing dissolution rate and/or apparent solubility and mucoadhesion owing to the significant increase in the surface area to volume ratio. Nanocrystals are unique, since the drug itself contributes the bulk of the product. They are unlike other delivery systems, where significant amounts of additional excipients are present [13, 31, 32].

Increase in dissolution kinetics and apparent solubility can be explained by the Noyes–Whitney equation and Kelvin–Ostwald–Freundlich equation, respectively. Nanocrystals offer versatility for administration through various routes such as oral, topical, parenteral, ocular, and pulmonary [13, 29].

3.4.3 Methods of Preparation

Nanocrystals are generated using bottom-up, top-down, and combination techniques. Among all these, top-down techniques are the preferred ones. Most of the marketed products of nanocrystals such as Emend®, Rapamune®, Tricor® are manufactured using top-down techniques. Table 3.6 captures techniques for nanocrystal formulation [13, 29].

3.4.4 Need for Stabilization

Small size of nanocrystals offers a drug delivery advantage but also contributes to physical instability. Reduction in particle size is associated with increase in the surface free energy. Hence, the system tends to aggregate in an attempt to reduce overall free energy of the system [33].

Literature describes three mechanisms for aggregation, namely prekinetic aggregation, differential sedimentation, and orthokinetic aggregation. In any dispersion, the particles are in continuous Brownian motion, and they may collide and stick together as a result of the attractive forces between them. Such a mechanism for aggregation is known as prekinetic aggregation. The rate of aggregation in such cases would be determined by the frequency of collisions as well as the by chances of cohesion during collision. Differential sedimentation assumes importance when particles of varying sizes and density settle in a suspension. Faster settling particles collide with slower moving ones, thereby resulting in aggregation. Lastly, the orthokinetic aggregation mechanism describes the aggregation brought about by increased particle collisions, arising out of particle transportation as a result of fluid motion. Orthokinetic aggregation depends on initial particle size and velocity gradient, but is independent of temperature [31].

TABLE 3.6 Methods of Preparation of Nanocrystals

S.No.	Technique	Technology	Advantages	Limitations
Bottom-up technologies				
1	Controlled crystallization	Freeze drying	• Simple method • Easy scale-up	• Requires strict process control • Control on crystal growth is required to prevent growth to micrometer size range
		Spray drying		• Requires drug solubility in at least one solvent • Problem of residual solvent
Top-down technologies				
2	Media milling	NanoCrystals® Nanomill®	• Simple method • Easy scale-up • Large number of products have been marketed	• Erosion from milling media • Large batches are difficult to produce
3	High-pressure homogenization	DissoCubes® Nanopure®	• Universally applicable • Amenable to production of large batches • Water free production is possible	• High-energy technique

Source: Rowe [17].

Not only aggregation, but also other resulting instabilities such as sedimentation and flocculation pose major challenges in development of successful nanocrystal dispersions. Sedimentation occurs when the force of gravity acting on the particle is greater than the buoyant force provided by the dispersion medium. It is generally considered as the extreme form of nanocrystal dispersion instability and is irreversible [34]. Flocculation is the process in which destabilized particles come together to form larger aggregates. It is a type of sedimentation, an out-of-equilibrium phenomena and is a consequence of the attractive forces between particles [31]. Another stability issue encountered in nanocrystal dispersions is crystal growth, also known as Ostwald ripening. Essentially, in Ostwald ripening, the larger particles grow at the expense of smaller ones [35].

3.4.5 Mechanisms of Stabilization

Primarily, there are two major mechanisms for the stabilization of nanosuspensions: (i) electrostatic stabilization and (ii) steric stabilization. These two can be achieved by adding ionic and nonionic stabilizers into the medium, respectively. It is also possible to achieve a combination of these two mechanisms by introducing chemical functionalities within the same molecule [31].

Electrostatic stabilization can be achieved by the addition of ionic excipients. Two types of forces act on the colloidal particle, that is, the repulsive forces and attractive van der Waals forces. Use of ionic stabilizers results in adsorption of charges onto the particle surface. An electrical double layer is created, and when the particles agglomerate, these electrical double layers overlap, leading to repulsion. This repulsion prevents agglomeration of the particles. The electrostatic repulsion has the advantage of simplicity and low cost, but at the same time, it is also very sensitive to the ionic strength of the medium [31, 35].

Steric stabilization is achieved by adding amphiphilic nonionic stabilizers and is governed by the solvation effect. These polymers are attached onto the particle via an anchor segment, while their well-solvated tail extends into the bulk medium. As the two particles approach each other, these "well-solvated" tails interpenetrate, thus squeezing the bulk medium of the interparticulate space. Since the tails are "well solvated," this event is thermodynamically unfavorable, and this helps to keep the particles at a distance from each other. The particles stabilized by this mechanism are redispersible, are not sensitive to electrolyte concentrations (below their "salting out concentrations"), and the method is suitable for multiple phase systems. However, the nanosuspensions stabilized by this mechanism are sensitive to temperature changes [31, 35].

3.4.6 Excipients Used as Stabilizers

A wide variety of stabilizers are used for the stabilization of nanosuspensions and are enlisted in Table 3.7.

3.4.7 FRCs of Excipients Used in Nanocrystal Formulations

Table 3.8 captures some of the critical functionality attributes of stabilizers [3, 31].

3.5 ORAL MODIFIED RELEASE DOSAGE FORMS

3.5.1 Introduction

An important concern of any formulation design is to maintain specified drug content at the site of action. Oral immediate release delivery systems are a popular choice, but they possess limitations such as frequent dosing and fluctuations in drug plasma levels. These can adversely affect patient compliance and treatment outcome [36, 37].

Some drugs have short half-life and require multiple dosing. Frequent dosing of such drugs can be reduced by designing MR drug delivery systems [36, 37]. MR

TABLE 3.7 List of Stabilizers Used for Nanocrystals

S.No.	Category	Excipient	Pharmacopoeial Status
1	Polymers		
	(a) Synthetic		
	(i) Linear	Polyvinyl alcohol (PVA)	USP-NF, EP
		Povidone (PVP)	USP-NF, EP
	(ii) Copolymeric	PVA-PEG graft copolymers	USP-NF, EP
	(b) Semisynthetic		
	(i) Ionic	Sodium CMC	USP-NF, EP
		Sodium alginate	USP-NF, EP
	(ii) Nonionic	HPMC	USP-NF, EP
		HPC	USP-NF, EP
		HEC	USP-NF, EP
2	Surfactants		
	(a) (i) Ionic	Docusate sodium	USP-NF, EP
		Sodium lauryl sulfate	USP-NF, EP
		Polyethylene imine	[a]
	(ii) Nonionic	Tweens	USP-NF, EP
		Poloxamers	USP-NF, EP
		D-α-Tocopheryl polyethylene glycol succinate (TPGS)	[a]
3	Miscellaneous	Food proteins	[a]
		Amino acids	
		Copolymers of PEO–PPO–PEO	[a]

[a]Not official in USP-NF or EP.

Source: Shete [13], https://doaj.org/toc/2150-2668/5. Used under CC BY-SA 4.0, https://creativecommons.org/licenses/by-sa/4.0/.

dosage forms were introduced in the market in 1952 and have been increasingly used to optimize drug delivery [38]. The "USP nomenclature guidelines 2014" define them as "those dosage forms which are formulated to modify the drug release." The definition includes two types of MR dosage products – namely, the delayed release and the extended release. The guidelines specifically mention to avoid the use of terms such as "prolonged release," "repeat action," and "sustained release" to refer to the MR dosage forms [39].

MR dosage forms encourage patient compliance by reducing the dosing frequency. They reduce the fluctuation of plasma drug levels and produce a more uniform therapeutic effect [36, 40].

3.5.2 Classification of MR Dosage Forms

Figure 3.4 captures the classification of MR systems that enables description of excipients used in their designing [41].

TABLE 3.8 FRCs and FRTs of Excipients Used

S.No.	Excipient	Usual Concentration Used	Chemical/Physical Property	Significance	FRT
1	Polymeric stabilizers	Up to 5%	Molecular weight, molecular weight distribution, degree of substitution	Used as steric stabilizers. Polymers with high molecular weight are preferred. The chain length should be long enough to overcome the van der Waals forces of attraction	SEC
			Hydrophobicity	Polymer adsorption onto the drug particle is a prerequisite for steric or electrostatic stabilization. Polymers with higher hydrophobicity produce stable nanocrystals of hydrophobic drugs. This can be attributed to the strong adsorption of hydrophobic polymer onto the hydrophobic drug surface	Contact angle determination
2	Surfactants	Up to 5%	Surface energy	Many studies have shown that stable nanocrystal dispersions have been prepared when the surface energy of polymer is similar to that of the drug	Contact angle studies
			Viscosity		Viscometers

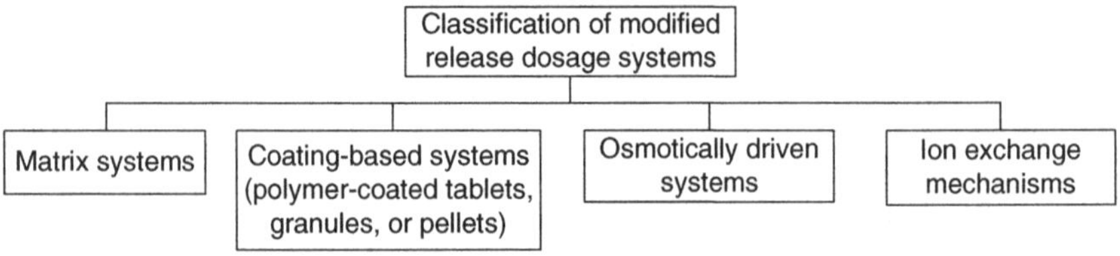

Figure 3.4 Classification of modified release dosage form.

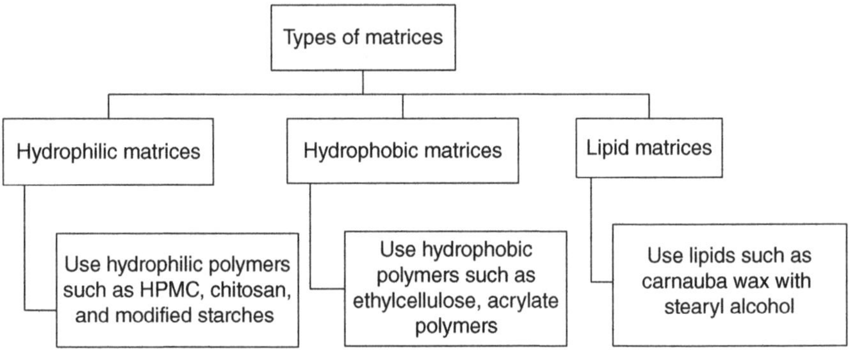

Figure 3.5 Classification of matrix MR systems.

3.5.3 Matrix-Based Systems

Matrix systems are one of the most widely used oral MR drug delivery systems. This can be attributed to their simplicity, ease of scale-up and manufacture [41]. The main advantage of matrix-based systems is that they preclude the use of complex production processes such as coating and pelletization [42]. The matrix system can be a hydrophilic system, erodible system, insoluble system, or a combination of any of these.

3.5.3.1 Types of Matrix Systems Based on the type of release control mechanism, the matrix systems can further be classified as given in Figure 3.5.

3.5.3.2 Hydrophilic Matrices Last four decades have seen widespread use of hydrophilic matrices for MR delivery. Their use to develop MR dosage forms has progressively widened, since they have a potential to control the release of diverse APIs and produce robust oral solid dosage forms [43].

The release of a drug from the hydrophilic matrix is governed by two simultaneously occurring mechanisms: (i) polymer swelling and (ii) polymer erosion. Hydrogen bonds are formed during the process of granulation and compaction of a drug with hydrogels or hydrophilic polymers. These hydrophilic matrices swell by interaction with dissolution media or biological fluids and the polymer chains eventually get disentangled, thus disrupting the hydrogen bonds. However, continuous entry of aqueous medium causes formation of hydrogen bonds between the polymeric chains and the liquid molecules. As a result, a "gel" layer is formed across the matrix tablet. Those polymeric chains that are on the surface tend to hydrate quickly than those deeper inside the core. The swelling step is followed by the erosion of the matrix. The kinetics of swelling and erosion of the polymer are pivotal in controlling the rate of drug release. The degree and kinetics of swelling, matrix erosion, and hence the rate of drug release is dependent on the concentration and viscosity of the polymer used. Highly soluble drugs by virtue of faster kinetics start diffusing through the gel layer before initiation of erosion of the matrix. Poorly water soluble can increase the matrix erosion by compromising the integrity of the gel layer. Hence, the solubility of the drug can also influence the mechanism of drug release from the matrix [43].

A wide variety of excipients are used for designing hydrophilic matrix-based MR systems. Table 3.9 lists the various excipients used in hydrophilic matrices. Mainly cellulosic polymers are used to prepare the hydrophilic matrices. They include methyl cellulose (MC), sodium carboxymethylcellulose, and HPMC. HPMC is widely used in the hydrophilic matrices because it has an excellent safety profile. HPMC is nonionic and it works well with acidic, basic, or other electrolytic systems. HPMC also works well with soluble and insoluble drugs, either at low or high doses [40]. Once the original protective gel layer is formed, it controls the ingression of additional water into the tablet.

3.5.3.3 Hydrophobic Matrices Release controlling matrices can also be prepared by compacting the drug with a hydrophobic polymer. Release control is achieved by virtue of the fact that the drug has to diffuse through a network of channels that exist between the compacted polymer chains [42]. Tablets prepared with hydrophilic polymer dissolve or erode away with time and in contrast, tablets prepared with hydrophobic polymers (Table 3.10) remain intact and maintain their integrity for longer times [46].

Drug release from hydrophobic matrices takes place by dissolution and diffusion of the drug through water-filled capillaries within the pore network of matrix. The release kinetics of the drug from the hydrophobic matrix can be explained by the concept of percolation theory. The "percolation threshold" of a component is defined as the critical concentration required to form a coherent network. Percolation threshold should be exceeded to obtain an integrated matrix and retard drug release. Similarly, the concepts of "bond percolation threshold" and "site percolation threshold" have also been proposed. The former involves a connection of the particles of the same species through a network of interparticulate bonds, whereas the latter is perceptible by the measured cohesion. Below the site percolation threshold, the matrix tablet would erode and below the bond percolation threshold it would disintegrate, thus resulting in a faster liberation of the drug [46].

3.5.3.4 Lipid Matrices Lipid matrices are prepared by using lipid waxes together with related materials. Lipids provide a hydrophobic environment and modulate the ingress of aqueous medium into the tablet matrix. In addition, diffusion of dissolved

TABLE 3.9 Excipients Used in MR Systems

S.No.	Polymer Type	Excipients	Pharmacopoeial Status	Usual Concentration Used
1	Cellulosic polymers	Hypromellose (HPMC)	USP-NF, EP	10–80%
		Methyl cellulose	USP-NF, EP	5–75%
		Hydroxypropyl cellulose (HPC)	USP-NF, EP	15–35%
		Hydroxy ethyl cellulose (HEC)	USP-NF, EP	Concentration depends on the solvent and molecular weight of the grade
		Sodium carboxymethyl cellulose (CMC)	USP-NF, EP	3–6%
2	Noncellulosic polymers	Polyvinyl pyrrolidine (PVP)	USP-NF, EP	10–25%
		Polyethylene oxide (PEO)	EP	5–40%
		Guar gum	USP-NF, EP	40–80%
		Xanthan gum	USP-NF, EP	
		Chitosan	USP-NF, EP	Up to 50%
3	Combination of two polymers			
	(a) Two nonionic polymers	Hydroxypropyl methyl cellulose–ethyl cellulose (HPMC–EC) EC is hydrophobic	[a]	2–40%
		HPMC–HPC	[a]	5–40%
	(b) Nonionic and anionic polymer	HPMC–CMC	[a]	5–40%
		PEO–CMC	[a]	2–20%
	(c) Cationic and anionic polymer	Chitosan (it contains both hydrophilic and hydrophobic groups)-sodium alginate	[a]	Up to 60%
		Chitosan–xanthan gum	[a]	

[a]Not official in USP-NF or EP.

Source: Rowe [17]. Patel, 2011 [42]. Jain [44]. Li [45].

TABLE 3.10 Excipients Used in Hydrophobic Matrices

S.No.	Excipient	Pharmacopoeial Status	Usual Concentration Used (%)
1	Ethyl cellulose (EC)	USP-NF, EP	3–20
2	Polymethacrylates (Eudragits)	USP-NF, EP	5–20
3	Hypromellose acetate succinate	USP-NF	5–10
4	Cellulose acetate	USP-NF, EP	5–10

Source: Rowe [17]. Jain [44].

TABLE 3.11 Excipients Used in Lipid Matrices [17, 44]

S.No.	Excipient	Pharmacopoeial Status	Usual Concentration Used (%)
1	Glyceryl behenate	USP-NF, EP	>10
2	Hydrogenated vegetable oils		5–10
	(a) Hydrogenated castor oil	USP-NF, EP	
	(b) Hydrogenated palm oil	USP-NF	
	(c) Hydrogenated cottonseed oil	USP-NF, EP	
3	Waxes		5–25
	(a) Paraffin wax	USP-NF, EP	
	(b) Carnauba wax	USP-NF, EP	
	(c) Candelilla wax	USP-NF	
4	Cetyl alcohol	USP-NF, EP	2–10
5	Stearyl alcohol	USP-NF, EP	
6	Glyceryl palmitostearate	[a]	10–25
7	Glyceryl monostearate	USP-NF, EP	
8	Glyceryl monooleate	USP-NF, EP	

[a]Not official in USP-NF or EP
Source: Rowe [17]. Jain [44].

drug through water-filled pores and erosion of lipid matrix also affect overall drug release [42].

Table 3.11 lists the excipients used in lipid matrices.

3.5.3.5 FRCs of Excipients Used in Matrix Systems Hydrophilic polymer matrix systems are the most common type of MR delivery systems [47]. Critical attributes for the polymers used in MR delivery systems are rate of hydration, viscosity, polymer molecular weight [48], polymer composition, substitution of polymer side chain [48], and particle size. Various chemical or physical material properties of the polymeric excipients that govern the release of drug from the matrix-based systems are captured in Table 3.12.

TABLE 3.12 FRCs and FRTs of Polymers Used in Hydrophilic Matrix Systems [3, 40, 49, 50]

S.No.	Chemical/Physical Property	Significance	FRT
1	Polymer molecular weight	Polymer molecular weight is directly related to gel strength that governs diffusion kinetics of dissolved drug and erosion capacity of polymer. The higher the molecular weight, the greater is the gel strength, leading to slower diffusion, lesser erosion, and slower drug release	SEC
2	Polymer viscosity	Viscosity of polymers is due to hydration of polymer chains and hydrogen bonding. This bonding causes the polymer chains to extend and form open random coils. These open coils may further get entangled and this contributes to an effective larger size and slowing of drug diffusion/polymer erosion	Ubbelohde viscometer, capillary viscometer, rotational viscometer
3	Hydration of polymer	The more hydrated a polymer is, more is the moisture gradient between surface and inner dry core. This moisture gradient drives the ingression of moisture into the deeper layers of matrix. Upon complete hydration, the chains start to detangle and polymer erosion starts	Dynamic vapor sorption
4	Substituents of the polymer side chain		Chemical analysis as given in respective monographs in EP (8th edition)
	(a) Structure of substituents or type of substitution	Substituent of the polymer side chain may change its polarity and melting point. If we substitute the side chain of polymer with more polar groups, it leads to decrease in crystallinity and a consequent decrease in melting point. The decrease in melting point affects the aqueous solubility, gel strength, and erosion of the polymer	

	(b) Extent of substitution	The longer the side chains are, the greater is hydrophobicity. This affects polymer hydration and diffusion kinetics of dissolved drug	
	(c) Uniformity of substitution	The greater the uniformity in degree and type of substitution, more is the homogeneity of the gel formed. As the gel formed is homogeneous, the ingression of dissolution media will be more, and this will hinder diffusion of drug toward the dissolution medium	
	(d) Pattern of substitution	Heterogeneously substituted polymers lead to slower release rates	
5	Thermal gelation	Increase in temperature causes polymer molecules to lose their hydration that leads to decrease in viscosity. Eventually sufficient dehydration causes polymer–polymer association, which is reflected by sharp increase in viscosity	DSC
6	Particle size and specific surface area	Small particle size of the polymer leads to faster formation of the gel and a slower release rate of the drug. Larger particles take more time to hydrate and swell to form the gel. Too high polymer particle size may lead to disintegration of matrix before a gel is formed	Laser diffraction
7	Powder flow	Powder flow affects processability of the excipient	Angle of repose

Source: France [3]. [40]. Viridén [49]. Maderuelo [50].

3.5.4 Oral Osmotically Driven Systems

Although the controlled drug delivery has witnessed a predominant usage of the polymer-based systems, various alternatives have also emerged. One such oral MR technology that has developed over the last 30 years is the oral osmotically driven systems [51].

The osmotically driven systems were primarily developed as device concepts for the delivery of veterinary medicines using Rose Nelson, Higuchi Leeper, and Higuchi Theeuwes pumps. Further evolution of these devices led to the development of oral osmotically driven systems [51].

Oral osmotic pumps utilize osmotic pressure as the energy source to control the drug delivery [52]. In the simplest form, they consist of a tablet core containing drug, osmotic agent, and a swellable polymer. The tablet is coated with a semipermeable membrane coating. The coating has one or more delivery orifices, through which a solution or a suspension of the drug is released over a period of time. The tablet, after coming in contact with the aqueous fluids, imbibes water at a rate governed by the osmotic pressure of the core formulation and fluid permeability of the membrane. This causes formation of a saturated solution of drug within the core, which is then released at a controlled rate via the delivery orifices on the membrane. All the oral osmotic delivery technologies have one thing in common, that is, the semipermeable membrane [52, 53].

A characteristic feature of osmotic drug delivery systems is that the rate of drug release is independent of the pH and hydrodynamics of the external dissolution medium. This results in a robust dosage form, in which the *in vivo* drug release rate is comparable to the *in vitro* rate thus providing an excellent *in vitro–in vivo* correlation. Another important advantage of osmotic systems is their applicability to drugs with a broad range of aqueous solubilities. A delivery rate of zero order is achievable, and furthermore, the delivery may be pulsed or delayed as desired [53].

3.5.4.1 Basic Components of an Osmotic Pump Osmotic pumps have many components, such as osmotic agents, wicking agents, solubilizing agents, semipermeable membranes, plasticizers, flux regulators, and pore formers [53].

Osmotic Agents/Osmagents/Osmotic Driving Agents The release rate of a drug from any osmotic system is principally governed by the osmotic pressure of the core formulation. Thus, it is critical to optimize the pressure gradient between the inside compartment and the external environment. In order to achieve and maintain a constant osmotic pressure, a saturated solution of osmotic agent must be maintained in the core compartment. In cases where the drug solution does not possess sufficient osmotic pressure, osmotic agents are added into the formulation [52]. They generate a driving force for the uptake of water and also aid in maintaining drug uniformity in the hydrated dosage form [53]. The amount of osmotic agents used varies in different products.

Some of the widely used osmotic agents are listed in Table 3.13.

TABLE 3.13 Osmotic Agents Used in Osmotic Pumps [17, 52]

S.No.	Category	Excipients	Pharmacopoeial Status
1	Water-soluble salts of inorganic acids		
	(a) Chlorides	Magnesium chloride	USP-NF, EP
		Lithium chloride	USP-NF
		Sodium chloride	USP-NF, EP
		Potassium chloride	USP-NF, EP
	(b) Sulfates	Magnesium sulfate	USP-NF, EP
		Lithium sulfate	USP-NF
		Sodium sulfate	USP-NF, EP
		Potassium sulfate	EP
	(c) Hydrogen phosphates	Sodium hydrogen phosphates	EP
		Potassium hydrogen phosphate	EP
2	Water-soluble salts of organic acids	Sodium acetate	USP-NF, EP
		Potassium acetate	USP-NF, EP
		Magnesium succinate	[a]
		Sodium benzoate	USP-NF, EP
		Sodium citrate	USP-NF, EP
		Sodium ascorbate	USP-NF, EP
3	Carbohydrates	Arabinose	[a]
		Ribose	[a]
		Xylose	USP-NF, EP
		Glucose	USP-NF, EP
		Fructose	USP-NF, EP
		Galactose	USP-NF, EP
		Mannose	[a]
		Sucrose	USP-NF, EP
		Maltose	USP-NF
		Lactose	USP-NF, EP
		Raffinose	[a]
4	Water-soluble amino acids	Glycine	USP-NF, EP
		Leucine	USP-NF, EP
		Alanine	USP-NF, EP
		Methionine	USP-NF, EP
5	Organic polymeric osmotic agents	Sodium CMC	USP-NF, EP
		Methylcellulose	USP-NF, EP
		Hydroxyethylmethylcellulose	EP
		HPMC	USP-NF, EP
		Cross-linked PVP	USP-NF, EP
		Polyethylene oxide	USP-NF
		Carbomers	USP-NF, EP
		Polyacrylamides	[a]

[a]Not official in USP-NF or EP.
Source: Rowe [17]. Verma [52].

FUNCTIONALITY-RELATED CHARACTERISTICS Osmotic agents such as salts, carbohydrates, and amino acids are highly water-soluble components. Their intended functionality in the dosage form is dependent on their degree and kinetics of solubilization. Being highly water soluble, their degree of solubilization is a noncritical attribute. However, particle size distribution can affect kinetics of dissolution and generation of osmotic pressure.

Hydration and swelling kinetics is a critical functionality attribute for polymeric osmotic agents. FRCs of hydrophilic polymers, as discussed in Section 3.5.3.5 dealing with oral MR systems, are also applicable for their function as osmotic agent. In addition, pH of the medium and ionic strength may be critical for functioning of ionizable polymers such as carbomers.

Semipermeable Membranes Semipermeable membranes are an essential part of the osmotic drug delivery systems. Hence, a choice of the rate controlling membrane is a key aspect in the formulation development of oral osmotic systems. The membrane must be permeable to water but impermeable to the solute(s). Because of this selectivity, the drug release from oral osmotic systems is largely independent of pH and agitation intensity of the GIT [52–54].

The semipermeable membrane must conform to some performance criteria to ensure success of the osmotic delivery system. The material used to make the membrane should retain its integrity throughout the operational lifetime of the device, so as to provide a constant osmotic driving force. It should remain impermeable to drug and other ingredients present in the core compartment. In addition, the membrane should be biocompatible [52–54].

Various polymers that are used for the semipermeable membrane are enlisted in Table 3.14.

FUNCTIONALITY-RELATED CHARACTERISTICS The pore size achieved in the coating of the semipermeable polymer is one of the most critical attribute for the functioning of the osmotic delivery system. Pore size of the coating is responsible for permitting entry of water into the core and controlling release of solute(s) either through "orifice"

TABLE 3.14 FRCs and FRTs of Semipermeable Membranes Used in Osmotic Pumps [17, 52]

S.No.	Polymer Category	Excipients/Polymers	Pharmacopoeial Status	Usual Concentration Used (%)
1	Cellulosic polymers	Cellulose acetate	USP-NF, EP	5–8
		Cellulose acetate butyrate	EP	
		Ethyl cellulose	USP-NF, EP	3–20
2	Eudragit	Eudragit RS 30D	USP-NF, EP	5–20
		Eudragit RL30D	USP-NF, EP	

Source: Rowe [17]. Verma [52].

or through "pores" in the coating. It is also essential that semipermeable membrane maintains its integrity throughout the functioning of delivery system. Failure to do so may cause "dose dumping" and toxic effects of the drug. Hence, permeability and thermomechanical properties of the film are critical functionality attributes for semipermeable films.

Reports linking chemical and physical properties of cellulose acetate to its permeability have been published. A decrease in acetyl content was found to increase the permeability of coating films. A direct correlation between acetyl content and glass transition temperature (T_g) of cellulose acetate has also been reported. T_g by affecting the thermomechanical properties has a significant influence on the permeability of the films. Hence, important FRC for semipermeable films could be molecular weight, uniformity of molecular weight, degree of substitution, T_g, and moisture content. Other coating ingredients such as plasticizers, coating solvent, drying temperature, and residual moisture content would also affect the properties of coating films [55].

Wicking Agents A wicking agent is a material that has the ability to draw water into the porous network of a delivery device. It has the ability to undergo physisorption with water. The role of a wicking agent is to act like a carrier and facilitate the entry of water to the inner surfaces of the core of the tablet, thereby leading to formation of channels [53, 56]. A wicking agent is usually dispersed throughout the composition and enhances the contact surface area of the drug with the incoming aqueous fluid. Various wicking agents used in osmotic systems are enlisted in Table 3.15. Various wicking agents used in osmotic systems are enlisted in Table 3.16. The usual concentration of wicking agents is around 2–5%. Kinetics of interaction with water and strength of interactions is a critical attribute for functionality of wicking agents. Particle properties such as particle size distribution and compressibility could be critical FRCs for wicking agents.

Pore Forming Agents Pore formers are mainly used for "controlled porosity" osmotic pumps (CPOP). The difference between CPOPs and elementary osmotic

TABLE 3.15 Excipients Used as Wicking Agents [17, 53]

S.No.	Excipient	Pharmacopoeial Status
1	Colloidal silica	USP-NF, EP
2	Povidone	USP-NF, EP
3	Sodium lauryl sulfate	USP-NF, EP
4	Kaolin	USP-NF, EP
5	Titanium dioxide	USP-NF, EP
6	Alumina	USP-NF
7	Bentonite	USP-NF, EP
8	Magnesium aluminum silicate	USP-NF, EP

Source: Rowe [17]. Ahuja [53].

TABLE 3.16 Excipients Used as Pore Formers [17, 52]

S.No.	Excipient	Pharmacopoeial Status
1	Dimethyl sulfone	USP-NF
2	Nicotinamide	EP
3	Saccharides	[a]
4	Amino acids	[a]
5	Sorbitol	USP-NF, EP
6	Pentaerythritol	[b]
7	Mannitol	USP-NF, EP
8	Organic aliphatic and aromatic acids	[a]
9	PLA	[b]
10	PGA	[b]

[a]These are broad categories of excipients; for specific examples, refer the official compendia.
[b]Not official in USP-NF or EP.
Source: Rowe [17]. Verma [52].

pumps (EOP) is the mechanism of drug release. In the latter, the drug is delivered through a laser-drilled orifice, whereas in the former, the drug is released through pores in the membrane. Pore formers are usually water-soluble additives that are incorporated in the membrane. Upon coming in contact with water, they dissolve, leaving behind pores in the membrane through which the drug release takes place [52].

Various excipients used as pore formers in oral osmotic dosage forms are enlisted in Table 3.16. The usual concentration of pore formers is around 2–5%. Particle properties such as particle size distributional would be an important FRC for pore former, if they are present as solid particles in the coated tablet.

Flux Regulators Flux regulators are added so as to regulate the permeability of the membrane. They may serve both purposes – hydrophilic materials may be added to improve the flux and hydrophobic materials may be added to decrease the flux [53]. Various agents used as flux regulators are mentioned in Table 3.17. The usual concentration of flux regulators is 2–5%.

Plasticizers Plasticizers play a critical role in the formation of polymeric films. They facilitate process of polymer particle coalescence by increasing the mobility of polymer chains. They decrease the intermolecular as well as intramolecular forces of attraction between polymer chains. Ideally, the plasticizer must remain in the polymeric film and have little or no tendency to volatilize [52, 53]. They can significantly change the viscoelastic behavior of the polymer and can also affect the permeability of the film formed. Some of the plasticizers used are listed in Table 3.18. Plasticizers are used in the concentration range of 0.1–0.5%. Plasticizers decrease the T_g of the polymers thus modifying mechanical properties, such as spreadability and brittleness,

TABLE 3.17 Excipients Used as Flux Regulators [17, 53]

S.No.	Category	Excipients	Pharmacopoeial Status
1	Hydrophilic materials	Polyethylene glycols	USP-NF, EP
		Polyhydric alcohols	[a]
		Polyalkylene glycols	[a]
2	Hydrophobic materials	Diethyl phthalate	USP-NF, EP
		Dimethoxy ethyl phthalate	[b]

[a]These are broad categories; for specific examples, refer the official compendia.
[b]Not official in USP-NF or EP.
Source: Rowe [17]. Ahuja [53].

TABLE 3.18 Excipients Used as Plasticizers [17, 53]

S.No.	Excipient	Pharmacopoeial Status
1	Polyethylene glycols	USP-NF, EP
2	Ethylene glycol monoacetate	[a]
3	Triethyl citrate	USP-NF, EP
4	Diethyl tartrate	[a]

[a]Not official in USP-NF or EP.
Source: Rowe [17]. Ahuja [53].

of the films. Some of the FRCs for plasticizers are degree of esterification (as many of the plasticizers are esters), molecular weight (typically plasticizers have a molecular weight of < 500 D), residual solvents, moisture content, T_g, and viscosity.

FRCs of Excipients Used in Osmotic Pumps Table 3.19 compiles FRCs of all categories of excipients used in osmotic pumps, which have been discussed in preceding sections.

3.5.5 Multiparticulate Systems

"Multiparticulate drug delivery systems" or "multiunit dosage forms" are those dosage forms that consist of numerous small discrete units [57]. A multiparticulate system for oral administration may consist of many "mini-depots" such as pellets or microencapsulated crystals contained in a capsule or tablet [58]. These "mini-depots" get distributed to varying degree in the gastrointestinal tract, after disintegration of the dosage form. These individual subunits may further be divided into fractions, based upon their specific characteristics such as size, coating, release properties, and drug content. Hence, the multiparticulates offer a wide array of possibilities for modulating drug release [57].

TABLE 3.19 FRCs and FRTs of Excipients Used in Osmotic Pumps [3]

S.No.	Excipient	Chemical/Physical Property	Significance	FRT
1	Osmotic agents	Particle size distribution, hydration and swelling kinetics, pH of medium, ionic strength	Helps in generating the osmotic pressure	Laser diffraction
2	Semipermeable polymers	Pore size, T_g of polymer, molecular weight, degree of substitution, coating solvent, drying temperature, moisture content	Pore size controls entry of water into the core. T_g affects thermomechanical properties and thus influences permeability of the film	DSC
3	Wicking agents	Particle size distribution, compressibility	Kinetics of water uptake and processability	Laser diffraction, dynamic vapor sorption (DVS)
4	Pore formers	Particle size distribution (if present as solid particles in the coated tablet)	Controls drug release	Laser diffraction
5	Plasticizers	Degree of esterification (since many plasticizers are esters), molecular weight, residual solvents, T_g, viscosity	Controls film properties	SEC, DSC, viscometer

Source: France [3].

The "pellets" used in multiparticulate systems may be defined as aggregates that are obtained from a wide variety of starting materials such as sucrose, starch, cellulose, and microcrystalline cellulose. Different drugs can be loaded onto these pellets and these pellets are further formulated as a single dosage form. This also allows simultaneous administration of two or more chemically incompatible drugs. Moreover, these pellets may have different release rates, and their combined effect may enable target release profile of the drug. Multiparticulate systems offer a variety of advantages over single-unit systems. Table 3.20 depicts a comparison of multiparticulates and single-unit systems [59].

Two basic approaches are used to achieve controlled release from the pellets: (i) matrix-coated pellets and (ii) reservoir-coated pellets. The former consist of pellets that are coated with a drug–polymer solution or dispersion, while the latter consist of a drug-loaded pellet, further having a polymer coating [59]. The practice of coating the pellets provides flexibility in modulating release profiles by varying the thickness of the coating. Pellets with variable coating thickness can be included in a unit dosage form to tailor-make release profiles. However, uncoated pellets are inferior to compacted hydrophilic matrix systems as the variable surface area of the pellets makes it difficult to achieve reproducible release profiles.

The drug and polymer are dissolved or dispersed in a common solvent for preparing matrix systems. A solid solution or a solid dispersion is obtained, upon evaporation of the solvent, which is coated onto the pellets. In the case where the drug concentration is less than its solubility in the polymer (i.e., solid solution), the drug release is mainly governed by the drug diffusivity in the polymer. In the case of a solid dispersion, the drug release can be approximated by Higuchi kinetics [59].

A reservoir-coated system on the other hand, consists of a drug-loaded pellet, which is coated with the polymer. Major advantages of this system are high drug loadings, and possibility of achieving variable release profiles by changing the type of coating polymer used [59].

3.5.5.1 Excipients Used in Multiparticulate Systems The formulation of multiparticulate systems involves the use of polymeric coatings, plasticizers, and pore formers. The polymers used for coating are essentially the same as those captured in Table 3.14. Plasticizers and pore formers have been discussed in sections "Plasticizers" and "pore forming agents", respectively.

3.6 ORODISPERSIBLE TABLETS

3.6.1 Introduction

Oral route of administration is the most preferred route owing to safety, convenience, and good patient compliance [32]. However, oral solid dosage forms sometimes pose problem of swallowing in patient populations such as psychiatric, geriatric, and pediatric. Orodispersible tablets (ODTs) are helpful for such patients as the tablets disperse readily in oral cavity thus obviating need for swallowing. They are also suitable for patients during traveling, as water is not required for ingestion of ODTs [60].

TABLE 3.20 Comparison of Multiparticulate Systems with Single-Unit Dosage Forms [58]

S.No.		Multiparticulates	Single-Unit Dosage System
1	Influence of gastric emptying	Use of multiparticulate systems reduces the dependence of drug effect on gastric emptying. The "mini-depots" are small enough to pass the actual opening of pylorus	Variability in the gastric emptying may influence reproducibility of therapeutic effects of single-unit dosage forms
2	Local GIT irritation	Wider distribution of mini-depots in the GIT decreases the incidence of local GIT irritation	Localized drug release at one site only increases the chances of local GIT irritation
3	Effect of food, intestinal transit time, and consequent variation in bioavailability	Multiparticulate systems are less affected by food and intestinal transit time. Hence, the bioavailability from these systems is more consistent	These systems are more prone to food effect for their oral bioavailability
4	Division of tablets	Multiparticulate systems can be divided into smaller doses, without the loss of the CR activity	Dividing a single-unit system usually destroys its CR activity

Source: Bechgaard [58].

EP defines ODTs as "uncovered tablet for buccal cavity, where it disperses before ingestion" [61]. These tablets, upon contact with saliva, form dispersion extemporaneously due to their highly porous nature. Orally disintegrating tablets, melt-in-mouth tablets, and fast disintegrating tablets are some of the terms that are synonymously used for this drug delivery system [62]. Disintegration time for ODTs can range from a few seconds to a minute, and according to EP, tablet that disperses or disintegrates within 3 minutes is called ODT [62]. ODTs offer formulation challenges such as rapid dispersion and taste masking of bitter APIs.

3.6.2 Methods of Preparation

Direct compression, tablet molding, freeze drying, spray drying, and sublimation are commonly used methods for the preparation of ODTs.

3.6.2.1 Direct Compression Direct compression is a process in which a mixture of drug and excipients is compressed to form tablets without any prior processing. Direct compression requires excipients that possess optimum cohesiveness for compaction and consolidation [63].Various technologies such as DuraSolv®, OraSolv®, WOWTAB®, and Flashtab® are based on direct compression method.

3.6.2.2 Tablet Molding In this process, a wet mass of drug and excipient blend is put into molds. This is followed by air drying of wetting solvent. This method produces less compact and more porous tablet structure that provides fast dissolution benefits. However, advanced packaging techniques are required for shipment of these products due to their poor mechanical strength [64].

3.6.2.3 Spray Drying Spray drying provides very rapid evaporation of solvent and porous powder. All the excipients can be solubilized in appropriate solvent to form solution for spray drying. Tablets made by compression of spray-dried powders show rapid disintegration in contact with water or saliva [64].

3.6.2.4 Freeze Drying Freeze drying is traditionally used for drying of thermolabile drugs. Proprietary techniques for ODTs such as Zydis® and Lyoc® use freeze drying to produce highly porous soft compacts that disperse rapidly in oral cavity [62].

3.6.2.5 Sublimation In this process, highly volatile excipients are compressed along with the other excipients and the tablets are subjected to sublimation. Sublimation of volatile excipients provides highly porous structure having rapid dispersion, suitable for ODTs [64].

3.6.3 Excipients Used in ODTs

Excipients in ODTs must contribute functionalities such as rapid dispersion in the presence of limited volume of water, pleasant mouth feel, taste masking, and sufficient mechanical strength. Rapid dispersion can be contributed by formation of highly porous hydrophilic structures that have ability to rapidly absorb water [61]. Excipients used must have high solubility and leave no or minimum amount of residue after the formation of dispersion [63]. Hygroscopicity can be a major challenge for ODTs, owing to the presence of porous matrix of hydrophilic excipients. Special packaging is sometimes required to protect ODTs from moisture gain [62].

A wide range of excipients is used for the preparation of ODTs and may include fillers, superdisintegrants, polymers, and other excipients such as collapse protectants, flocculating agents, preservatives, flavors and sweeteners, based on the need of product or process. Table 3.21 captures various excipients used in preparation of ODTs [64].

Mannitol and sorbitol used in ODTs provide bulk of the tablet. They are preferred owing to their negative heat of solution that contributes a pleasant mouth feel. Sorbitol is hygroscopic at humidity above 65% while mannitol is nonhygroscopic in nature [17].

Superdisintegrants contribute fast disintegration after contact with saliva and work by mechanisms such as swelling, wicking, or a combination of both. Croscarmellose sodium, crospovidone, and sodium starch glycolate are the most widely used superdisintegrants. Superdisintegrants can swell up to 40 times of their weight in contact with water. Generated swelling stress within the mechanical structure of tablet triggers disintegration into smaller granules or particles [65].

These swelling agents are made up of cross-linked polymeric chains that are highly hydrophilic in nature and form three-dimensional networks. Osmotic pressure, electrostatic forces, and entropy-based forces are three forces involved in the process of swelling and water uptake [65].

3.6.4 FRCs of Excipients Used in ODTs

Rapid dispersion, nonhygroscopicity, and pleasant mouth feel are some of the critical functionality attributes for ODTs. Rapid dispersion is to be balanced with optimal mechanical strength of the tablet. Optimization of the formulation is required to achieve a fine balance between all these criteria. Out of the functionalities mentioned here, rapid dispersion is the most critical functionality criteria for ODTs.

Rapid dispersion in ODTs can be achieved by inclusion of superdisintegrants. Superdisintegrants are cross-linked materials that swell many times of their weight upon contact with water. Density of cross-linking is the most important factor for swelling. High degree of cross-linking is desirable to prevent collapse of chains and reduce interchain intermolecular forces. This helps in sorption of water and subsequent swelling. Polymers that are not cross-linked tend to hydrate and form gel-like structure. Hence, the rate of swelling of highly cross-linked polymers is higher compared to noncross-linked polymers [65].

Table 3.22 captures the chemical and physical parameters that act as FRCs of various excipients used in ODTs. Corresponding FRTs for these FRCs are also captured in the table.

TABLE 3.21 Excipients Used in ODTs [17]

S.No.	Excipient		Function	Usual Concentration Used (%)	Preferred Processing Method	Pharmacopoeial Status
General excipients						
1	Fillers	Mannitol, sorbitol	Contribute bulk, mechanical strength, and pleasant mouth feel	10–90	Direct compression	USP-NF, EP
2	Superdisintegrants	Croscarmellose sodium Crospovidone Sodium starch glycolate	Help in rapid dispersion by way of swelling and/or wicking action upon contact with water	2–8	Direct compression	USP-NF, EP
3	Collapse protectant[a]	Dextran	Used during freeze drying to prevent collapse of cake	2–5	Lyophilization	USP-NF, EP
4	Preservatives	Methyl paraben, propyl paraben	Prevent microbial growth	0.01–0.02	All processes	USP-NF, EP
5	Sweeteners	Saccharin sodium, aspartame, acesulfame potassium	Taste masking	0.04–0.25	All processes	USP-NF, EP

(continued)

TABLE 3.21 (*Continued*)

S.No.	Excipient		Function	Usual Concentration Used (%)	Preferred Processing Method	Pharmacopoeial Status
Coprocessed excipients						
6	Disintegrant + Filler	Modified chitosan with silicon dioxide	Helps in disintegration	5–10	Direct compression	[b]
7	Filler	Spray-dried mannitol (Mannogem® EZ), spheronized granulated mannitol (Pearlitol® SD)	Filler and sweetener	50–70	Direct compression	[b]
8	Binder + Filler	Spheronized mannitol (Orocell® 200 and 400), Isomalt (Galen® IQ)	Filler and binder	2–5	Direct compression	[b]
9	Filler + disintegrant	Mannitol with crospovidone and polyvinyl acetate (Ludiflash®)	Filler and disintegrating agent	10–40	Direct compression	[b]

[a]Used in freeze-drying-based technology such as Zydis.
[b]Not official in USP-NF or EP.
Source: Rowe [17].

TABLE 3.22 FRCs and FRTs of Excipients Used in ODTs [65]

S.No.	Excipient	Physical/Chemical Property	Significance	FRTs
1	Fillers	Water capturing capacity	Disintegration and ease of dispersibility	Gravimetric techniques such as dynamic vapor sorption
		Bulk density	Compressibility of powder blend	Carr's index, Hausner ratio
		Flow properties	It will decide the need of glidants, which sometimes are having inverse effect on wettability	Angle of repose studies
2	Superdisintegrants	Degree of cross-linking	High degree of cross-linking is required for swelling	DSC and dynamic mechanical analysis (DMA)
		Degree of ionization	Increase in ionic content in structure of superdisintegrants causes increase in osmotic and electrostatic forces inside the system thus facilitating swelling	Chemical analysis
		Hydrophilic content	Higher hydrophilic content in the structure of superdisintegrants promotes swelling and disintegration process	Chemical analysis
3	Collapse protectants	Stabilization of mechanical structure	Prevents collapse of mechanical structure during lyophilization process	Thermal property such as glass transition temperature

Source: Omidian [65].

3.7 FUTURE DIRECTIONS

Advanced drug delivery systems shall play a vital role in future, as more and more "difficult-to-deliver" molecules are expected to enter the drug development pipeline. In contrast to conventional dosage forms, excipients have a greater enabling role in the processing and performance of these drug delivery systems.

Contribution of excipient variability to designing, manufacturing, and performance of drug delivery systems is critical for meaningful application of "QbD" approach. More vigorous efforts are required to understand excipient functionality and its relationship to CQAs of drug products. This would call for pharmaceutical manufacturers to work in tandem with excipient manufacturers, from early stages of development, to achieve quality targets.

Efforts have been initiated by professional bodies such as IPEC and pharmacopoeial agencies such as EP, USP, and JP to address the issue of excipient functionality. FRCs and FRT of excipients shall evolve further to achieve their intended functionality in the drug product. It is already recognized that one excipient may have different functionality in different types of drug delivery systems. Convincing evidence exists that material attributes of excipient can drastically affect performance of drug delivery systems. These material attributes emanate from differences in chemistry, solid-state properties (e.g., crystallinity, polymorphism, and amorphous form), particle properties (e.g., size, surface area, surface free energy, and surface roughness), and bulk properties (e.g., powder flow). A thorough understanding of physical and chemical material properties of the excipient and its impact on CQAs shall facilitate greater understanding of "CMAs" of an excipient.

It is beyond doubt that future shall witness intense developments in the area of excipient functionality, excipient variability, and contribution of excipients to the objectives of QbD. It is imperative that FRT keeps pace with these expectations and reliable, fast, and cheap analytical methods become available to support these initiatives.

REFERENCES

1. Pifferi, G. and Restani, P. (**2003**) The safety of pharmaceutical excipients. *Il Farmaco*, **58**, 541–550.
2. Moreton, R. C. (**2006**) Functionality and performance of excipients. *Pharmaceutical Technology,* **34**, S4, http://www.pharmtech.com/functionality-and-performance-excipients.
3. European Pharmacopoeia (8.0), Directorate for Quality of Medicines of the Council of Europe (EDQM), Strasbourg Cedex, France, **2014**.
4. Kristensen, H. G. (**2007**) Functionality-related characteristics of excipients. *Pharmaceutical Technology,* **31**, 134, http://www.pharmtech.com/functionality-related-characteristics-excipients.
5. http://ipeceurope.org/UPLOADS/09.07.21USPEuPharmacopoeiaPerspectiveOn FunctionalityRelatedCharac.-A.Gayot.pdf. Accessed on 24-12-2014
6. http:/www.ipec.org. Accessed on 24-12-2014.

7. Rios, M. (**2006**) Debating excipient functionality. *Pharmaceutical Technology,* **30**, 50, http://www.pharmtech.com/debating-excipient-functionality.
8. http://www.elementoorganika.ru/files/hpmcp.pdf. Accessed on 26/03/2015.
9. Srinarong, P., de Waard, H., Frijlink, H.W., and Hinrichs, W.L. (**2011**) Improved dissolution behavior of lipophilic drugs by solid dispersions: The production process as starting point for formulation considerations. *Expert Opinion on Drug Delivery*, **8**, 1121–1140.
10. Leuner, C. and Dressman, J. (**2000**) Improving drug solubility for oral delivery using solid dispersions. *European Journal of Pharmaceutics and Biopharmaceutics*, **50**, 47–60.
11. Wairkar, S. and Gaud, R. (**2013**) Solid dispersions: Solubility enhancement technique for poorly soluble drugs. *International Journal of Research in Pharmaceutical and Biomedical Sciences*, **4**, 847–854.
12. Vo, C.L.N., Park, C., and Lee, B.J. (**2013**) Current trends and future perspectives of solid dispersions containing poorly water-soluble drugs. *European Journal of Pharmaceutics and Biopharmaceutics*, **85**, 799–813.
13. Shegokar, R. and Müller, R.H. (**2010**) Nanocrystals: Industrially feasible multifunctional formulation technology for poorly soluble actives. *International Journal of Pharmaceutics*, **399**, 129–139.
14. Brough, C. and Williams, R. (**2013**) Amorphous solid dispersions and nano-crystal technologies for poorly water-soluble drug delivery. *International Journal of Pharmaceutics*, **453**, 157–166.
15. Vasconcelos, T., Sarmento, B., and Costa, P. (**2007**) Solid dispersions as strategy to improve oral bioavailability of poor water soluble drugs. *Drug Discovery Today*, **12**, 1068–1075.
16. Surikutchi, B.T., Patil, S.P., Shete, G. *et al.* (**2013**) Drug-excipient behavior in polymeric amorphous solid dispersions. *Journal of Excipients and Food Chemicals*, **4**, 70–94.
17. Rowe, C.R., Sheskey, P.J., and Quinn, M.E. (**2009**) *Handbook of Pharmaceutical Excipients*, Pharmaceutical Press and American Pharmacists Association, London.
18. Kakumanu, V.K. and Bansal, A.K. (**2002**) Enthalpy relaxation studies of celecoxib amorphous mixtures. *Pharmaceutical Research*, **19**, 1873–1878.
19. Ali, W., Williams, A.C., and Rawlinson, C.F. (**2010**) Stoichiometrically governed molecular interactions in drug: Poloxamer solid dispersions. *International Journal of Pharmaceutics*, **391**, 162–168.
20. Pouton, C.W. and Porter, C.J. (**2008**) Formulation of lipid-based delivery systems for oral administration: Materials, methods and strategies. *Advanced Drug Delivery Reviews*, **60**, 625–637.
21. Pouton, C.W. (**2006**) Formulation of poorly water-soluble drugs for oral administration: Physicochemical and physiological issues and the lipid formulation classification system. *European Journal of Pharmaceutical Sciences*, **29**, 278–287.
22. Pouton, C.W. (**2000**) Lipid formulations for oral administration of drugs: Non-emulsifying, self-emulsifying and 'self-microemulsifying' drug delivery systems. *European Journal of Pharmaceutical Sciences*, **11**, S93–S98.
23. Sarpal, K., Pawar, Y.B., and Bansal, A.K. (**2010**) Self-emulsifying drug delivery systems: A strategy to improve oral bioavailability. *Current Research and Pharmaceuticals Sciences*, **11**, 342–349.
24. Tarate, B., Chavan, R., and Bansal, K.A. (**2014**) Oral solid self-emulsifying formulations: A patent review. *Recent Patents on Drug Delivery & Formulation*, **8**, 126–143.

25. Patel, M., Patel, S., Patel, N., and Patel, M. (**2011**) A review: Novel oral lipid based formulation for poorly water soluble drugs. *International Journal of Pharmaceutical Sciences and Nanotechnology*, **3**, 1182–1192.
26. Zara, S. and Nabila, M. (**2014**) Optimizing oral drug delivery using lipid based formulations. *International Research Journal of Pharmacy*, **5**, 514–522.
27. Kalepu, S., Manthina, M., and Padavala, V. (**2013**) Oral lipid-based drug delivery systems – An overview. *Acta Pharmaceutica Sinica B*, **3**, 361–372.
28. Porter, C.J., Trevaskis, N.L., and Charman, W.N. (**2007**) Lipids and lipid-based formulations: Optimizing the oral delivery of lipophilic drugs. *Nature Reviews Drug Discovery*, **6**, 231–248.
29. Junghanns, J.-U.A. and Müller, R.H. (**2008**) Nanocrystal technology, drug delivery and clinical applications. *International Journal of Nanomedicine*, **3**, 295–310.
30. Müller, R., Jacobs, C., and Kayser, O. (**2001**) Nanosuspensions as particulate drug formulations in therapy: Rationale for development and what we can expect for the future. *Advanced Drug Delivery Reviews*, **47**, 3–19.
31. Shete, G., Jain, H., Punj, D. *et al.* (**2014**) Stabilizers used in nanocrystal based drug delivery systems. *Journal of Excipients and Food Chemicals*, **5**, 184–209.
32. Thanki, K., Gangwal, R.P., Sangamwar, A.T., and Jain, S. (**2013**) Oral delivery of anticancer drugs: Challenges and opportunities. *Journal of Controlled Release*, **170**, 15–40.
33. Sun, W., Tian, W., Zhang, Y. *et al.* (**2012**) Effect of novel stabilizers – Cationic polymers on the particle size and physical stability of poorly soluble drug nanocrystals. *Nanomedicine: Nanotechnology, Biology and Medicine*, **8**, 460–467.
34. Wang, Y., Zheng, Y., Zhang, L. *et al.* (**2013**) Stability of nanosuspensions in drug delivery. *Journal of Controlled Release*, **172**, 1126–1141.
35. Wu, L., Zhang, J., and Watanabe, W. (**2011**) Physical and chemical stability of drug nanoparticles. *Advanced Drug Delivery Reviews*, **63**, 456–469.
36. Chandana, N., Gopinath, H., Bhowmik, D. *et al.* (**2013**) Modified release dosage forms. *Journal of Chemical and Pharmaceutical Sciences*, **6**, 13–21.
37. Gaur, P.K., Mishra, S., Kumar, A., and Panda, B.P. (**2013**) Development and optimization of gastroretentive mucoadhesive microspheres of gabapentin by Box–Behnken design. *Artificial Cells, Nanomedicine, and Biotechnology*, 1–11.
38. Park, K. (**2014**) Controlled drug delivery systems: Past forward and future back. *Journal of Controlled Release*, **190**, 3–8.
39. United States Pharmacopeia, The United States Pharmacopeial Convention, Rockville, MD, 2014.
40. http://msdssearch.dow.com/PublishedLiteratureDOWCOM/dh_0379/0901b803803797ad.pdf?filepath=methocel/pdfs/noreg/198-02075.pdf&fromPage=GetDoc (accessed 23 January 2015).
41. Rathbone, M.J., Hadgraft, J., and Roberts, M.S. (**2003**) *Modified Release Drug Delivery Technology*, Marcel Dekker, Inc., pp. 4–8.
42. Patel, H., Panchal, D.R., Patel, U. *et al.* (**2011**) Matrix type drug delivery system: A review. *Journal of Pharmaceutical Science and Bioscientific Research*, **1**, 143–151.
43. Ghori, M.U., Ginting, G., Smith, A.M., and Conway, B.R. (**2014**) Simultaneous quantification of drug release and erosion from hypromellose hydrophilic matrices. *International Journal of Pharmaceutics*, **465**, 405–412.
44. Jain, K.K. (**2008**) *Drug Delivery Systems*, Humana Press, pp. 220–221.

45. Li, L., Wang, L., Li, J. *et al.* (**2014**) Insights into the mechanisms of chitosan–anionic polymers-based matrix tablets for extended drug release. *International Journal of Pharmaceutics*, **476**, 253–265.
46. Grund, J., Koerber, M., Walther, M., and Bodmeier, R. (**2014**) The effect of polymer properties on direct compression and drug release from water-insoluble controlled release matrix tablets. *International Journal of Pharmaceutics*, **469**, 94–101.
47. Miranda, A., Millán, M., and Caraballo, I. (**2006**) Study of the critical points of HPMC hydrophilic matrices for controlled drug delivery. *International Journal of Pharmaceutics*, **311**, 75–81.
48. Jain, A.K., Söderlind, E., Viridén, A. *et al.* (**2014**) The influence of hydroxypropyl methylcellulose (HPMC) molecular weight, concentration and effect of food on *in vivo* erosion behavior of HPMC matrix tablets. *Journal of Controlled Release*, **187**, 50–58.
49. Viridén, A., Wittgren, B., and Larsson, A. (**2009**) Investigation of critical polymer properties for polymer release and swelling of HPMC matrix tablets. *European Journal of Pharmaceutical Sciences*, **36**, 297–309.
50. Maderuelo, C., Zarzuelo, A., and Lanao, J.M. (**2011**) Critical factors in the release of drugs from sustained release hydrophilic matrices. *Journal of Controlled Release*, **154**, 2–19.
51. Malaterre, V., Ogorka, J., Loggia, N., and Gurny, R. (**2009**) Oral osmotically driven systems: 30 years of development and clinical use. *European Journal of Pharmaceutics and Biopharmaceutics*, **73**, 311–323.
52. Verma, R.K., Krishna, D.M., and Garg, S. (**2002**) Formulation aspects in the development of osmotically controlled oral drug delivery systems. *Journal of Controlled Release*, **79**, 7–27.
53. Ahuja, N., Kumar, V., and Rathee, P. (**2012**) Osmotic-controlled release oral delivery system: An advanced oral delivery form. *The Pharma Innovation*, **1**, 7–16.
54. Singla, D., Kumar, S.H., and Nirmala (**2012**) Osmotic pump drug delivery – A novel approach. *International Journal of Research in Pharmacy and Chemistry*, **2**, 661–670.
55. Yuan, J., Dunn, D., Clipse, N. M., Newton (Jr), R. J. (**2009**) Characterization of cellulose acetate films: Formulation effects on the thermomechanical properties and permeability of free films and coating films. *Pharmaceutical Technology*, **33**, 88–100.
56. Singh, K., Walia, M.K., Agarwal, G., and Harikumar, S. (**2013**) Osmotic pump drug delivery system: A novel approach. *Journal of Drug Delivery and Therapeutics*, **3**, 156–162.
57. Nikowitz, K., Kása, P. Jr.,, Pintye-Hódi, K., and Regdon, G. Jr., (**2011**) Study of the preparation of a multiparticulate drug delivery system with a layering technique. *Powder Technology*, **205**, 155–159.
58. Bechgaard, H. and Nielsen, G.H. (**1978**) Controlled-release multiple-units and single-unit doses a literature review. *Drug Development and Industrial Pharmacy*, **4**, 53–67.
59. http://www.diss.fu-berlin.de/diss/servlets/MCRFileNodeServlet/FUDISS_derivate_000000007987/thesis.pdf (accessed 27 January 2015).
60. Schiermeier, S. and Schmidt, P.C. (**2002**) Fast dispersible ibuprofen tablets. *European Journal of Pharmaceutical Sciences*, **15**, 295–305.
61. Fini, A., Bergamante, V., Ceschel, G.C. *et al.* (**2008**) Fast dispersible/slow releasing ibuprofen tablets. *European Journal of Pharmaceutics and Biopharmaceutics*, **69**, 335–341.
62. Fu, Y., Yang, S., Jeong, S.H. *et al.* (**2004**) Orally fast disintegrating tablets: Developments, technologies, taste-masking and clinical studies. *Critical Reviews in Therapeutic Drug Carrier Systems*, **21**, 433–475.

63. Chotaliya, M.K.B. and Chakraborty, S. (**2012**) Overview of oral dispersible tablets. *International Journal of PharmTech Research*, **4**, 1712–1720.

64. Kumar, S. and Garg, S.K. (**2014**) Fast dissolving tablets (FDTs): Current status, new market opportunities, recent advances in manufacturing technologies and future prospects. *International Journal of Pharmacy and Pharmaceutical Sciences*, **6**, 22–35.

65. Omidian, H. and Park, K. (**2008**) Swelling agents and devices in oral drug delivery. *Journal of Drug Delivery Science and Technology*, **18**, 83–93.

4

EXCIPIENTS USED IN BIOTECHNOLOGY PRODUCTS

Eva Y. Chi

Department of Chemical and Biological Engineering and Center for Biomedical Engineering, University of New Mexico, Albuquerque, NM, USA

Protein-based therapeutics, including vaccines and antigens, has found a prominent place in the pharmaceutical industry. They offer some of the most effective clinical methods to prevent and treat a wide range of diseases and disorders, including cancer, infectious diseases, autoimmune diseases, and HIV/AIDS. The protein therapeutics market holds tremendous growth potential and is estimated to reach USD 156 billion by 2018 (Global Protein Therapeutics Market Outlook 2018). However, the proper stabilization of peptide and protein pharmaceuticals remains a major challenge for the industry as these molecules are only marginally stable. Both chemical and physical stabilities of biopharmaceuticals need to be optimized during formulation in order to preserve biological activity, sustain the release from a controlled released formulation, and avoid undesirable immunological reactions. A wide variety of excipients are often added to formulations and dosage forms to stabilize proteins and to optimize other desirable properties. The choice of excipients, aside from considerations of the safety, toxicity, and immunogenicity of the excipients themselves, needs to be guided by knowledge of the degradation pathways of the active pharmaceutical ingredient and mechanisms by which different excipients ameliorate those instabilities. This chapter gives an overview of the primary degradation pathways of protein-based therapeutics and the different classes of excipients used, with a focus on providing a fundamental understanding of excipient–protein interactions and the mechanisms by which excipients stabilize protein therapeutics in different dosage forms.

Pharmaceutical Excipients: Properties, Functionality, and Applications in Research and Industry, First Edition. Edited by Otilia M.Y. Koo.

4.1 UNIQUE CHALLENGES IN THE FORMULATION DEVELOPMENT OF BIOTECHNOLOGY PRODUCTS

Recombinant technology in the past few decades has led to a significant increase in the number of approved biotechnology medicines and a shift away from producing biologically active materials such as proteins from animals or humans toward cloning and fermentation [1]. The increasing use of recombinantly expressed therapeutic proteins has highlighted issues such as their stability during manufacturing and long-term storage and methods of efficacious delivery that avoid adverse immunogenic side effects. In many respects, the efficacy and safety requirements of biotechnology products are similar to those for small molecule therapeutics. However, owing to the macromolecular structure of biologics, the formulation of protein drugs faces unique challenges, including conformational (or physical) instability in addition to chemical instability, immunogenicity, and delivery route and dosage form challenges.

Proteins are very large molecules with molecular weights that can range from a few thousand to a million Daltons (e.g., 6 kDa for insulin and 1000 kDa for glutamate dehydrogenase). Proteins are made of linear chains of covalently linked amino acids and each protein has a unique amino acid sequence [2]. Most proteins fold into specific globular conformations, stabilized by noncovalent interactions such as hydrogen bonds and salt bridges, and their biological functions depend on their three-dimensional structures. The thermodynamic stability of the folded protein conformation, or conformational stability, is only about 20–80 kJ/mol in free energy more stable than unfolded, biologically inactive conformations [3–7] and is much weaker than covalent bonds (~450 kJ/mol) [3]. The small conformational stability of protein results from a delicate balance between large stabilizing forces (e.g., hydrophobic interactions, hydrogen bonding, van der Waals forces, and electrostatic forces) and large destabilizing forces (e.g., conformational entropy) [3].

Because the native protein conformation is only marginally stable, small changes of the protein environment (e.g., temperature, pH, salts) and/or of the proteins themselves (e.g., chemical modifications, mutations) can destabilize the protein structure, leading to adverse events such as denaturation, aggregation, and precipitation [8, 9]. In fact, aggregation is often observed even under physiological conditions where the protein native state is highly thermodynamically favored and in the absence of any stresses (e.g., neutral pH and 37 °C) [10]. As such, protein instability is commonly encountered during all steps of the manufacture of protein therapeutics, including cell culture, purification, formulation, fill-finish, labeling, packaging, storage, transport, and delivery, which leads to low yield and unstable product [11, 12].

Aside from protein instability, immunogenicity is another major roadblock to the clinical success of novel protein therapeutics [13]. All exogenous proteins have the potential to cause antibody formation, possibly leading to severe allergic response, reduce (or abolish) therapeutic efficacy, or even induce autoimmunity to the patients' own endogenous proteins [14–16]. Adverse antibody-mediated immune responses in treatment with early therapeutic protein products such as intravenous immune globulin (IVIG) and human growth hormone have been linked to the presence of aggregation of administered proteins for well over a half century [17]. These early case studies have shown that among the many factors that contribute to the immunogenicity of protein pharmaceuticals [14–16], aggregates of administered proteins, even at minute levels, can cause severe allergic responses, leading to anaphylactic shock and even death [14–16, 18–20]. Without costly and time-consuming clinical trials, whether aggregates of a given protein product can induce adverse responses in patients cannot be predicted, nor can the minimal level of aggregates required for safety be determined without costly and time-consuming clinical trials [21]. Moreover, the link between protein aggregates and immunogenicity is often not discovered until side effects surface following either long-term administration or increases in patient population after the drug has been approved. Thus, a major goal of protein therapeutic process development and formulation is to minimize protein aggregation.

Another challenge to the formulation and excipient selection of protein therapeutics is the special delivery routes and dosage forms that are required and/or preferred. Oral administration of medicines is the most widely used route of administration. However, this route is generally not feasible for the delivery of proteins. The inherent instability of proteins in the gastrointestinal tract, as well as low permeability through biological barriers such as the lipid membrane due to high molecular weight and hydrophilicity, implies that proteins need to be administrated parenterally or through alternative routes such as nasal or pulmonary delivery. Liquid formulations are convenient to manufacture and use and indeed has been a common dosage form. However, some proteins may not be stable enough for handling and storage in liquid formulations. Various dried (e.g., lyophilized or spray dried) and suspension formulations (e.g., insulin zinc suspensions) have been developed and successfully used. In addition, improvements in devices designed for the easier use of lyophilized products, for example, dual-chamber syringes, dual-chamber cartridges, and convenient reconstitution devices, have helped the pharmaceutical industry to develop lyophilized products without too many concerns surrounding patient compliance issues [22]. Multidosage forms of protein therapeutics are also being increasingly used when the dose needs to be split (e.g., dose titration or dose combination). The various delivery routes and dosage forms impose additional product properties that need to be achieved with formulation and excipient selection. For example, in developing spray-dried formulations for pulmonary delivery, in addition to protein stability, properties pertinent to powder particle size, flowability, hygroscopicity, agglomeration, and the density and crystallization of excipients also need to be taken into consideration [23].

For protein pharmaceuticals, the marginal stability of the drug product means that both their chemical and physical stabilities need to be optimized during development and formulation in order to preserve biological activity, sustain the release from a controlled delivery formulation, and avoid undesirable immunological reactions. Our knowledge of protein stability is increasing. However, optimizing formulation conditions, including the proper selection and use of excipients, to completely suppress aggregation and ensure 18–24 months of shelf lives remains a major challenge [1, 24–28]. The successful formulation of a protein therapeutic requires knowledge of causes and mechanisms of protein instability as well as how formulation conditions, excipients being an important component, affect protein stability.

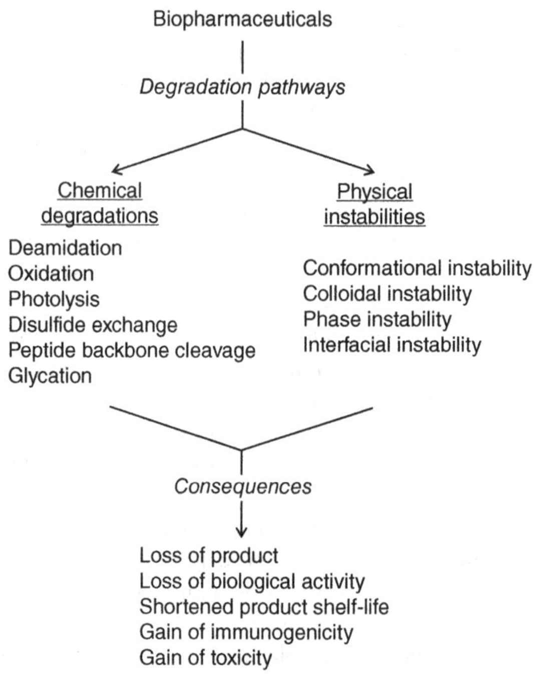

Figure 4.1 Overview of the chemical and physical instability processes observed in biopharmaceuticals and their consequences on the drug product.

4.2 DEGRADATION PATHWAYS OF PROTEINS

Degradation pathways of proteins can be separated into two general categories: chemical instability and physical instability (Figure 4.1). Chemical instability refers to processes that break or form covalent bonds, generating new chemical entities. Commonly observed chemical modifications of protein therapeutics occurring during their *in vitro* purification, storage, and handling include deamidation, oxidation, photolysis, disulfide exchange, cleavage of peptide bonds, and glycation. Physical instability refers to processes where the physical state of the protein, that is, secondary and higher order structures, changes while the chemical composition of the protein remains unaltered. This includes unfolding/denaturation, undesirable adsorption to surfaces and interfaces, misfolding, aggregation, and precipitation. A summary of the current understanding of each of these processes is presented as follows. In addition, the interrelationship between chemical and physical instabilities is briefly discussed.

4.2.1 Chemical Degradations

4.2.1.1 Deamidation Deamidation is the most commonly encountered chemical modifications of proteins and involves the cleavage of the —NH_2 group from the amine containing asparagine and glutamine residues. Deamidation can occur under acidic, neutral, or alkaline conditions. Under acidic conditions ($pH < 3$), deamidation proceeds through the direct hydrolysis of the amide linkage to form a free carboxylic acid where asparagine and glutamine are converted into aspartic acid and glutamic acid, respectively. Hydrolysis is strongly pH dependent as the reaction is catalyzed by acid. The more prevalent and faster deamination reaction occurs at neutral to alkaline conditions ($pH > 6$), where the reaction proceeds through the formation of a cyclic imide intermediate, followed by the formation of two degradation products, aspartic acid and isoaspartic acid, for example, from asparagine. This deamidation pathway is also strongly pH dependent as it is base catalyzed. The deamidation rate of asparagine under neutral or alkaline condition is also sequence dependent; only those followed by small or hydrogen bond-donating residues (e.g., glycine, serine, asparagine, or aspartic acid) are found to undergo deamidation on a timescale relevant to pharmaceutical products, with asparagine–glycine sequence being the most reactive (see Manning et al. and references therein [29]). In addition, deamidation rate is also dependent on protein secondary and tertiary structures where the flexibility of the polypeptide chain in the region of susceptible asparagine side chains can either inhibit or enhance succinimide formation [30]. Conformationally rigid regions (e.g., ordered secondary structures such as α-helices, β-sheets, β-turns) of a protein may inhibit deamination at labile asparagine sites whereas the flexible regions may enhance the susceptibility of this site to deamination.

For biopharmaceuticals, the major concern is the alteration of protein function upon deamidation. Irreversible inactivation of enzyme activity, decrease in biological activity, and slower rates of protein refolding have all been observed due to deamidation (see Manning et al. and references therein [29]). From a regulatory perspective, deamidation generates process-related impurities and degradation products that may contribute to increased immunogenicity.

A number of formulation approaches have been used to slow the rate of deamidation. The most effective approach is controlling the pH to be in the range of 3–6, where deamidation rate for a reactive asparagine has been shown to be lowest [29]. Low processing and storage temperatures also slow the rate of deamidation, as the reaction displays the typical Arrhenius behavior. As flexibility of the polypeptide chain can impact deamidation rate, decreasing chain flexibility by the inclusion of excluded solutes, such as sugars and polyols that compact the structures of proteins, has been shown to slow deamidation [31, 32]. It has been long known that most buffers exhibit some degree of catalysis. Although the mechanistic aspects of this buffer-catalyzed deamidation are not clear, it is prudent to reduce ionic strength when possible, especially if the drug product is prone to deamidation. In addition, it has been reported that deamidation rate is highest in phosphate buffer and lowest in citrate buffer at $pH < 5$ [29].

4.2.1.2 Oxidation Oxidation is another major chemical degradation pathway for biopharmaceuticals. Any protein comprised of amino acids that contain a sulfur atom (methionine and cysteine) or an aromatic ring (histidine, tryptophan, and tyrosine) can be potentially damaged by oxidation due to the high reactivity of these amino acids with various reactive oxygen species (ROS) [27, 29, 33]. Oxidation of the reactive amino acid side chains in a protein can occur during any stage of protein production, purification, formulation, and storage and can occur through multiple mechanisms. Oxidation can be induced by contaminating oxidants, catalyzed by the presence of redox active metal ions (metal-catalyzed oxidation that is site specific), or by exposure to light (photooxidation or photolysis that is nonsite specific).

Contaminating oxidants may derive from various sources in the system including formulation excipients such as polyethylene glycol (PEG) and surfactants (see Li et al. and references therein) [33]. Residual peroxides in polysorbate 80 from the bleaching step during purification have been reported to increase the amount of oxidation products in an interleukin-2 formulation. Trace amounts of metal ion contamination that originate from reagents (e.g., buffer salts) used in protein processing and formulation have also been found to catalyze oxidation.

Redox active metal ions (e.g., Fe(II) and Cu(I)) catalyze oxidative reactions in a number of ways. First, metal ions can form high-affinity complexes with proteins through binding to the side chains of certain amino acids (e.g., glycine, aspartic acid, histidine, and cysteine), generating ROS at or near the metal binding sites and react predominantly with labile amino acids (e.g., histidine and cysteine) that are in close proximity. As such, metal-catalyzed oxidation is generally considered a site-specific mechanism. Second, they may complex and react with molecular oxygen (O_2) to produce a variety of ROS such as superoxide radical ($O_2^{\bullet -}$), hydrogen peroxide (H_2O_2), or hydroxyl radical ($^{\bullet}OH$).

The oxidation of methionine and cysteine resides has been well documented. The chemical stability of methionine has been shown to be important for protein conformation and function as its oxidation has been associated with the loss of biological activity for many proteins and can be caused by a wide range of ROS. Even molecular oxygen is potent enough to convert methionine side chain into its corresponding sulfoxide. Methionine oxidation is nearly pH independent. It is also known that different methionine residues can oxidize at different rates depending on the degree of solvent accessibility of the particular residue, where more solvent-exposed residues exhibit higher oxidation rates than less solvent-exposed residues. As such, the rate of methionine oxidation has also been observed to correlate with protein conformational stability, where non-Arrhenius kinetics has been observed near the melting temperature of the protein due to increased solvent exposure of previously buried methionine residues caused by protein unfolding.

The thiol group in cysteine can be oxidized in successive steps to generate a number of oxidation products. Formation of disulfide linkages is one of the primary oxidation products if the spatial positioning of the thiol groups in the protein is such that they are close and contact between them unhindered. Similarly, intermolecular disulfide bonds can form under favorable spatial and steric conditions, causing the formation of irreversible, covalent protein aggregates. Rate of thiol oxidation depends on pH, temperature, buffer, the type of catalyst (e.g., metal ions), and the oxygen tension. Oxidation of the thiol groups occurs not only in the presence of oxidizing agents (e.g., (H_2O_2)) but can also occur spontaneously, or autoxidation, by oxygen from the air, and this reaction can take place at an appreciable rate in the presence of trace quantities of metal ions, such as iron or copper.

Photooxidation, or the photolytic degradation of proteins, has been recognized as a potential source of chemical degradation. Both ionizing radiation (e.g., γ-rays, X-rays, electrons, and α-particles encountered during sterilization processes) and nonionizing radiation (e.g., exposure to UV and visible light) can induce photooxidation, potentially leading to irreversible damages to proteins. Tryptophan, tyrosine, phenylalanine, and cysteine residues are particularly susceptible to UVA (320–400 nm) and UVB (250–320) photolysis. The absorption of photons leads to the ionization and the formation of photolysis products through either direct interaction with amino acids or indirectly via various sensitizing agents, such as dyes, riboflavin, or oxygen. In addition, the side chains of histidine and methionine can also be oxidized by visible light in the presence of dyes and susceptibility is sensitive to pH. Photooxidation has been shown to cause loss of protein activity, discoloration, changes in protein structure, and increase in aggregate level. In addition to contaminant oxidants, formulation excipients themselves such as polysorbates (or Tweens®) have also been found to facilitate photooxidation by acting as photoenhancers, leading to more facile production of singlet oxygen [34].

Similar to other chemical degradations, oxidation may lead to modifications of the physiochemical properties of the proteins, including polarity, net charge, hydrophobicity, and isoelectric point of the protein. These changes may lead to conformational modifications to the proteins that have the potential to induce undesirable immunogenic response, diminished biological activity, or altered biological half-life [33]. To limit oxidation, a number of approaches can be used, including (i) designing packaging to avoid exposure to UV, (ii) reducing headspace to minimize exposure to oxygen, (iii) adding preferentially excluded solutes, such as polyols and sugars, to limit solvent accessibility of oxidation-sensitive side chains, (iv) adding sacrificial additives such as free methionine that will be oxidized instead of the drug product, (v) adding antioxidants, (vi) adding chelating agents to remove redox-active metal ions [29], and (vii) adopting a solid formulation, for example, lyophilized solid forms, where oxidation is much reduced compared to liquid formulation. Although the addition of excipients can be useful in mitigating oxidation, it is important to keep in mind that many excipients carry oxidative impurities [35], including polysorbates and PEG, and/or can act as photoenhancers themselves.

4.2.1.3 Peptide Backbone Cleavage In addition to covalent changes from deamidation and oxidation reactions, the peptide backbone can also be cleaved via three major mechanisms: (i) preferential hydrolysis of peptide bonds at aspartic acid residues under acidic condition, (ii) C-terminal succinimide formation at asparagine residues under physiological pH conditions, and (iii) enzymatic proteolysis including autolysis. The first mechanism, the selective hydrolysis at aspartic acid residues, can occur by heating for 15–18 hours at pH 2 and 110 °C and results in the cleavage of the aspartic acid-X peptide bond, where X can be any amino acid. In particular, the aspartic acid-proline bond is the most labile and can be hydrolyzed under conditions where other aspartic acid bonds are stable. Cleavage of the C-terminal aspartic acid bond has been observed in acidic to physiological pH.

More commonly, contaminating proteases are often found to cleave recombinant proteins during both fermentation and purification. Addition of protease inhibitors, selection of protease negative hosts, sequence modifications of susceptible sites, and optimizing fermentation and purification conditions have been useful strategies to limit proteolysis. The cleavage of peptide bonds obviously disrupts the linear sequence of the amino acid chain, but may or may not affect protein higher order structure or activity. Nonetheless, peptide bond cleavage should be minimized as it chemically modifies the drug product and can yield side products that are potentially immunogenic.

4.2.1.4 Glycation Glycation of proteins occur when a protein is stored in the presence of a reducing sugar such as glucose, lactose, fructose, and maltose. These sugars can react with protein amino groups, for example, side chain of lysine and N-terminus of polypeptide chain, leading to the formation of a Schiff base, which can undergo rearrangement to more stable products [29, 36]. These associated reactions are referred to Maillard reaction or nonenzymatic browning after the color that evolves. The Maillard reaction can occur in the solid state as well as in aqueous solution [37]. Avoiding using reducing sugars is by far the most effective formulation strategy at minimizing glycation. However, reducing sugars can even be generated *in situ* by the hydrolysis of sucrose [38, 39]. Fortunately, sucrose hydrolysis requires elevated temperatures and acidic pH and trehalose has been found to be much more stable against this type of decomposition that yields reducing sugars.

4.2.2 Physical Degradations

Physical degradations refer to processes where the physical state of the protein, that is, secondary and higher order structures, changes while the chemical composition of the protein remains unaltered. These include unfolding, denaturation, undesirable adsorption to surfaces and interfaces, misfolding, aggregation, precipitation, and phase separation. Because the protein native conformation is only marginally stable, physical degradations are routinely encountered and particularly problematic, with nonnative protein aggregation (or the assembly from native, folded proteins to form aggregates containing nonnative protein structures) being the most common. As each protein is both chemically and physically unique, each will exhibit unique stability profiles, for example, temperature, pH, and excipients for adequate stability. However, a formulation scientist can rationally approach formulation design and excipient selection if she/he has an understanding of the fundamental forces that drive protein physical instability and the pathway by which proteins undergo physical degradations. The remainder of this section summarizes general protein aggregation pathways and the different causes known to drive the aggregation of therapeutic proteins. The review by Chi et al. [8, 9] has been extensively used for the following sections and the readers are referred to the review and references therein, for further information.

4.2.2.1 General Protein Aggregation Pathway Nonnative protein aggregation is the most common and persistent physical degradation encountered during the processing and storage of biopharmaceutical. Many other changes, such as unfolding, denaturation, adsorption to surfaces/interfaces, and misfolding, also lead to the formation of nonnative aggregates as the final degradation product. Aggregation is often irreversible and accompanied by the loss of native protein structures. One common feature of protein aggregates is an increased level of nonnative intermolecular β-sheets [40]. This structural change occurs regardless of the initial structures of the native protein [40], type of stress (thermal, chemical, or physical), or the morphology of the aggregate (amorphous or fibrillar) [10, 41–45].

In order to transform protein molecules from natively folded monomers (or higher order native assemblies such as native dimers) to structurally perturbed, higher order aggregates, protein molecules in the native state must undergo both structural changes and assembly processes. The well-known Lumry–Eyring framework has been used as a starting point to analyze the aggregation pathway of many proteins [46, 47]. One representation of this framework, shown in Scheme 4.1, involves reversible conformational change of a protein (Scheme 4.1a) followed by irreversible aggregation of the nonnative species to form aggregates (Scheme 4.1b) [46–48].

$$\mathrm{N} \leftrightarrow \mathrm{TS}^{*} \rightarrow \mathrm{A}_{I} \qquad (1a)$$

$$\mathrm{A}_{I} + \mathrm{A}_{m} \rightarrow \mathrm{A}_{m+I} \qquad (1b)$$

Scheme 4.1 Lumry–Eyring framework of protein aggregation.

In Scheme 4.1, N is the native protein, TS^* represents a transition state preceding the irreversible formation of an intermediate A_I, that is, aggregation competent. A_m and A_{m+I} are aggregates containing m and $m + I$ protein molecules, respectively. Figure 4.2 shows the schematic reaction coordinate diagram of the protein aggregation pathway depicted in Scheme 4.1. By definition, TS^* is the highest energy state in the aggregation process and the free energy difference between N and TS^* is the activation free energy $\Delta G^{\ddagger}$. For a multiple step reaction, such as protein aggregation, the step that has the highest $\Delta G^{\ddagger}$ is the rate-limiting step, where increasing $\Delta G^{\ddagger}$ exponentially decreases the rate of reaction and vice versa.

Because of the importance of the transition state in controlling both the rate, where the reaction rate constant decreases exponentially with increasing $\Delta G^{\ddagger}$ according to the Eyring equation, and the order, for example, unimolecular versus bimolecular, of aggregation, it is critical to have an understanding of the physical state of TS^*.

4.2.2.2 Conformational Instability Protein can aggregate from fully or partially unfolded states, for example, aggregation of proteins during chaotrope refolding. More often, and more problematic, is the aggregation of biopharmaceuticals under physiological conditions and without applied stresses or perturbations (i.e., physiological buffers, low temperatures, and without denaturants). It has been demonstrated

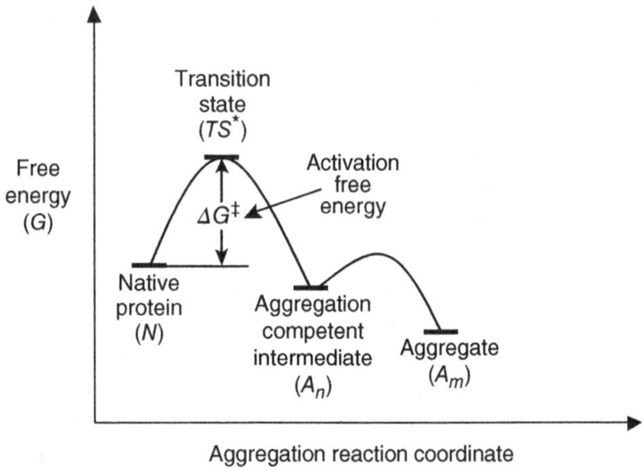

Figure 4.2 Schematic reaction coordinate diagram of protein aggregation depicted in Scheme 4.1 on an arbitrary free energy scale. Curved lines indicate kinetic energy barriers. Used with permission from Chi et al. [8, 9].

through the careful investigation of the aggregation of several pharmaceutically relevant proteins, including recombinant human interferon-γ (rhIFN- γ) [42] and recombinant human granulocyte colony stimulating factor (rhGCSF) [10], that TS* is a slightly conformationally expanded state within the native state ensemble. TS* is only approximately 15–30% expanded in the surface area compared to that for the complete unfolding of the protein [10, 42, 49]. Because the native protein conformation is not a static structure, but rather is an ensemble of native substates with a distribution of structural expansion and contraction, the interpretation from these studies is that the TS* for protein aggregation is an expanded species within the native state ensemble. As such, complete unfolding, or even partial unfolding, is not required to cause protein aggregation. Small perturbations that shift the native state ensemble toward structurally expanded species may be sufficient to induce aggregation. Furthermore, the surface area increase to form the structurally expanded species that precedes rhIFN-γ aggregation is independent of denaturant concentration, pressure, or temperature, suggesting a common intermediate for aggregation under these various stresses (Webb, [49]).

It is apparent that the intrinsic conformational stability of the protein native state plays an important role in aggregation. First, aggregation is often accompanied by the loss of native protein structures. Second, partially unfolded proteins are especially prone to aggregation. Third, the aggregation transition state of some proteins has been identified as a structurally expanded species within the protein native state ensemble. Hence, aggregation is governed by the conformational stability of the protein native state (*N*) relative to that of the aggregation transition state (TS*) (Figure 4.2).

Conformational stability of proteins is defined as the free energy difference between the unfolded state (*U*) and the natively folded state (*N*) ($\Delta G_{\text{unf}} = G_U - G_N$) and can be experimentally determined by a number of methods, such as chaotrope-induced or thermally induced unfolding experiments [50, 51]. It has been demonstrated for a few proteins, including rhIFN- γ [42] and rhGCSF [10], that increasing ΔG_{unf} values correlated with decreasing rates of aggregation (Figure 4.3). This effect is attributed to increase in $\Delta G^{\ddagger}$ as ΔG_{unf} is increased. Excipients that cause increase in the values of ΔG_{unf}, such as the preferentially excluded cosolute sucrose used in experiments depicted in Figure 4.2, shift the native state ensemble toward more structurally compact states. As a result, the cosolute is expected to be effective at reducing the rate of aggregation of proteins that proceed through a structurally expanded transition state. On the contrary, conditions that destabilize the native state, or shift the native state ensemble toward more structurally open states, can drive aggregation.

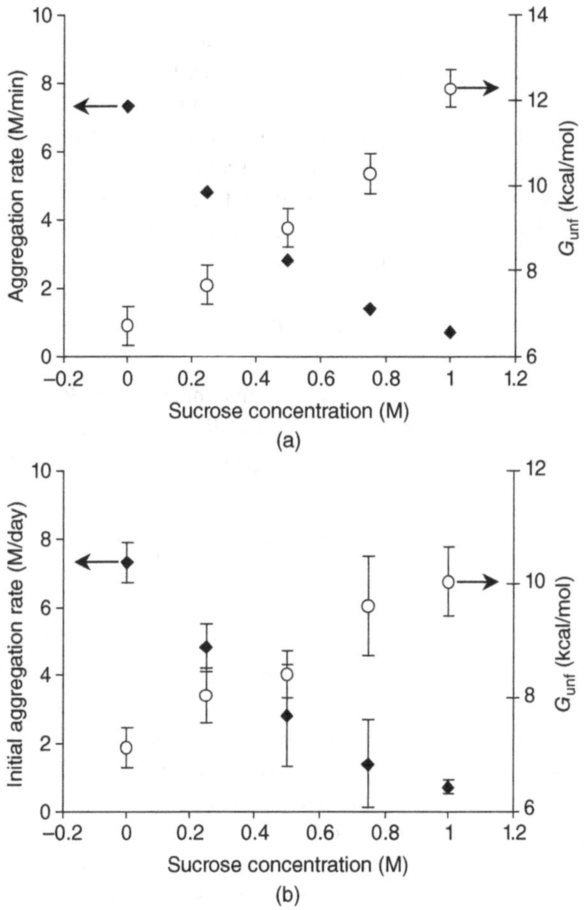

Figure 4.3 The effect of conformational stability on the aggregation rates of (a) recombinant human interferon-g (rhIFN-g) [43, 49] and (b) recombinant human granulocyte colony stimulating factor (rhGCSF) [10]. Increasing the free energy of unfolding (ΔG_{unf}) by the addition of sucrose decreased protein aggregation rates. Used with permission from Chi et al. [8, 9].

4.2.2.3 Colloidal Instability Conformational stability, however, is not the only factor influencing protein aggregation. A lack of correlation between the rate of protein aggregation and values of ΔG_{unf} has also been observed. For example, rhGCSF is stable in low salt, acidic condition (e.g., pH 3.5 HCl) but aggregates irreversibly in pH 6.1 (the protein's p*I*) phosphate buffer containing 150 mM NaCl (PBS). However, ΔG_{unf} values of rhGCSF under these two solution conditions are nearly identical [10]. Moreover, the addition of 150 mM NaCl was sufficient to induce rhGCSF aggregation in pH 3.5 HCl, although the addition of NaCl caused a negligible change in the value of ΔG_{unf} [8, 9]. Thus, the different aggregation behaviors of rhGCSF in different pH and salt conditions are not due to differences in its conformational stability. It was found that in these cases colloidal stability, which describes the propensity for self-assembly in solution, is the dominating effect in controlling the aggregation of rhGCSF [8, 9].

The role of colloidal stability in protein aggregation can be understood when one considers that in addition to the structural changes that occur during aggregation, protein molecules also need to assemble to form higher order aggregates. Molecular assembly processes occur as a result of attractive intermolecular interactions. Thus, understanding protein aggregation also requires information about the nature and magnitude of these interactions. The osmotic second virial coefficient (B_{22}) is a measure of nonideal solution behaviors that arise from two-body interactions and expressed from the osmotic virial expansion:

$$\frac{\pi}{cRT} = 1 + B_{22}c + \cdots \tag{4.1}$$

where π is the osmotic pressure, c is the protein concentration, R is the gas constant, and T is the absolute temperature. Importantly, B_{22} can be derived from statistical mechanics in terms of the pairwise, spherically symmetric interaction forces between identical proteins [52]:

$$B_{22} = \frac{2\pi}{M^2}\int_0^{\infty} r^2\left(1 - e^{-u(r)/kT}\right) dr \tag{4.2}$$

where M is the protein molecular weight, r is the intermolecular separation distance, $u(r)$ is the potential of mean force, and k is the Boltzmann constant. The interaction potential, $u(r)$, describes all of the interaction forces between two protein molecules, which include hard-sphere, electrostatic, van der Waals, and all other short-range and orientational interactions. Positive B_{22} values indicate the overall dominance of repulsive forces between protein molecules, where protein–solvent interactions are favored over protein–protein interactions [53] (i.e., the protein solution is colloidally stable). Negative B_{22} values reflect overall attractive forces between proteins, with protein–protein interactions favored over protein–solvent interactions (i.e., the protein solution is colloidally unstable).

Experimentally, the intermolecular interactions characterized by B_{22} are typically too weak to be measured by methods that probe stronger protein interactions and binding events, for example, surface plasmon resonance or fluorescence polarization. However, traditional colloidal characterization techniques, such as static light scattering [8, 9, 54–56], small angle X-ray scattering [57, 58], small angle neutron scattering [59, 60], membrane osmometry [56, 61, 62], and sedimentation equilibrium [63], are suitable for measuring B_{22} values. Measurement of B_{22} using these traditional methods, however, is not trivial and often laborious. Recent development in instrumentation, such as the flow cell-based light scattering instrument from Wyatt, vastly simplified B_{22} measurement. Furthermore, faster and potentially less expensive methods for measuring B_{22} values based on protein immobilization have been developed, namely, self-interaction chromatography [64–69] and self-interaction nanoparticle spectroscopy [70, 71]. These techniques can potentially be adopted for parallel, high-throughput formats. Alternatively, it has been shown that for monoclonal antibodies the interaction parameter, k_D, extrapolated from dynamic light scattering measurement of diffusion coefficients correlates well with B_{22} values [72]. As dynamic light scattering is amendable to parallelized, high-throughput formats and require small amounts of protein, k_D can be a useful alternative to assessing protein–protein interactions for formulation studies.While conformational stability, or ΔG_{unf} values, was found to not correlate with the rate of rhGCSF aggregation in the different pH and salt conditions, colloidal stability, as reflected in B_{22} values, was predictive of the different aggregation behaviors. In pH 3.5 HCl, B_{22} value was large and positive for the rhGCSF solution, whereas in pH 6.1 PBS, the B_{22} value was negative for the protein solution. Thus, the protein in the acidic and low salt solution is colloidally stable, exhibiting an overall strong and repulsive interaction between the protein molecules. In contrast, the protein in pH 6.1 PBS is colloidally unstable, exhibiting an overall attractive interaction. This difference in protein colloidal stability in the two different solution conditions can be understood in terms of the different electrostatic interactions that arise from the net charge of the protein at the different pH conditions. At pH 3.5, rhGCSF has a net charge of +14. Thus, the electrostatic interactions between proteins are strongly repulsive. At pH 6.1, the protein's p*I*, the next charge is zero. Thus, there is no electrostatic interaction between the proteins due to the net charge. However, the interactions are slightly attractive, as reflected in the small negative B_{22} value. The attractive interaction may arise from other contributions to protein–protein interactions, such as dipole–dipole interactions due to asymmetric charge distributions on the protein, van der Waals interactions, or any specific interactions between the protein molecules. The addition of NaCl to the low pH 3.5 HCl solution sufficiently screened the repulsive electrostatic interactions among rhGCSF molecules that aggregation occurred. The screening effect in protein–protein interactions is reflected in the decreasing values of B_{22} with increasing ionic strength. In the presence of 150 mM of NaCl, the electrostatic repulsion between protein molecules are sufficiently screened such that self-assembly was energetically favorable. For assembly-controlled protein aggregation processes, that is, aggregation is of second or higher order where the (TS*) is a multimeric state, conditions and excipients that increase the colloidal stability of the protein solution is expected to be effective at reducing the aggregation of proteins. These conditions can range from the obvious, such as pH and salt, to the subtle, such as adding preferentially excluded cosolutes [8, 9].

4.2.2.4 Phase Instability Phase instability is another driving force for undesired physical changes that can occur with biopharmaceuticals, particularly in high protein concentration liquid formulations such as monoclonal antibodies that require high dosing (~150 mg/ml) [73]. In these formulations, opalescence and high viscosity are often observed [74, 75]. Opalescence (i.e., milky white in appearance with a slight blue-tinge to human eyes) is due to Rayleigh scattering of visible light [56, 76]. Proteins, by virtue of their size, are Rayleigh scatters (particles with diameters less than 30 nm). Opalescence can therefore arise in solutions that do not contain protein aggregates or particulates. Nonetheless, the phenomenon is problematic because it can be easily confused with turbidity that results from protein aggregation or other particulate formation. High viscosity, on the other hand, can pose challenges in manufacturing processes, for example, tangential flow filtration that concentrates and formulates the protein drug, as well as in the administering of the drug by injection, particularly for subcutaneous delivery [74].

Opalescence has been correlated to attractive interactions between proteins and linked to liquid–liquid phase separation of the protein solutions. Attractive protein–protein interactions in a concentrated protein solution can lead to opalescence even in the absence of any significant association between the protein molecules [77]. In a solution containing high concentrations (e.g., 90 mg/ml antibody) of weakly but favorably interacting proteins (e.g., at a pH close to p*I*), extensive Rayleigh scattering due to critical density fluctuations can occur if the system is near the critical point, as defined by a critical temperature (T_c) and a critical concentration (C_c). The extent of opalescence, that is, how cloudy the solution appears, has been found to be dependent on proximity to the critical point [56].

Liquid–liquid phase segregation has also been reported for several antibody solutions, where a light (protein-poor) phase and a heavy (protein-rich) phase are in coexistence [78]. In such a phase-separated solution, the heavy (opalescent) phase can settle to the bottom of the vial and become completely segregated over time from the light phase [78]. Self-association of antibodies mediated mainly by favorable electrostatic interactions in the heavy phase was found to induce the liquid–liquid phase segregation.

Since opalescence is linked to the phase behavior of the protein solution, it is a reversible phenomenon. Moving the system away from the critical point or the liquid–liquid phase boundary by either increasing temperature or lowering protein concentration usually results in the disappearance of opalescence. Moreover, as opalescence is correlated with protein–protein interactions, conditions such as pH, salty type, and ionic strength can be chosen to increase repulsive protein interactions, thereby moving the critical point to temperatures below storage temperature and concentrations above dosage requirements. For example, the further away the pH is from the p*I* of an antibody, at which electrostatic repulsion due to antibody net charge is minimized, the less opalescence the solution is [76]. For formulation near the p*I*, the addition of salts generally decreases opalescence, as salts weaken attractive intermolecular electrostatic interactions. Enhancing protein–protein repulsive interactions, or reducing attractive protein–protein interactions, can reduce the propensity for phase separation and density fluctuations.

A related, but different, phenomenon in high-concentration antibody formulations is high viscosity. Antibody solutions often exhibit sharp exponential increases in solution viscosity with increasing protein concentration [56, 72–74, 79, 80]. At high antibody concentrations, viscosity has been found to depend on shear rate, or behaves as a non-Newtonian fluid. Contrary to Newtonian fluids that are comprised of noninteracting monomeric proteins, it has been suggested that concentrated antibody solutions, due to the small intermolecular separation distances that are on the order of protein molecular dimensions, may form long-range networks mediated by weak but favorable protein–protein interactions [72]. Attractive protein–protein interactions, reflected in negative values of B_{22} or the diffusion interaction parameter k_D, have been found to generally correlate with high solution viscosity, whereas repulsive protein–protein interactions, reflected in positive values of B_{22} or k_D, correlate with low solution viscosity. In low ionic strength solutions, however, the correlation between protein–protein interactions with solution viscosity is weaker. For example, a humanized monoclonal antibody of IgG_1 exhibits high viscosity in low ionic strength solution conditions where protein–protein interactions have been found to be overall repulsive. It is hypothesized that in this case, electroviscous effects from long-range charge repulsion and the related electric double layer give rise to high solution viscosity [56]. Increasing ionic strength, which diminishes the electroviscous effect, has been found to significantly decrease solution viscosity.

Since attractive protein–protcin interactions can lead to high viscosities in concentrated protein solutions, modulating these interactions by changing pH or ionic strength has been found to be effective at reducing solution viscosity. In addition, excipients such as arginine salts have been shown to be effective in reducing monoclonal antibody solution viscosity [74].

4.2.2.5 Interfacial Instability Aside from conformational, colloidal, and phase instabilities, it has also been well documented that the interfacial instability of proteins is another cause for irreversible and undesirable physical changes to the protein drug product. Proteins are inherently surface active. Compounded by the fact that interfaces and surfaces are ubiquitous in protein drug products, interfacial instability can emerge to become a potent driving force for irreversible denaturation and aggregation even if bulk solution stability, that is, conformational and colloidal, and phase stability, is optimized through formulation and excipient selection. Thus, behaviors of proteins at interfaces and the effect of the interactions with interfaces on protein physical and chemical stability, biological activity, and immunogenicity are of interest to the formulation scientists.

Peptides and protein are inherently amphiphilic as they are linear chains of nonpolar, polar uncharged, and charged amino acids. Although a polypeptide chain generally folds into a conformation that shields hydrophobic residues from aqueous solvent by positioning them in the interior of the folded structure, while exposing hydrophilic and charged residues to the aqueous solvent by placing them on the surface of the folded structure, it has long been known that proteins are surface active, spontaneously partition to air/water, liquid/liquid, and liquid/solid interfaces. The surface activity of proteins, or the propensity of proteins to go to interfaces, stems from the inhomogeneity of the protein surface, where hydrophobic patches are interspersed among hydrophilic and charged patches. Such a surface can favorably interact with a hydrophobic surface (e.g., air/water interface and silicon oil droplets) or hydrophilic surface (e.g., glass, stainless steel, cellulose filters). Upon partitioning or binding to these interfaces and surfaces, the inherent structural flexibility of proteins, as evidenced by their low conformational stability as discussed earlier in this section, allows the proteins to undergo structural reorganization to maximize favorable contacts with the interface and with other surface-adsorbed proteins. Such structural changes of the adsorbed proteins essentially cause the absorption/adsorption process to be irreversible. For natively folded proteins, such as pharmaceutical proteins, adsorption to hydrophobic surfaces usually induces large perturbations to the natively folded conformation as the proteins tend to unfold to expose the hydrophobic core in order to maximize hydrophobic interaction with the surface. Interaction between a protein molecule and a hydrophilic surface is largely mediated by electrostatic interactions and may not lead to perturbations to the protein structure. However, even under conditions where the adsorption is reversible and not accompanied by perturbation to the protein secondary or tertiary structures, partitioning to interfaces can induce aggregation and particulate formation [81].

It is clear that the adsorption of proteins to surfaces is a complex process that is driven primarily by a combination of electrostatic forces, hydrophobic interactions, and entropic changes due to contributions from both water and protein [82], although changes in hydrogen bonds and van der Waals interactions should also be considered [83]. Adsorption may be reversible or irreversible and may or may not lead to unfolding or partial unfolding of the adsorbed protein. Therapeutic proteins are exposed to various contacting surfaces, particles, and leachables during manufacturing, shipping, storage, and delivery. The extent and significance of protein adsorption within a formulation and the potential impact on product shelf life, efficacy, and safety are hard to predict. Figure 4.4 depicts some of the processes of how solid and liquid contact surfaces and leachables have been found to cause instabilities in protein products (see Bee et al. and references therein) [82]. The figure uses a vial as an example. It is important to recognize that these processes may also occur in other upstream operations and in other containers and delivery devices [82]. Surfaces and interfaces that are often encountered by protein therapeutics and known interface-induced aggregation pathways of therapeutic proteins are briefly reviewed in the following text.

Protein adsorption is affected by characteristics of the protein, solvent, and surface/interface that affect protein adsorption are summarized in Table 4.1. In liquid formulations, including manufacturing steps that involve handling of the protein in a liquid solution, a protein will encounter the air/water interface. The air/water interface can be thought of as an ideal hydrophobic interface and as such, adsorption of proteins to this interface is often accompanied by structural perturbations to the protein native state to maximize hydrophobic interactions at the interface. Denaturation or misfolding (e.g., nonnative β-sheet formation [84–86] and subsequent aggregation

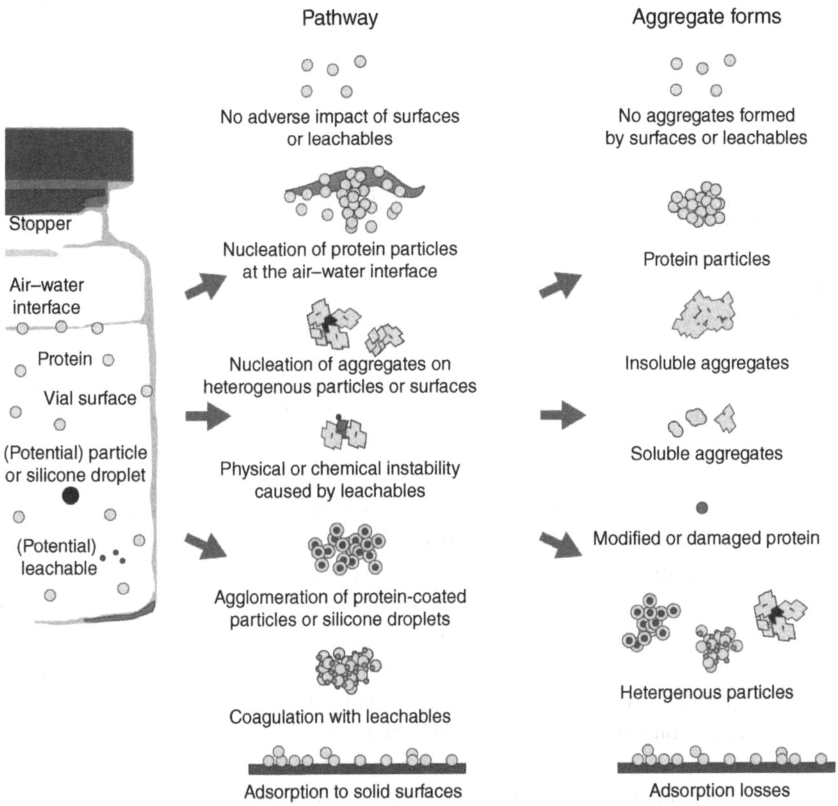

Figure 4.4 Possible physical degradation pathways and aggregate forms of proteins caused by interfaces, foreign particulates, and leachables. The figure shows a vial as an example. These aggregation processes may also occur in other upstream operations and in other containers, closures, and delivery devices. Used with permission from Bee et al. [82].

TABLE 4.1 Factors that Influence Protein Adsorption to Surfaces and Interfaces

Factors	Properties
Problem	Size, shape, charge (positive, negative, and net), charge distribution, p*I*, conformational stability (ΔG_{unf})
Solution	pH, ionic strength, specific interaction with excipients
Surface	Charge, hydrophobicity, morphology, roughness, flexibility
Common	Temperature, polarity, shear

Adapted with permission from Pinhold et al. [83].

are often observed when dissolved proteins spontaneously partition to the air/water interface. This detrimental effect on the physical stability of the protein solution is greatly amplified by agitation from movement during transportation of the drug product, for example, as it introduces dissolved gas/water interface in the bulk solution. In addition, other processing steps such as diafiltration operations have also been cited as causes of air/water interface-induced aggregation as the operations lead to air bubble entrainment and/or microcavitation (reviewed in Bee et al. [82]). Not all proteins, however, are susceptible to air/water interface-induced degradations. The susceptibility of a particular protein to air/water interface-induced denaturation and aggregation may be assessed from the surface activity of the protein, which can be experimentally determined by measuring decreases of the air/water surface tension (or increases in surface pressure) from the adsorption of a protein to the air/water interface using a force tensiometer (e.g., Wilhelmy plate tensiometer) [85].

Perhaps the most common interface encountered by therapeutic proteins during production, purification, storage, and usage is the solid–liquid interface. Some of these interfaces are obvious, including purification membranes and column matrices, a wide variety of container walls and closures, syringes, tubing, and bags. There are also liquid/solid interfaces that are not so obvious, but have been found to cause protein aggregation and/or particulate formation. Stainless steel nanoparticles shed from positive displacement piston filling pump's solution contacting surfaces have been found to nucleate the formation of IgG particulates [87]. Particles shed from various brands of syringe filters have been shown to accelerate the rate of protein aggregation and particle formation during agitation [88]. Tungsten nanoparticles introduced into some prefilled syringes during the creation of the needle hole have been found to precipitate monoclonal antibodies in formulations below pH 6 where tungsten polyanions are formed [89]. Silicon oil used as lubricant on barrels and stoppers in prefilled syringes that sloths off into the solution as droplets have also been shown to have the ability to greatly accelerate the aggregation and particulate formation of proteins [90]. Borosilicate glass has been widely used as primary container material for biopharmaceuticals. Although extremely rare, glass flakes from pitting and delamination of vial surfaces have been reported (see Bee et al. and references therein [82]. Adsorption of proteins to these glass flakes can potentially nucleate protein aggregation and particulate formation [81].

The cases described above, as well as many others as reviewed in Carpenter et al. [91] and Bee et al. [82], highlight the importance of taking interfacial instability into consideration as a cause for physical instability of pharmaceutical proteins so that exposure to detrimental interfaces can be minimized and/or the formulation can be designed to be more resistant to interface-induced denaturation and aggregation. For example, in response to the potential tungsten-induced protein degradation, syringe manufacturers have developed proprietary manufacturing processes and engineering changes that control or effectively eliminate residual tungsten contamination in glass-staked needle syringes [82]. The use of a siloxane coating in glass vials has been shown to reduce protein adsorption. By far the most convenient, commonly used, and effective mean for suppressing agitation- or interface-induced aggregation is the addition of nonionic surfactants to formulations. This category of excipients is discussed in detail in the following section. Aside from surfactants, other formulation components can also modulate protein–interface interactions. Different buffering ions have been shown to change the relative affinity of the protein adsorption to the interface and the degree of protein unfolding at interfaces. Cyclodextrin (CD) has been reported to stabilize proteins exposed to agitation and air/water interface turnover. Human serum albumin (HSA) has been used as an effective excipient for preventing adsorptive losses of low-concentration protein formulations and also as a stabilizer for very hydrophobic proteins.

While the various chemical and physical instabilities discussed in this section can each lead to undesired changes to a protein therapeutic, another important consideration is interrelationship between chemical and physical instabilities. Perturbation of the native protein structure often leads to the exposure of previously buried, or solvent inaccessible, amino acid residues, facilitating their chemical reactivity. In fact, partial unfolding is often observed prior to the onset of irreversible chemical modifications. Conversely, chemical changes to the protein backbone or amino acid side chains may lead to loss of protein conformation. For example, the reduction of disulfide bonds or the oxidation of cysteine residues can induce protein unfolding and aggregation. Deamidation has been found to produce species that are less conformationally stable, more aggregation prone, and less soluble (see Manning et al. and references therein [29]. Oxidation has also been observed to reduce conformational stability and increase aggregation propensity. Coupling between interfacial instability and chemical instability has also been observed in antibody formulations. For example, the adsorption of a mAb to stainless steel particulates is believed to cause the exposure of buried residues, which can then be oxidized and ultimately lead to aggregation [92]. The coupled interaction between chemical and physical instabilities thus has the potential to significantly complicate efforts in formulation and excipient selection.

4.3 COMMON CLASSES OF EXCIPIENTS USED FOR BIOTECHNOLOGY PRODUCTS

Stability during manufacturing and long-term storage and means of efficacious delivery that avoid adverse immunogenic side effects are the challenges that must be met for recombinantly expressed proteins to be successfully used as therapeutics. Various chemical and physical instabilities, as outlined earlier in the chapter, are responsible for protein denaturation, aggregation, and loss of activity. Stability proteins can be optimized by different approaches, including modifying the inherent properties of the protein or by changing the protein external environment. For example, inherent properties can be modified by site-directed mutagenesis to substitute labile amino acids with ones not prone to chemical degradations or chemical modifications such as PEGylation that stabilize proteins against denaturation and aggregation. However, such modifications are complex and may compromise the biological activity of the protein. Thus, the simplest and most common method of protein stabilization is to change the nature of the environment surrounding the protein. As most protein therapeutics are still formulated as suspension, aqueous solutions either in a ready-to-use form or as a lyophilized product for reconstitution, formulation, and excipient selection remain the most important and versatile tools for achieving optimized stability and biological activity of protein pharmaceuticals.

A variety of excipients have been added to formulations to stabilize proteins, act as antimicrobials, aid in the manufacture of the dosage form, control or target drug delivery, and minimize pain upon injection (see reviews from Kamerzell et al. and Ohtake et al. [93, 94] and references therein) (Table 4.2). Examples include buffering agents to control pH, surfactants to inhibit protein adsorption to interfaces, preservatives to prevent microbial growth, carbohydrates as bulking agents for lyophilization, polymers to increase solution viscosity, and salts or sugars to stabilize proteins and to obtain physiological tonicity and osmolality. The choice of excipients, aside from considerations of the safety, toxicity, and immunogenicity of the excipients themselves, needs to be guided by knowledge of the degradation pathways of the active drug product and mechanisms by which different excipients ameliorate those instabilities. A well-tested general approach is to first preserve the native structure of a protein by optimizing the solution conditions and excipient selection to minimize structural changes and to minimize intermolecular interactions. Subsequently, if phase or interfacial instabilities are contributors of protein instability, manufacturing and formulation conditions need to be modified and chosen to eliminate these instabilities. This section gives an overview of the different classes of excipients used, with a focus on excipient–protein interactions and the mechanisms by which excipients stabilize protein therapeutics in different dosage forms.

Excipients can be broadly divided into eight categories based on their modes of action: buffering agents, tonicity modifiers, sugars and polyols, polymers and proteins, amino acids, salts, surfactants, and preservatives. Several excellent reviews have been published on different excipients used in biopharmaceuticals and the readers are referred to these reviews as well as references therein for more information [93, 94, 96].

4.3.1 Buffer Agents

Buffering agents are routinely used in formulations to control and stabilize solution pH as proteins are often stable only over narrow pH ranges and may degrade rapidly outside these ranges (see Chi et al. [8, 9] and Wang et al. [96] and references therein). pH has a strong influence on both the chemical and physical stabilities of proteins. As described earlier in the chapter, chemical integrity of the amino acids (e.g., asparagine deamidation, and methionine oxidation) is highly dependent on solution pH. As pH determines the type (positive or negative) and net charge on the protein, it affects electrostatic interactions, both intra- and intermolecular. At pH conditions far removed from its isoelectric point (either more acidic or more basic), increased net charge on the protein, which increases charge repulsion within the protein, can destabilize the protein native structure to result in the unfolding of the protein, which reduces charge density. In addition, pH may affect specific ionic interactions, such as salt bridges that contribute significantly to the stability of natively folded protein structures, and may cause the loss of the native structure. Net charge on the protein also

TABLE 4.2 Summary of Excipients Used in Biopharmaceutical Formulations and Their Effects

Excipient Class	Representative Examples	Formulation Effects and Cautions
Buffering agents	Acetate, citrate, tartrate, histidine, glutamate, phosphate, Tris, glycine, bicarbonate, sulfate, nitrate	• Maintain solution pH • Ion-specific interactions with proteins that can be stabilizing or destabilizing • pH may change with temperature • Decomposition during storage • Crystallization during freezing
Tonicity modifiers	Mannitol, sorbitol, lactose, dextrose, trehalose, sodium chloride, glycerol, and glycerin	• Maintain solution tonicity
Sugars and polyols	Sucrose, trehalose, glucose, lactose, sorbitol, mannitol, glycerol	• Nonspecific stabilizers in liquid and lyophilized formulations • High concentrations often required
Amino acids	Histidine, arginine, and glycine, methionine, alanine, arginine, aspartic acid, proline, lysine	• Antioxidants (histidine, methionine) • Buffering and tonicity modifiers • Increase protein solubility • Decrease solution viscosity
Salts	Sodium chloride, potassium chloride, sodium sulfate, potassium phosphate	• Tonicity modifiers • Stabilizing or destabilizing to proteins (Hofmeister series)
Surfactants	Polysorbates 20 and 80 (Tween 20 and Tween 80), poloxamer (Pluronic F68 and F127), Triton X-100, Brij 30 and Brij 35	• Competitive inhibitor of protein adsorption and interface/surface denaturation • Assist in cell culture and protein refolding • Specific interaction with proteins that can be stabilizing or destabilizing • Peroxides can cause oxidation • May degrade during storage • Complex interactions and behaviors during membrane filtration
Polymers and proteins	Gelatin, PVP, PLGA, PEG, dextran, cyclodextrin and derivatives, starch derivatives HSA, BSA	• Competitive inhibitor of protein adsorption • Lyophilization bulking agent • Drug delivery vehicles

(continued)

TABLE 4.2 (*Continued*)

Excipient Class	Representative Examples	Formulation Effects and Cautions
Preservatives		
Antioxidants	Amino acids (His, Met), ascorbic acid, glutathione, vitamin E, poly(ethylenimine)	• Free radical scavengers
Chelators	EDTA, DTPA, citric acid, hexaphosphate, thioglycolic acid	• Metal binding compounds
Antimicrobials	Benzyl alcohol, *m*-cresol, phenol	• Prevent microbial growth in multidose formulations

Adapted with permission from Kamerzell et al. [93] and Jorgensen et al. [95].

affects intermolecular electrostatic interactions. As protein net charge increases when pH moves away from its isoelectric point, repulsive electrostatic interactions between protein molecules also increases, which increases the protein colloidal stability and solubility as well as enhances protein solution phase stability.

Because of the multifaceted effects of pH on the chemical and physical stabilities of proteins, the choice of optimal pH for formulation is not straightforward. For example, a pH value that is optimal for physical stability and solubility may not coincide with a pH value where the rate of deamidation is lowest. Optimizing the pH is a major issue and one that is usually studied early on in the protein formulation process to establish pH stability and solubility profiles typically in the pH 3–10 range. A buffer system is subsequently selected to maintain the pH at an optimal level such that pH-sensitive chemical modifications (e.g., deamination) and physical instabilities are minimized. Common buffering agents include acetate, citrate, tartrate, histidine, glutamate, phosphate, Tris, and glycine that cover the pH range of approximately 3–10.

Aside from stabilizing solution pH, which is the primary effect of a buffer salt, it is also important to keep in mind that different buffering ions can have specific effects on the chemical and physical stabilities of proteins. These effects can be stabilizing or destabilizing and thus can greatly impact the selection of buffering agents. For example, the rate of deamidation appears to be faster in phosphate and bicarbonate buffers than in sulfate, nitrate, acetate, chloride, and pyruvate buffers (see Jorgensen et al. and references therein) [95]. Multivalent carboxylate buffers such as citrate have been shown to decrease solubility and cause gelation of monoclonal antibodies formulated at high concentrations. It is also well known that the pH of solutions of certain buffers, such as phosphate and Tris, change upon freezing or with temperature, respectively. In addition, some buffer ions can decompose during storage and their degradation products can interact with the protein and destabilize the drug product, for example, Tris buffers at elevated temperatures and citrate buffers upon exposure to trace metals and light.

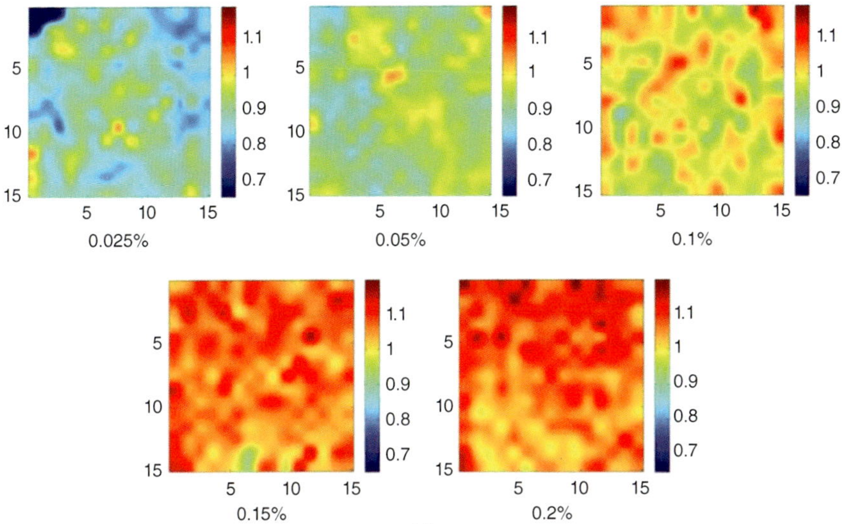

Figure 1.4 Chemical imaging of drug product stability showing (a) surface-enhanced Raman chemical imaging of between 0.025% and 0.2% 4-aminophenol (degradant/impurity) versus the pixel position in tablets of acetaminophen and PVP. Images were obtained from plotting the median intensity of the principal band of 4-aminophenol normalized butanethiol peak. Source: De Bleye [9]. Reproduced with permission of Elsevier. (b) Identification of prodrug (top) to parent (bottom) conversion in a prototype BMS tablet formulation. Images and data courtesy of Boyong Wan and Christopher Levins (Bristol-Myers Squibb, 2015).

Pharmaceutical Excipients: Properties, Functionality, and Applications in Research and Industry,
First Edition. Edited by Otilia M.Y. Koo.

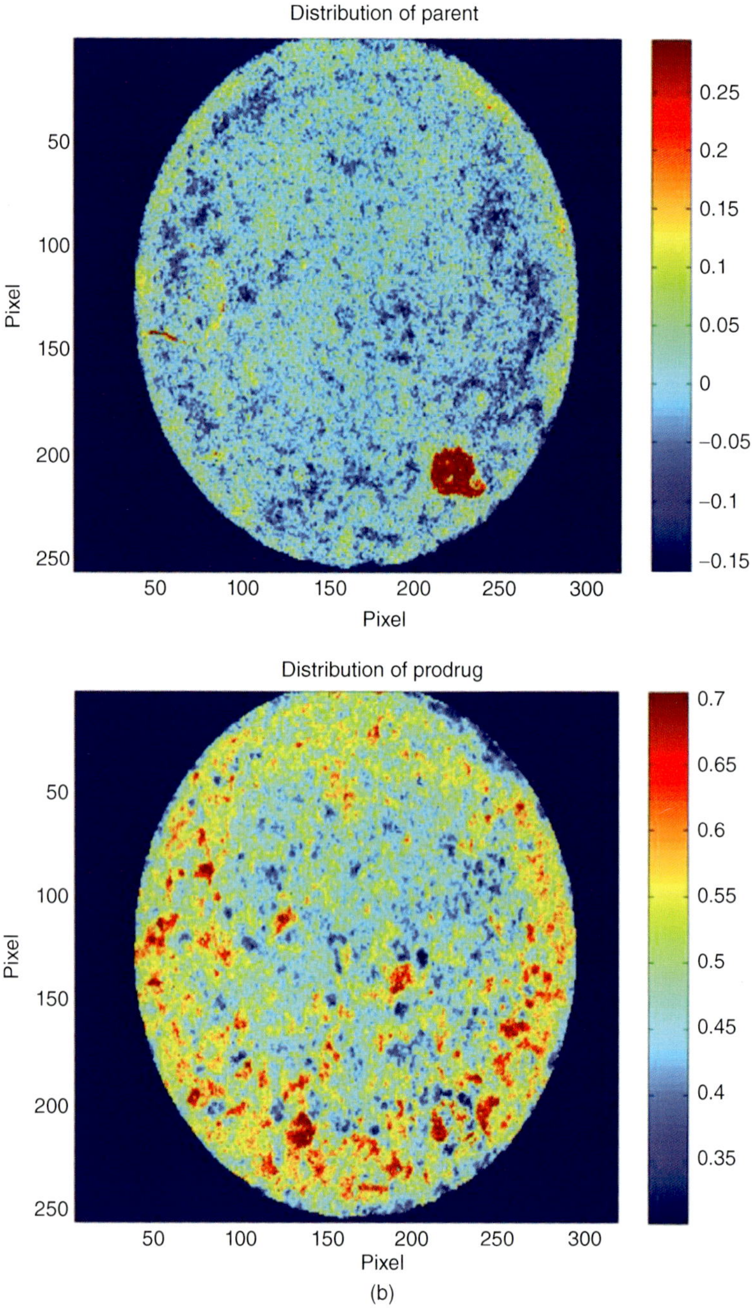

Figure 1.4 *Continued*

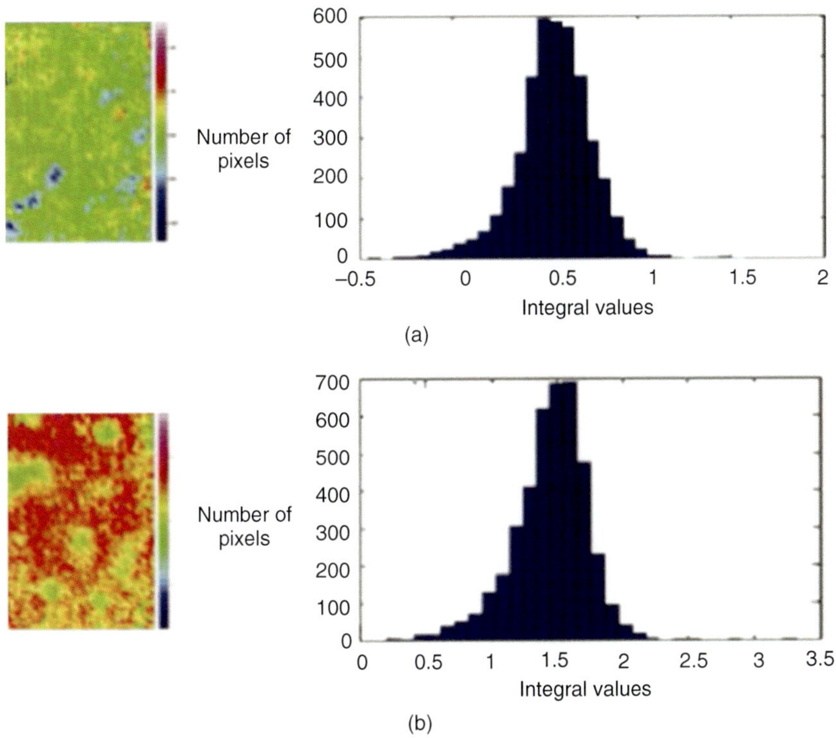

Figure 1.5 FTIR images and histograms of HPMC ibuprofen tablets using blends stored at two RH conditions and compressed at two forces: (a) 60% RH blend compressed at 80 cN m; (b) 80% RH blend compressed at 120 cN m. Source: Elkhider [11]. Reproduced with permission of Elsevier.

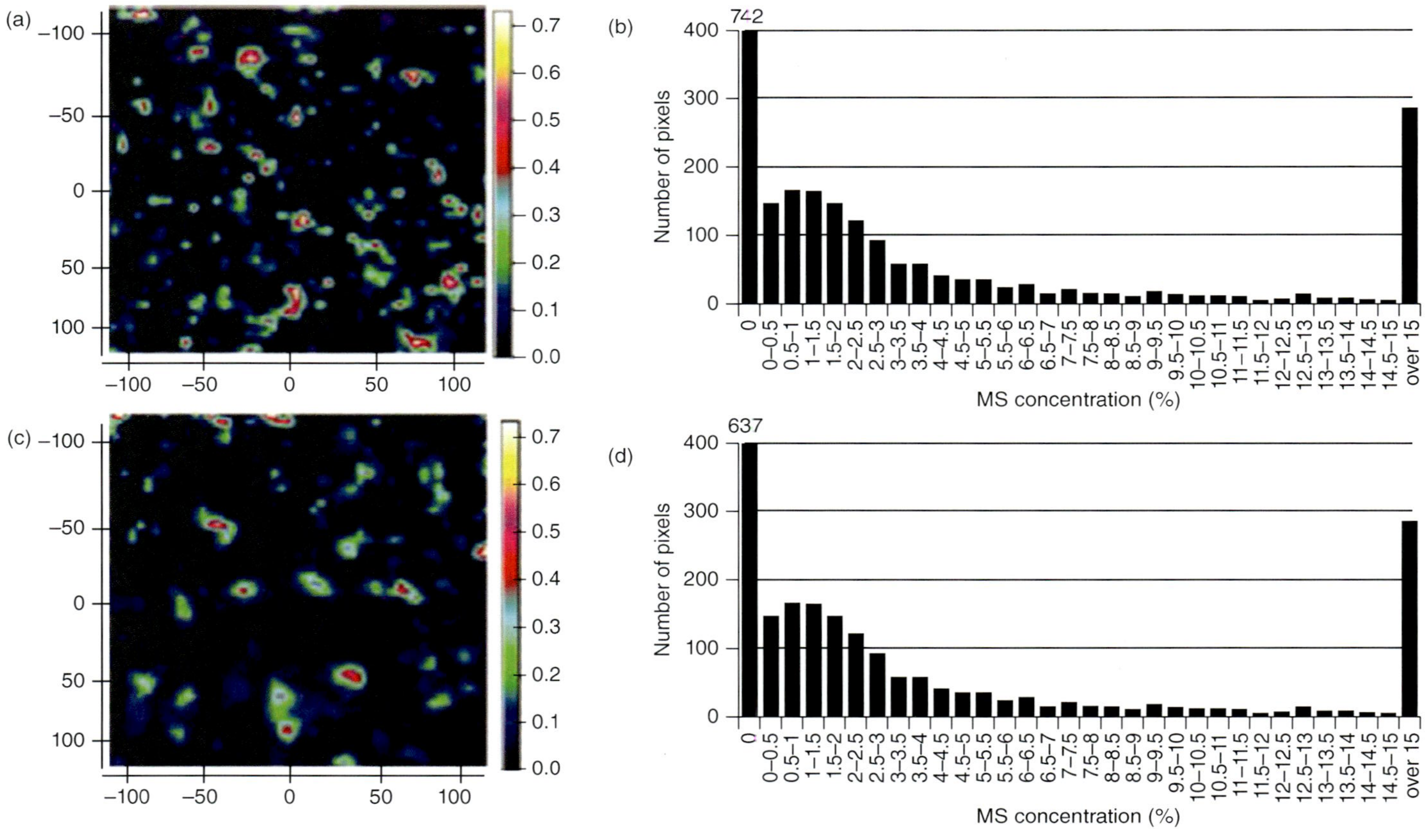

Figure 1.6 Dispersion of magnesium stearate (MS) lubricant particles in physical blends analyzed by Raman chemical imaging. Quantification of domain size, number, and localization is provided. Blending time increases from 2 to 60 minutes from the top to bottom tablet images. Source: Lakio

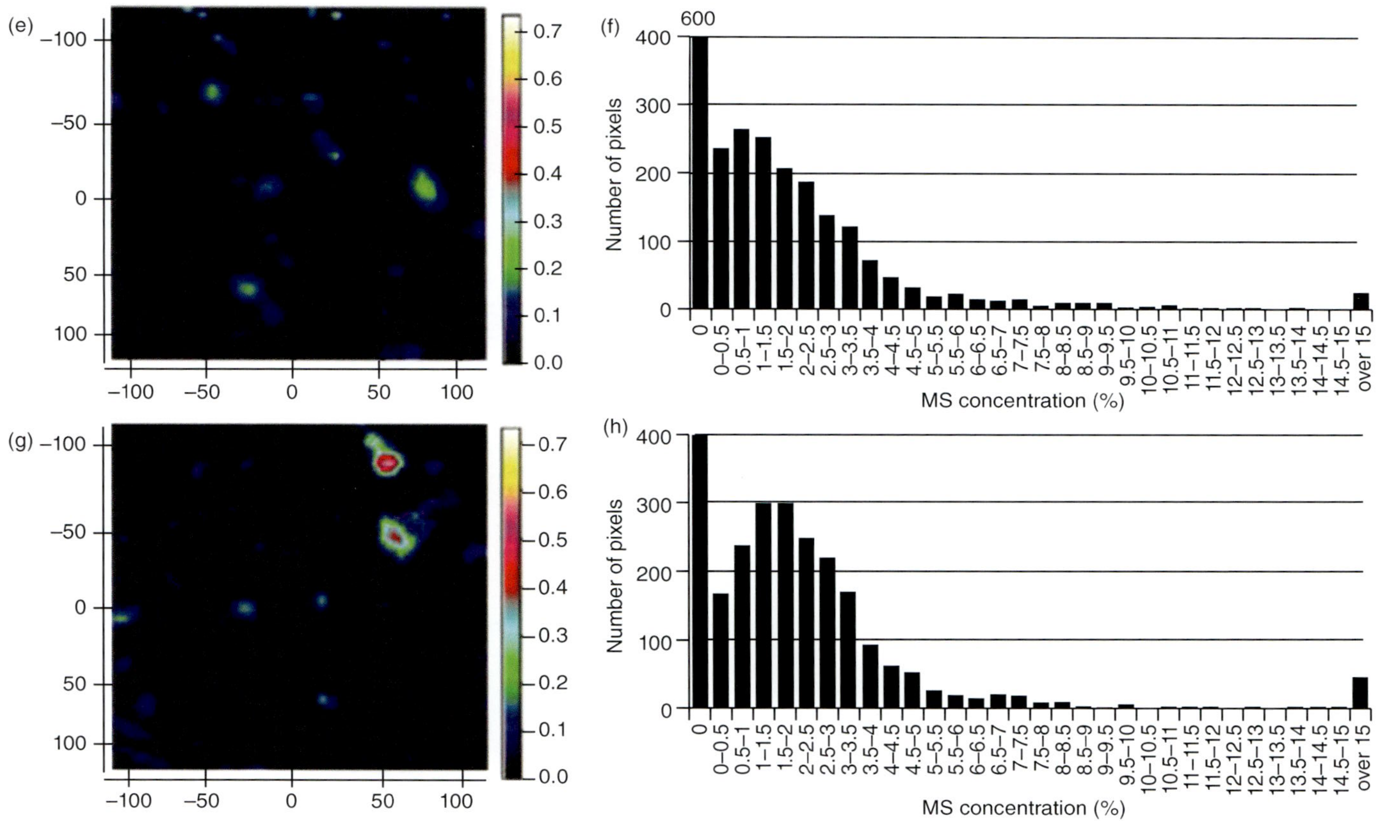

Figure 1.6 *Continued*

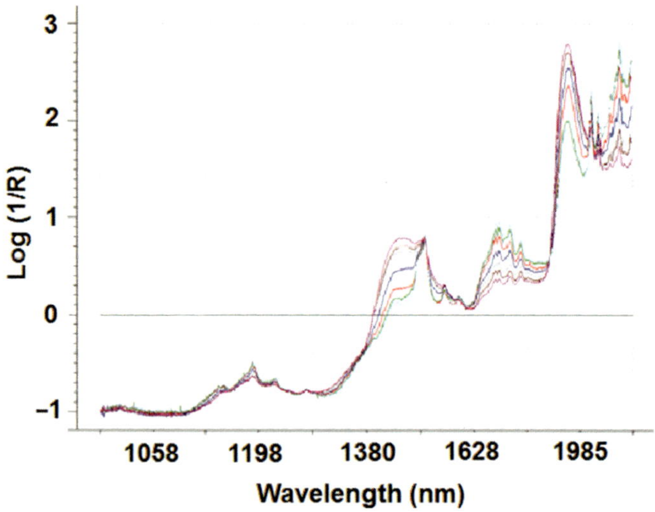

Figure 8.5 NIR spectra of calibration blends during fluid bed drying showing the main absorption bands at 1470–1408 nm and 1960–1890 nm.

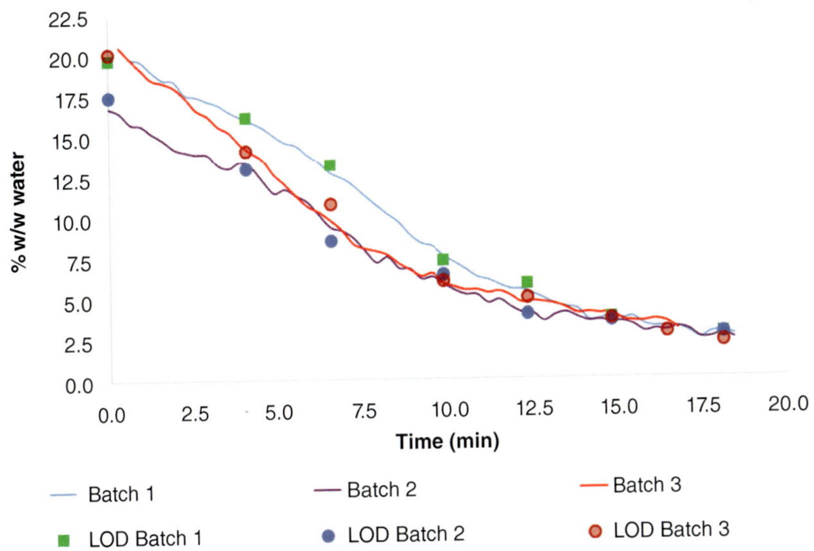

Figure 8.6 % w/w water versus time for three different batches of a wet granulated product and LOD reference values.

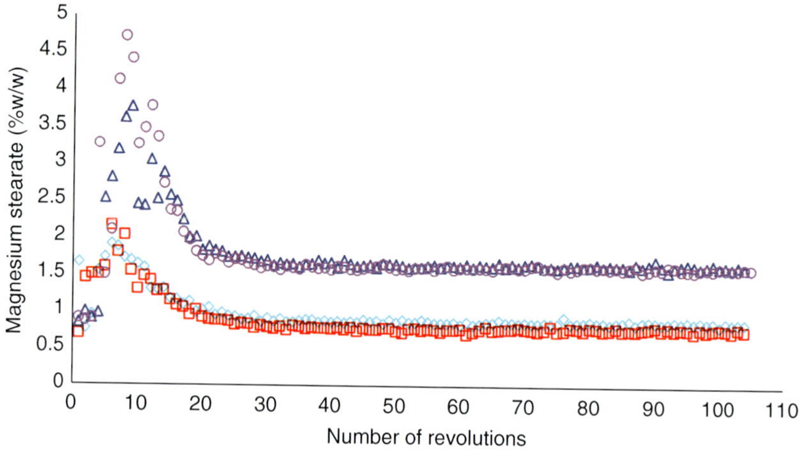

Figure 8.8 Typical blending profile of magnesium stearate in PB (□: Batch 1, ◇: Batch 2) and FB (Δ: Batch 1, ○: Batch 2).

Depending on the dosage form of the biopharmaceutical product, for example, solution or lyophilized drug product, additional factors will need to be taken into consideration in the selection of buffering agents. For example, the buffering agent should have a high collapse temperature, be nonvolatile, and have a high glass transition temperature [97, 98]. A high collapse temperature would facilitate a faster primary drying, the nonvolatile nature would prevent pH drifts, and a high glass transition temperature ensures stability of the lyophilized product during storage. Based on these criteria, acetate buffer is not used lyophilized formulations due to its volatile nature. Sodium and potassium phosphate salts are not often used because pH of their solutions can decrease by 4 units during cooling and freezing. Comparing the crystallization behaviors of citrate, tartrate, and succinate buffers, citrate buffer was found to be the most preferred as it remained amorphous, whereas tartrate and succinate buffers crystallized during lyophilization [97, 98].

4.3.2 Tonicity Modifiers

Tonicity is a measure of the effective osmotic pressure difference between two solutions separated by a semipermeable membrane. Defined relative to the red blood cell membrane, the normal saline (0.9% or 154 mM sodium chloride) is considered to be isotonic to human plasma. Parenteral formulation for subcutaneous or intramuscular administration should be isotonic with human plasma so as to avoid pain, irritation, or tissue damage at the site of administration. Protein drugs at their recommended dosage that are not isotonic with blood thus require the addition of a tonicity-adjusting agent to the formulation. Commonly used isotonicizing agents include mannitol, sorbitol, lactose, dextrose, trehalose, sodium chloride, glycerol, and glycerin. Note that many of these of these agents also serve other roles as excipients in a protein formulation.

4.3.3 Sugars and Polyols

The use of various sugars and polyols as protein-stabilizing excipients has a long history with many documented cases (see the review Wang [96], Ohtakc et al. [94], Kamerzell et al. [93], and Jorgensen et al. [95] and references therein). These excipients have been reported to stabilize the structure of the native proteins at moderate (0.1 M) to high concentrations (1 M) [93, 94, 96, 99], hence reducing aggregation and chemical degradations via the pathways reviewed earlier in the chapter. Because of the compatibility of these excipients with proteins and solvent (i.e., water), they are often referred to as stabilizing cosolvents. In fact, the use of these stabilizers is partly motivated by the large body of work that has been done in the past decades on the nature and mechanism of action of intracellular solutes (osmolytes) that stabilize microorganisms under stress conditions such as elevated temperature and desiccation [100]. Naturally occurring osmolytes, including methylamines, polyols, and amino acid derivatives, play a critical role in preserving macromolecular function and maintaining cell viability.

Sugars and polyols are the most commonly used excipients that nonspecifically stabilize proteins and their stabilizing effect increases with increasing concentration. Among sugars, sucrose and trehalose are the most frequently used. They are also osmolytes used in nature to stabilize microorganisms under harsh environmental conditions such as high temperatures and low water environment. Sugars and polyols have been shown to be highly effective at increasing the protein melting temperature, preserving enzyme activity, and reducing protein aggregation. Generally, a correlation can be drawn between excipients that stabilize the protein against thermal stress and those that stabilize the proteins during processing and storage. Increasing melting temperature typically translates to a shift in the equilibrium constant of unfolding toward the native state. Thus, there will be a reduction in the population of unfolded and structurally perturbed proteins that are aggregate competent. The polyol glycerol has been found to protect several enzymes from thermal inactivation; however, its stabilizing effect is also protein dependent. In general, large polyols (e.g., sorbitol) have been to confer greater stability to the protein therapeutic than smaller polyols (e.g., glycerol and erythritol) [94]. Other polyols that have shown protective effects against heat-induced denaturation include caprylate, tryptophanate, sorbitol, sarcosine, and glycine.

Extensive studies of the interactions of the protein-solvent system have given us insights into the mechanism of protein stabilization by cosolvents [101–103]. Stabilization of the protein native structure by cosolvents does not stem from the binding of the cosolvents to the protein, but from several interrelated, somewhat indirect, stabilization mechanisms described here.

Protein-stabilizing cosolvents, likely without exception, increase the surface tension of water [94]. As such, the work that is required to create a cavity, for example, the volume that a protein occupies, increases with increasing surface area of the cavity. Thus, increasing surface tension of the solvent by the addition of cosolvents favors more compact protein conformations that minimize protein surface area.

Another mechanism by which cosolutes, particularly at high concentrations, stabilize the protein native state is the excluded volume effect. The excluded volume of a molecule, which equals four times the molecular volume, describes the volume that is inaccessible to other molecules in the system. Thus, at a high cosolvent concentration, volume of solvent (e.g., water) available to proteins is reduced, increasing the effective concentration (or chemical potential) of the protein. Consequently, changes that reduce system volume, for example, a shift in the ensemble of native protein structures towards more compact structures, are favored.

Unfavorable interaction between the cosolvent and the peptide bonds in proteins has also been shown to contribute to the stabilizing effect of cosolvents. Transfer free energy values of the polypeptide backbone calculated from experimental measurements showed that in stabilizing cosolvents (e.g., osmolytes) the exposure of the polypeptide backbone to the solvent is highly unfavorable [104, 105]. This effect favors the natively folded structure of proteins.

Perhaps the best-known mechanism of cosolvent stabilization of proteins is the preferential exclusion mechanism established by the seminal work from Timasheff et al. From equilibrium dialysis experiments, water has been shown to be present in excess in the vicinity of the protein surface than that in the bulk phase. At the same time, the cosolvents are depleted from the protein surface compared to the bulk phase. This effect is called "preferential hydration" or "preferential exclusion" of the cosolvent. This exclusion can be thought of as negative binding of the cosolvent to the protein and is thermodynamically unfavorable. The degree of exclusion is proportional to the solvent-exposed surface area of the protein. During denaturation, or unfolding, protein surface area increases, resulting in a larger extent of preferential exclusion. The net effect of increased thermodynamically unfavorable preferential exclusion is to favor the native, compact, and less solvent-exposed state over an unfolded state. Many sugars, polyols, and certain salts, which are known to stabilize proteins and decrease their solubility, are all preferentially excluded from the vicinity of the protein.

In light of the mechanisms reviewed above, stabilizing cosolvents such as sugars and polyols stabilize the natively folded state of protein with respect to the unfolded state by a combination of interrelated mechanisms. The cosolvents raise the surface tension of water and are preferentially excluded from the protein surface, which is in accordance with the repulsive interactions of these cosolvents with proteins. At high concentrations, for example 1 M, or 30 mass%, sucrose, the cosolvents also exert an excluded volume effect. All these effects are thermodynamically unfavorable and raise the free energy of the protein native state that scales with solvent-exposed surface area. Critically, the free energy of the unfolded state, which has a larger solvent-exposed surface area, is raised even further. This leads to greater free energy difference between the unfolded and folded states of the protein in the presence of cosolvents, resulting in a larger conformational stability of the protein native state. As discussed earlier in the chapter, increased conformational stability decreases the propensity and rate of nonnative aggregation, whether aggregation is driven by conformational instability or colloidal instability. A dominant factor that causes protein aggregation is the reduced exposure of hydrophobic patches on proteins. This process is entropically driven since the system entropy increases when structured water molecules around hydrophobic surfaces are released into the bulk water that is far more disordered. The unfolded or partially unfolded states are more aggregation prone due to the exposure of hydrophobic residues that are normally buried in the natively folded state. Excipients that promote a compact, native state of the protein such as sugars and polyols by mechanism outlined above will inhibit aggregation by reducing the equilibrium concentration of the structurally expanded aggregation-competent states.

The stabilizing effect of sugars and polyols depends on their concentration. A minimum concentration of 0.3 M (or ~5%) has been suggested to achieve significant stabilization. As high as 1 M sucrose or 10% glycerol have been routinely used to protect the activity of proteins. Sorbitol levels between 20% and 33% have been found to stabilize proteins and suppress aggregation.

Sugars and polyols not only stabilize proteins against physical degradations, but they have also been shown to stabilize proteins against chemical degradations such as oxidation. The inhibitory effect arises from complexation, although weak, of the cosolvents with metal ions that catalyze oxidation reactions. It is also worthwhile to note that while sugars and polyols are effective at stabilizing proteins whose degradation is driven by conformational instability, they are not effective against degradations that proceed through different pathways, for example, interface-induced unfolding and aggregation. Thus, it remains important to have an understanding of the primary pathways by which a particular protein therapeutic undergoes degradation.

4.3.4 Amino Acids

The use of amino acids as excipients in biopharmaceutical formulations has been gaining more attention due to their safety in humans, highly beneficial effects in protein processing and formulation and recent advances in the molecular level understanding of their interactions with proteins. The most commonly used amino acids are histidine, arginine, and glycine. Other amino acids used as formulation excipients include methionine, alanine, arginine, aspartic acid, proline, lysine, and mixtures such as glutamic acid and arginine (see Kamerzell et al. [93] and references therein).

The effects of the amino acids on protein stability and formulation are multifaceted. For example, histidine has been used extensively as a buffering agent for antibodies to not only control solution pH (p*I* of 7.59) but also to provide stabilizing noncovalent interactions with antibodies in the solid state. Histidine is also an antioxidant that scavenges hydroxyl radicals in solution. Glycine has been used as a buffering agent in solution (p*I* = 5.97). In addition, the amino acid is used as a bulking agent during lyophilization of proteins.

Arginine has been widely used as a solubilizing and aggregation-suppressing agent in protein purification steps such as inclusion body recovery and as a component of mobile phases in liquid chromatography. Inclusion of arginine during refolding led to increased recovery of the protein by suppressing the aggregation of folding intermediates without imparting any stabilizing effect on the protein native structure [94]. High concentrations of arginine (2 M) have been shown to result in high recovery of antibodies from Protein-A columns above pH 4.0 [106]. Arginine is also frequently used in protein formulations to enhance shelf life. The arginine salt, arginine sulfate, at 0.4 M, has been shown to increase the solubility of fibroblast growth factor 20 (FGF20) by up to 1000-fold, those solubility is otherwise too low (e.g., 0.25 mg/ml at pH far removed from p*I*) for processing. It is believed that for FGF20, which binds polyanions, the sulfate and arginine synergistically increase the solubility of the protein [107]. A similar effect of arginine on the solubility of recombinant plasminogen activator (rPA) has been observed. This protein has an extremely low solubility in aqueous solutions, less than 1 mg/ml, and its solubility was found to increase with increasing arginine hydrochloride concentration, leading to >50 mg/ml in 1 M arginine. In contrast, 1 M NaCl or a combination of 0.5 M NaCl and 0.5 M glycine marginally improved solubility of the protein. These findings clearly demonstrate the unique nature of arginine in affecting protein solubility that is neither purely ionic nor just a concentration effect [108]. In addition, it has been shown recently that the addition of arginine hydrochloride at >0.15 M significantly reduced the viscosity of high-concentration antibody formulations [74, 109], but not globular proteins such as bovine serum albumin (BSA) and α-amylase [109].

Since the same amino acid can serve different roles as an excipient in a therapeutic protein formulation, it is not surprising that the mechanism of amino acid stabilization is multifaceted. The antioxidant effect arises from the ability of histidine to scavenge radicals and methionine acting as the sacrificial excipient that becomes oxidized instead of the protein active ingredient. Mechanism of protein-stabilizing effects from amino acids, however, remains an active area of investigation.

Despite the frequent use of arginine in biopharmaceutical processing and formulation, the mechanism of arginine's solubilizing and aggregation-suppressing effects remains to be fully elucidated. Although arginine raises the surface tension of water, it is neither a protein-stabilizing agent nor is it used in osmotolerant organisms. Thus, unlike sugars and polyols, arginine does not stabilize the natively folded protein conformation and is not an osmolyte. In fact, the effect of arginine on protein solution is more similar to the protein denaturant guanidine hydrochloride. Arginine has been shown to increase the solubility of most amino acids, especially the two aromatic amino acids tyrosine and tryptophan. Similar to guanidine hydrochloride, increase in amino acid solubility is believed to result from favorable interactions between arginine and amino acid side chains, particularly aromatic side chains. Molecular dynamic simulations have shown that arginine is capable of interacting with proteins through multiple types of interactions [110]. Arginine forms both hydrogen bond and electrostatic interaction with charged residues. The guanidinium group of arginine can interact directly with the protein backbone and the methylene group of arginine can associate with hydrophobic patches on the protein. Finally, arginine can also form cation-π interactions with aromatic amino acids, particularly tryptophan [110]. These interactions are believed to be responsible for suppressing aggregation during refolding by stabilizing folding intermediates.

Unlike guanidine hydrochloride, which preferentially binds to proteins and unfolds proteins, preferential interaction measurements show that arginine's binding to proteins is much weaker compared to that of guanidine hydrochloride [110, 111]. Thus, consistent with the observation that arginine has little effect on protein stability, the compound does not bind strongly to proteins nor is it strongly excluded from the protein surface. Such cosolutes have been termed "neutral crowders" and have been proposed to slow protein association by being preferentially excluded from protein–protein contacting surfaces [112]. Such an effect impacts only reactions that involve multiple proteins (e.g., protein association reactions such as aggregation), while having little or no effect on unimolecular reactions such as protein folding and unfolding. Recent investigations have further revealed the nuanced and complex nature of arginine self-interaction and arginine–protein interactions that contribute to its protein solubilizing and aggregation-suppressing effects [113]. Nonetheless, arginine remains a useful excipient in stabilizing protein formulations and reducing viscosity of high-concentration antibody formulations.

4.3.5 Salts

Salts are common components in protein formulations. Aside from buffering salts used to maintain solution pH as discussed earlier, common physiological salts such as sodium chloride have been used as tonicity-modifying agents. The effect of salts on protein is complex, partly because of the complex ionic intra- and intermolecular interactions. In general, salts may stabilize, destabilize, or have no effects on protein conformation and stability depending on the type and the concentration of salts, net charge and charge distribution of the protein, charged amino acids on the protein, and the nature of the ionic interactions of the salts with the protein [96]. The net effect is a balance among nonspecific (Debye–Hückel) electrostatic screening and specific interaction with the protein.

At low concentrations, salts weaken ionic interactions, both repulsive (from like charges) and attractive (from opposite charges) by acting as counterions. This electrostatic screening may either be stabilizing when there are major intramolecular repulsive interactions that can lead to protein unfolding, or destabilizing when there are salt bridges or ion pairs that contribute significantly to the folded conformation of the protein.

At high concentrations, the salt effects are believed to stem from their effects on the solvent and from specific salt–protein interactions. Salt–protein interactions have been studied for over a century and the effects of salts, both anions and cations, exhibit a recurring trend in chemistry and biology called the Hofmeister series, proposed by Franz Hofmeister in 1888 [96, 114–116]:Cations: $(CH_3)_4N^+ > NH^{4+} > K^+,\ Na^+ > Mg^{2+} > Ca^{2+} > Ba^{2+} > GdnH^+$Anions: $CO_3^- > SO_4^{2-} > CH_3CO^{2-} > H_2PO_4^- > F^- > Cl^- > Br^- > NO^{3-} > ClO_4^- > SCN^-$Ions to the left of the series are the most stabilizing. Less is known about the cation series, but the anions in the series have been shown to have a more drastic effect on protein stability, ranging from those that typically precipitate proteins from solution by preventing protein unfolding and reducing solubility (carbonate CO_3^{2-} and sulfate SO_4^{2-} ions) to ions that increase solubility and induce protein unfolding (thiocyanate SCN^- and perchlorate ClO_4^- ions). Chloride (Cl^-) ions, the most commonly used in protein formulations in the form of sodium chloride (NaCl), are ranked in the middle of the Hofmeister series.

The molecular level detail of the Hofmeister effect continues to be unraveled. It has been believed that the primary mechanism of the Hofmeister effects arises from the ability of the anions to alter the hydrogen-bonding ability of water, either being "water structure makers" (kosmotropes) or "water structure breakers" (chaotropes). More recently, femtosecond mid-infrared pump-probe spectroscopy measurements that directly probed the structure of water outside the hydration shell of ions showed that the long-range hydrogen-bonding network of bulk water is not affected by the presence of kosmotropic or chaotropic ions [117]. Several lines of investigation point to direct and local interactions with the proteins and the first hydration shell being largely responsible for most aspects of the Hofmeister series [74].

Although the molecular origin of the Hofmeister series continues to be debated and studied, it is useful as a general guide for ion effects on proteins. Ions to the left, commonly referred to as kosmotropes, tend to precipitate proteins from solution and prevent protein unfolding, whereas ions to the right, commonly referred to as chaotropes, increase the solubility and promote the denaturation of proteins. However, behaviors that are not predicted by the Hofmeister series are also often observed. Both Na^+ and Cl^-, from the commonly used NaCl salt, rank in the middle of the Hofmeister series and are expected to exert very little effect on the stability and solubility of protein solutions. However, NaCl has been shown to be a powerful stabilizer for a number of proteins, including increasing denaturation temperature of BSA, recombinant Factor VIII (rFVIII), and RNase T1, and interleukin-1 receptor (IL-1R) [96]. More recently, NaCl has been shown to dramatically decrease viscosity of high-concentration monoclonal antibody formulations by affecting self-association of the antibodies [74]. These examples highlight the complexity of effects of ions on proteins stemming from complex ion–protein interactions and effects on solvent properties. It is thus difficult to determine *a priori* the effect of a particular salt on the stability and activity of a protein therapeutic. The choice of salts to use in formulations thus needs to be guided by prior examples and experimental assessment of the effect of a particular salt on a particular protein.

4.3.6 Surfactants

Surfactants, or surface-active agents, are now routinely added to protein solutions to prevent physical damages during purification, filtration, transportation, freeze drying, spray drying, and storage. They have been shown to stabilize proteins, suppress and prevent aggregation, and assist in protein refolding. Surfactants are amphiphilic, containing a polar head group, which can be anionic, cationic, uncharged, or amphoteric, and a nonpolar tail, which is often a long-chain aliphatic hydrocarbon group (Figure 4.5). This dual nature causes surfactants to adopt specific orientations at interfaces and in aqueous solutions and underlie the mechanisms by which surfactants affect the physical stability of proteins. Ionic surfactants such as the anionic surfactant sodium dodecyl sulfate (SDS) have been long known as effective protein denaturants and are commonly used for this purpose, for example, as a pretreatment for proteins in polyacrylamide gel electrophoresis (SDS-PAGE). In contrast, surfactants used as stabilizing excipients in pharmaceutical formulations are typically nonionic. These surfactants generally bind weakly to proteins and are largely nondestabilizing to the native protein conformation. In addition, nonionic surfactants exhibit low toxicity and show low sensitivity toward various salts and buffers in solution. The most commonly used nonionic surfactants are polysorbate 80 and polysorbate 20, also commercially known as Tween 80 and Tween 20, respectively (Figure 4.5). Other examples include poloxamers (poloxamer 88 and poloxamer 407, also commercially known as Pluronic F68 and Pluronic F127, respectively), PEG dodecyl ethers (commercially known as Brij 35 and Brij 30), and PEG *tert*-octylphenyl ether (commercially known as Triton X-100) (Figure 4.5).

There are two primary mechanisms by which nonionic surfactants exert their protective effects on protein physical stability, prevention of protein surface (or interface)-induced denaturation and direct interaction (or binding) to proteins. The first mechanism is generic to all surfactant excipients, whereas the second mechanism is specific to both the protein and the surfactant. Some surfactants exert their effects through only one of these mechanisms, while others may function through both.

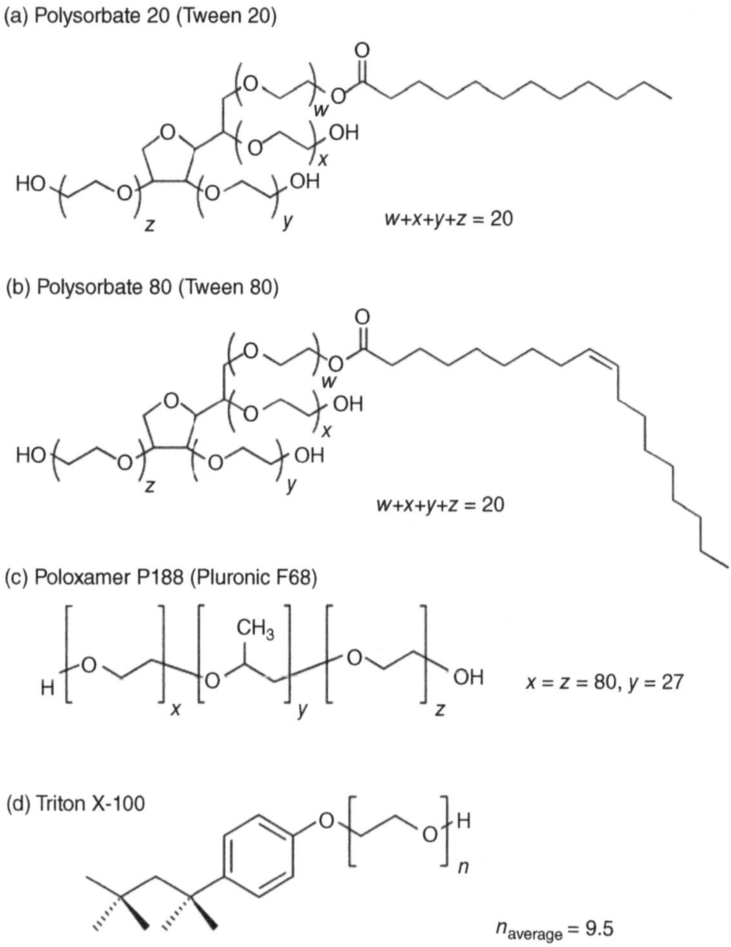

Figure 4.5 Structures of commonly used nonionic surfactants in biopharmaceutical formulations.

Because of their amphipathicity, surfactants in solution tend to orient themselves so that the exposure of the hydrophobic portion of the molecule to the aqueous solution is minimized. Thus, in systems containing air/water interface, surfactants will accumulate at the interface, forming a layer of surfactant molecules positioned in such a way that the hydrophobic tails are orientated toward the airside (i.e., the more "hydrophobic" side) of the interface and the hydrophilic ends are exposed to water. Such orientation also occurs at the solid/water interface including that found in vials, syringes, and other containers. Protein molecules are themselves amphipathic and as such also tend to adsorb and accumulate at interfaces. Protein adsorption to the air/water interfaces can be attributed to hydrophobic interactions, while electrostatic interactions often contribute to protein interactions at the solid–liquid interface. As reviewed earlier in the chapter, interface interactions often result in protein structural perturbations, leading to protein loss, chemical degradations, and aggregation directly on the interface or following desorption into the bulk solution. As interfaces and surfaces are routinely encountered by the drug product, interface-induced denaturation can be a major degradation pathway for protein therapeutics. The generic pathway by which surfactants protect and stabilize proteins against surface activity loss and/or surface-induced aggregation is by saturating the interface such that protein adsorption and subsequent denaturation are minimized. In this case, complete inhibition of protein adsorption can be achieved because generally smaller surfactant molecules diffuse much faster to the interface than much larger proteins. The adsorbed surfactants coat the interface and sterically prevent proteins from adsorbing to the interface. For this type of stabilization, surfactant concentrations near or above their critical micelle concentrations (CMCs) are needed.

In addition to altering protein interactions at interfaces, there is also ample evidence that nonionic surfactants also interact directly with proteins in solution. Although the interactions are considered weak, they can significantly alter protein behaviors. For example, aside from the complete inhibition of protein adsorption, surfactants have also been shown to partially prevent protein adsorption or even increase the amount of protein adsorbed at interfaces. These last two cases are attributed to the formation of surfactant–protein complexes with reduced or increased surface affinity (see Lee et al., and references therein [118]. Chou et al. showed that polysorbates 20 and 80 completely inhibited agitation-induced aggregation of recombinant fusion protein Albutropin solution at surfactant concentrations well below their CMC values [119]. The polysorbates exhibited saturable binding to Albutropin with a molar binding stoichiometry close to 10:1 (surfactant:protein). More importantly, this binding increased the free energy of unfolding by approximately 1 kcal/mol. Thus, at concentrations below CMC, the stabilizing effects of the polysorbates arise from their specific binding to the protein that increased the protein's conformational stability [119]. Joshi et al. investigated the stabilizing effect of polysorbate 80 on the aggregation of unagitated and agitated samples of recombinant human Factor VIII [120]. Association of the surfactant to the native conformation of Factor VIII was attributed to the stabilizing effect in unagitated samples as the binding provided an effective steric barrier to aggregation. In agitated samples, the stabilizing effect of polysorbate 80 against agitation-induced aggregation of Factor VIII is attributed to the rapid and competitive adsorption of polysorbate 80 to nascent air–water interfaces introduced to the protein solution by agitation [120]. Clearly, polysorbates can affect protein physical stability through both the nonspecific competitive adsorption mechanism and the protein-specific binding mechanism. Importantly, stabilization of the protein through surfactant binding is not universal. Stabilization occurs when binding of the surfactant ligand is greater to the native state than to a nonnative state (e.g., unfolded and partially unfolded states). In addition, the direct binding of surfactants to protein's hydrophobic surface covers and protects them from interacting with other proteins or surfaces. In this case, a surfactant's protective effect often correlates with surfactant:protein molar ratio rather than its CMC.

Nonionic surfactants generally bind to hydrophobic regions on a protein and these interactions are also attributed to their chaperone-like ability to assist in protein refolding. Interactions between the exposed hydrophobic regions of the partially unfolded proteins and hydrophobic tails of the surfactants can assist protein refolding by preventing misfolding and aggregation and allowing more time for the protein to refold to its native conformation. *In vitro*, surfactants such as polysorbates, PEG, Triton X-100, and lubrol have been shown to aid in protein refolding by acting as chemical chaperones (see Randolph and Jones and reference therein [121]).

Polysorbates 20 and 80 (Tween 20 and Tween 80) are the most commonly used nonionic surfactant excipients. The surfactants are composed of fatty acid esters of polyoxyethylenesorbitan, where polyoxyethylenesorbitanmonolaurate is polysorbate 20 and polyoxyethylenesorbitanmonooleate is polysorbate 80 [122] (Figure 4.5). Polysorbates used in biopharmaceutical formulations are mixtures of different fatty acid esters with the monolaurate fraction of polysorbate 20 making up only 40–60% of the mixture and the monooleate fraction of polysorbate 80 making up >58% of the mixture [122]. Compared to polysorbate 20, polysorbate 80 has a longer and monounsaturated aliphatic chain and as a result, is considerately more surface active and has a lower CMC (Figure 4.5). The two surfactants also exhibit different binding to proteins due to their different fatty acid chains. They have been widely reported to suppress aggregation upon agitation, shaking, freeze drying, and freeze–thawing processes, and can significantly reduce protein adsorption at solid surfaces (see Lee et al., and references therein [118]). The effectiveness of the polysorbates appears to be dependent on the stress involved, for example, stirring versus shaking [123]. Due to the interplay between surface adsorption inhibition and protein interaction, the effects of the surfactants are difficult to precisely predict, and need to be tested for a particular protein and formulation conditions in the presence of other excipients.

Although most nonionic surfactants can be considered as chemically inert, polysorbates are susceptible to autoxidation at moderate temperatures [122] and hydrolysis at higher temperatures [124]. Autoxidation leads to superoxide formation, side-chain cleavage, and eventually the formation of short chain acids such as formic acid [122]. The buildup of these autoxidation products from degraded polysorbates has been shown to increase the rate of oxidation of the protein drug product, compromising its storage stability (Wang, [125]). Singh et al. showed that polysorbate 80 is photosensitive, where exposure to light in aqueous solution resulted in peroxide generation that can lead to oxidation of susceptible amino acid residues in the protein drug product [126]. In addition, the authors tested an IgG1 formulation containing polysorbates of different grades from different vendors and found that photostability of the formulation indeed was affected by the quality of the polysorbate [126]. Residual peroxides were found in some commercial polysorbate 80 products. This study underscores the importance of carefully screening the quality and vendor of this critical excipient to ensure a robust, stable, and efficacious formulation delivered to the patients. In addition to oxidation, some enzymes have shown hydrolytic activity toward polysorbates, including several esterases [127]. Thus, particular caution should be used when formulating proteins with esterase activities using polysorbate solutions.

Poloxamers are triblock copolymers of the form polyethylene oxide–polypropylene oxide–polyethylene oxide (PEO–PPO–PEO) (commercially available as Pluronics or Synperonics) (Figure 4.5). They comprise another class of nonionic surfactants that are extensively used in a variety of pharmaceutical formulations. Poloxamers show complex aggregation behaviors in solution, including monomers, oligomers, various micelles, and larger clusters that strongly depend on temperature and concentration. The CMC and critical micelle temperature values of poloxamers have been found to vary over a wide range depending on the molecular weight and PPO/PPO ratio (see Lee et al., and references therein [118]). Poloxamer 188 (or Pluronic F68) is widely used in large-scale mammalian cell cultures and in bioreactors. It has been shown to be a shear-protective excipient that enhances cell yield in agitated culture and reduces cell adhesion in stationary cultures [128]. These protective effects have been attributed to the surfactant's ability to inhibit damages associated with cell–bubble interactions and shear stress. Poloxamer 188 has also been reported to facilitate refolding and suppress aggregation of a thermally denatured lysozyme and BSA [129]. Removal of poloxamer F68 during product purification and recovery can be difficult and may comprise product yield and stability [130].

Surfactants clearly represent a useful and important class of excipients for stabilizing protein formulations. Small amounts of surfactants often adequately stabilize proteins against interface-induced denaturation and aggregation by competitive adsorption. In this case, surfactant concentrations near or above CMC should be used. On the other hand, high concentrations of surfactants are known to destabilize proteins through hydrophobic binding to the protein. Thus, particular attention needs to be paid in selecting the proper surfactant concentration used in a formulation. In cases where no specific binding to the protein is observed, maximum level of protection is generally correlated to the CMC of the surfactant. It is recommended that surfactant levels slightly above the CMC values should be used [121]. The choice of surfactant is often dedicated by a trade-off. Lower concentrations are needed for surfactants with lower CMC values. However, these surfactants tend to be less soluble and more difficult to remove if needed [121]. In cases where surfactant binding stabilizes the native conformation of the protein, specific surfactant:protein stoichiometry needs to be maintained in order to provide optimal protection. A general recommendation for surfactants that show specific binding to the native state of the protein is to formulate so that the surfactant:protein ratio is slightly above the binding stoichiometry [121].

4.3.7 Polymers and Proteins

Naturally derived hydrolyzed gelatins and HSA were the earliest polymer- and protein-based stabilizers used in both protein drugs as well as live virus vaccines [93]. Concerns about potential infectious agents in animal-derived products have promoted restrictions in their usage as well as development of synthetic polymeric excipients and recombinant versions of HSA. Various kinds of polymers are now been used as excipients in both solution-based formulations and solid dosage forms. We briefly review the classes of polymers used and their stabilization mechanisms in the following text.

Hydrophilic polymers, such as PEGs, polysaccharides, and inert proteins, have been often used to nonspecifically stabilize proteins and enhance protein assembly [94]. For example, hydroxyethyl (heta) starch, the high-molecular-weight PEG4000, and gelatin were found to be effective, concentration-dependent inhibitors of thermal-induced aggregation of low-molecular-weight urokinase [131]. These polymers were found to be effective at stabilizing the enzyme by preferential exclusion and excluded volume (or molecular crowding) effects, which consequently suppressed protein aggregation [131]. In contrast, polyvinylpyrrolidone (PVP) and low-molecular-weight PEGs (e.g., PEG200, PEG300, and PEG1000) were found to be ineffective at stabilizing urokinase in solution [131]. It is believed that more hydrophobic nature of the polymers offsets their stabilizing effects. Specifically, PEG contains a small nonpolar moiety. The stabilizing volume exclusion effect exerted by small-molecular-weight PEGs, including PEG200 and PEG1000, appears to be overwhelmed by their hydrophobicity. In addition, as polymers are strong protein precipitants, they have been shown to enhance protein–protein self-interaction and protein–macromolecule interactions, possibly leading to undesirable protein self-association and aggregation [94].

Aside from stabilizing proteins via preferential exclusion and molecular crowding mechanisms, hydrophilic polymers that contain nonpolar moieties, such as PEGs and poloxamers, also stabilize proteins by suppressing surface adsorption-induced aggregation [94]. Because of their amphiphilic nature, these polymers have surfactant properties and been observed to decrease water surface tension. Poloxamers, also known by the trade names Synperonics, Pluronics, and Kolliphor, are nonionic triblock copolymers composed of a central hydrophobic chain of polyoxypropylene flanked by two hydrophilic chains of polyoxyethylene. Because the lengths of the polymer blocks can be customized, a wide variety of poloxomers with different properties exist and are commercially available. In this regard, amphiphilic proteins such as BSA have been used to stabilize proteins from surface adsorption. However, concerns regarding potential infectious agents in animal-derived products have prompted restricted use of BSA, and nonionic surfactants, as reviewed in the previous section, have been increasingly used as a replacement for BSA. As the polymers and proteins are competitive inhibitors of protein adsorption, typically only low concentrations (e.g., 0.1–1 wt%) are needed to cover protein adsorbing surfaces.

Charged polymers, such as sulfated polysaccharides, sulfated and phosphorylated polymers, and polyamino acids, can stabilize proteins via electrostatic interactions with multiple charged binding sites. Thus, in contrast to the nonspecific effects of hydrophilic polymers, effects from charged polymers are rather protein specific. For example, a variety of anionic polymers, including anionic heparin, dextran sulfate, pentosan sulfate, enoxaparin, phosvitin, and phytic acid, have been found to be effective at stabilizing acidic fibroblast growth factor, which has clusters of positively charged amino acids on the protein surface [132]. The cationic polymer polyethyleneimine (PEI) was found to improve the storage stability of porcine muscle lactate dehydrogenase (LDH) by suppressing its aggregation [133]. Unlike nonspecific protein stabilizers, PEI did not increase the denaturation temperature of LDH, but rather suppressed the oxidation of free sulfhydryl groups on the protein by chelating metal ions [133]. For the LDH enzyme, the putative inactivation pathway starts with the oxidation of the enzyme that modifies certain amino acids, followed by alterations in protein secondary structure and aggregation [133]. In this case, PEI was found to be even more effective than EDTA at stabilizing the protein against metal-catalyzed oxidation.

Another class of polymeric compounds that have been found to have significant potential in reducing protein aggregation is the CDs and their derivatives [1, 29]. CDs are circular polymers of typically five to seven sugar molecules that have an annulated, ring-shaped structure. They are now found in a number of approved pharmaceutical products for parenteral administration, including hydroxypropyl-β-CD for the peptides leucine enkephalin and a neuromedin B-receptor antagonist, and sulfobutylether-β-CD for small-molecule drugs ziprasidone (Geodon; Pfizer) and voriconazole (Vfend; Pfizer) (see [1] and references therein). CDs are hydrophobic on the inside of the ring and hydrophilic on the outside of the ring. CDs are known to form complexes with hydrophobic compounds, increasing their solubility. β-CDs have been shown to suppress the aggregation of several therapeutically relevant proteins, including insulin and growth hormone (see [1] and references therein). It is believed that CDs derive their aggregation suppression effect from binding to aromatic residues, which can lead to the preferential stabilization of the partially or completely unfolded proteins, reducing folding and aggregation rates. The ability of CDs to suppress aggregation has also led to their use in the refolding of proteins. In addition, CDs may also stabilize proteins from aggregation by acting as a surface-active agent. Hydroxypropyl-β-CD has been shown to protect porcine growth hormone and an IgG monoclonal antibody from agitation-induced damages [134].

A dosage form where polymers have been demonstrated to be a critical excipient component is the dry (or solid) state, for example, lyophilized or spray-dried products. Polymers have been successfully used as bulking agents and lyoprotectants in lyophilized products, including dextran, hydroxyethyl starch, polyvinyl alcohol, and polyvinylpyrrolidone (see Kamerzell et al., and references therein) [93]. Bulking agents are generally used to provide an adequate structure to the cake, particularly for low dose, high potency drugs (<2% solid content) that do not have sufficient bulk to support their own structure. Lyoprotectants stabilize and prevent the degradation of the active drug molecule during freeze drying, storage, and reconstitution. Polymers have been shown to suppress protein aggregation during lyophilization and to prevent the solubility decrease observed during reconstitution. Water-soluble polymers dextran, carboxymethyl cellulose, diethylaminoethyl-dextran, and PEG have been shown to reduce the aggregation of lyophilized BSA significantly during storage at 37 °C [135]. Hydroxypropyl-β-CD has been shown to improve the solubility and prevent the lyophilization-induced aggregation formation for many proteins, including growth hormone, interleukin-2, and insulin [136]. Hydroxypropyl-β-CD has also been shown to stabilize lyophilized mouse monoclonal antibody storage at 56 °C [137] and inhibit dimerization of lyophilized tumor necrosis factor during storage at 37 °C [138]. Polymers stabilize protein drug products in solid formulations through a variety of mechanism (see Ohtake et al. and references therein), including raising the glass transition temperature (T_g) and collapse temperature of the protein formulation, increasing protein solubility, and suppressing interface (e.g., air/water and ice/water)-induced protein denaturation and aggregation.

4.3.8 Preservatives

Three types of preservatives are commonly added to formulations to better maintain the stability and safety of protein therapeutics over the shelf life of the drug product. Antioxidants and chelators are used to prevent or minimize oxidation reactions of drug products and/or excipients and antimicrobial agents are used to prevent microbial growth.

As discussed previously, metal-catalyzed oxidation of certain amino acid residues in proteins (e.g., methionine, cysteine, histidine, and tryptophan) is a commonly encountered chemical degradation pathway during storage of protein pharmaceuticals. Trace impurities such as metal ions, as well as hydrogen peroxide, are present in many pharmaceutical excipients, which can initiate oxidative damages to proteins. In addition, proteins may inadvertently be exposed to trace levels of sanitizing agents, for example, hydrogen peroxide, during processing. Antioxidants directly inhibit oxidation of other molecules by acting as electron donors. The most commonly used antioxidants include ascorbic acid, glutathione, acetylcysteine, sulfurous acid salts (bisulfite, metabisulfite), and monothioglycerol [97, 98]. Chelating agents are molecules that can form two or more separate coordinate bonds with metal ions and are often used together with antioxidants to further reduce oxidation. Metal chelating agents such as edetate disodium (EDTA) and diethylenetriaminepentaacetic acid (DTPA) have been shown to be useful at inhibiting the oxidation of growth factors [139] and monoclonal antibodies [140]. These chelators have also been shown to enhance the stability of pharmaceutical preparations of plasmid DNA and adenovirus in development as vaccines, when used in combination with free radical scavengers such as ethanol [141–144]. Other strategies of preventing metal-catalyzed oxidation include protection from light, optimizing solution pH [145], and the addition of amino acids such as methionine and histidine [96].

Multidose, single-container formulations are useful when the dose needs to be split (e.g., dose titration and dose combination) [23]. In administering of these drugs, multiple disruptions of the container closure integrity increase the chance of bacterial contamination. Thus, antimicrobials are needed in the formulation as a preservative to prevent microbial growth during the product's shelf life. There are a limited number of regulatory approved antimicrobial preservatives that can be included in medicinal products and the number is constrained even further in parenteral protein products. Antimicrobial preservatives must be efficacious (i.e., broad-spectrum and exert sufficient toxicity), chemically and physically stable, and cause minimal adverse effects toward the active drug product. The commonly used antimicrobial agents in parenteral protein therapeutic formulations belong to a few classes of simple organic compounds: alkyl alcohols (e.g., benzyl alcohol), amino aryl acid esters (e.g., methyl, propyl, and butyl parabens), and phenols (e.g., phenol, meta-cresol, and chlorocresol) [93, 97, 98, 146].

Unlike most pharmaceutical excipients that can be considered essentially inert, antimicrobials exert broad-spectrum toxic biological activity through their interactions with multiple cellular targets, which has several important consequences in the choice and utility of these compounds in protein therapeutic formulations. First, lowest levels that achieve adequate preservation should be used to avoid toxicity to mammalian cells. Examples of typical levels used are 0.75–5% benzyl alcohol, 0.1–0.315% *m*-cresol, 0.15–0.5% phenol, and 0.005–0.1% paraben propyl [97, 98]. An ideal antimicrobial should be active only against opportunistic pathogens, with little or no effects on mammalian cells. However, antimicrobial preservatives are chosen for their toxicity against a wide range of pathogens, including bacteria (Gram-positive and Gram-negative), yeasts, fungi, and molds. The broad-spectrum antimicrobial activity is achieved by targeting multiple cellular targets, including the cell wall (e.g., phenols), cytoplasmic membrane (e.g., phenols, parabens, and benzyl alcohol), and the cytoplasm (e.g., benzyl alcohol and phenols), through multiple modes of action, including enzyme inhibition, membrane permeabilization, and metabolic activity inhibition. These toxic pathways also affect mammalian cells, especially at high doses. Hence, inclusion levels of antimicrobial preservatives should be minimal, consistent with adequate preservation.

Second, additional stabilizers might be needed to counter the destabilizing effect antimicrobial preservatives may exert on protein drug products. Because antimicrobials exert their toxic activity by interacting with various biomolecules in the cells, they are capable of interacting with protein therapeutics as well. In general, the addition of antimicrobial preservatives, regardless of the preservative used, significantly affects the stability of proteins [147–150]. In some cases, visible precipitation and aggregation have been reported. Zhang et al. investigated the mechanism of benzyl alcohol–induced aggregation and precipitation of recombinant human interleukin-1 receptor antagonist (rhIL-1ra) in aqueous solution and found that benzyl alcohol caused minor perturbation to the tertiary structure of the protein, without affecting protein secondary structures, through weak hydrophobic binding to the protein [149, 150]. This minor structural perturbation was sufficient to increase the level of partially unfolded, aggregation-competent species to induce aggregation and precipitation. Thus, formulation and excipient selection also need to take into consideration antimicrobial preservative-induced protein instability. In the same study, Zhang and coworkers found that the preferentially excluded cosolute sucrose partially inhibited benzyl alcohol–induced aggregation by reversing the tertiary structural changes induced by benzyl alcohol [149, 150].

In addition, the influence of pH on the activity of the antimicrobial preservatives also needs to be taken into consideration during formulation. For example, the pH of optimal activity for phenol, parabens, and benzyl alcohol are pH 4–9, pH 4–8, and pH <5.0 [146]. Formulation pH condition outside of the pH range can lead to diminished antimicrobial efficacy. Lastly, as the antimicrobial preservatives are no different than any other group of organic compounds, they are subject to chemical and physical instabilities that comprise their activity through interactions with the drug product, other excipients, containers/closures, delivery devices, and so on.

For example, surfactants such as polysorbate 80 have been known to reduce or compromise the antimicrobial activity of benzyl alcohol and *m*-cresol through complex formation that sequester the antimicrobial agent [146]. Parabens are susceptible to base-catalyzed ester hydrolysis. Phenols have been found to disrupt the crystallinity of insulin in zinc suspension formulations. Parabens, phenols, and benzyl alcohol are all volatile to different extents. This renders them susceptible to losses to sublimation and evaporation during manufacturing and throughout product life. They are not suitable as preservatives in formulations that need to be lyophilized or for storage in container/closure systems that are permeable to gases [146]. It is clear that possibility for antimicrobial preservative degradation and incompatibility area is manifold. Potential problems can be mitigated at the onset by thorough knowledge of all the formulation components and appropriate preformulation studies to determine interaction and degradation propensity at early stages of product design.

4.4 EXCIPIENTS USED IN SOLID DOSAGE FORMS OF BIOPHARMACEUTICALS

Discussion of the different classes of excipients used in biopharmaceuticals and their mechanisms of action in this chapter has so far focused on liquid formulations. For obvious reasons, aqueous liquid formulation is easy and economical to handle during manufacturing and is convenient for the end user for parenteral and subcutaneous delivery [151]. However, proteins are susceptible to chemical and physical degradations in liquid formulations. In addition, during shipping, precise control of conditions is not always feasible so that products can be subject to numerous stresses such as temperature fluctuations (e.g., high, low, and freezing) and agitation. Furthermore, long-term stability of liquid formulations during storage remains an obstacle even if the formulation has been optimized for manufacturing and shipping. Theoretically, all of these problems could be avoided with a properly formulated solid dosage form. The removal of water can improve the long-term storage stability of the protein drug product as well as easy shipping and storage (see Remmele et al. and references therein) [152]. In fact, pH- and/or temperature-induced hydrolysis and deamidation reactions have been reported to be reduced for proteins in the dry state. In addition, proteins in the dry state are much less prone to shear-induced denaturation and aggregation during shipping. Lyophilized (or freeze-dried) and spray-dried products are now widely used for biopharmaceuticals.

The formulation design and excipient selection for solid dosage form require formulation scientists to take into consideration the same stability issues that go into designing liquid formulations. However, as the processing of biopharmaceuticals into a solid dosage form exposes additional stresses to the drug product, formulation needs to be designed to additionally protect the product from those stresses. For example, during freeze drying, the protein is exposed to supercooling, freeze concentration, selective crystallization, or precipitation of important excipients such as buffers and salts, water–ice interface, amorphous solid phase, and dehydration, all of which can adversely affect protein conformation and the stability of the drug product [152]. Thus, the development of a stable freeze-dried biopharmaceutical dosage forms requires an understanding of the freezing and drying processes. In addition, excipient selection also needs to take into account the desired physical property of the solid dosage form, for example, the ruggedness of the lyophilized cake and particle size, size distribution, and flowability of the spray-dried powder [23]. A number of excellent reviews have been published on the development of stable lyophilized protein drug products [151, 152]. Next, stresses encountered by the protein drug product during freezing and dehydration are briefly discussed, followed by a review of the common categories of excipients used in solid dosage forms with an emphasis on lyophilized products, which represent the most common solid dosage form for biopharmaceuticals.

4.4.1 Stresses Encountered During Lyophilization

For a lyophilized product, every step from vial filling to reconstitution of the dried product has the potential to damage the protein and require formulation excipients to inhibit degradation. During lyophilization, the liquid protein drug product is first frozen, followed by the removal of water by sublimation under vacuum. Proteins experience various physiochemical stresses during the freezing and drying steps [153]. During freezing, temperature decreases and water ice crystals form. As a result, the unfrozen portion of the formulation becomes concentrated in protein and excipients. Solutes may undergo crystallization or precipitation, and/or undergo liquid–liquid phase separation. Thus, the protein drug product is potentially exposed to low temperatures, ice–water interface, high protein concentrations, high excipient concentrations, solid–liquid interface, and liquid–liquid interface. Cold denaturation of the protein can also occur, which could lead to aggregation upon reconstitution if the protein does not properly refold upon rehydration. In addition, it is well known that certain buffer salts selectively precipitate, resulting in pH shifts. Phase separation of stabilizing excipients can also result in undesirable structural changes in proteins. For example, although stabilizers such as sugars, amino acids, salts, and polyols have been shown to effectively stabilize proteins during the freeze–thaw process, some of these stabilizers may lose efficacy when they precipitate or crystallize into a frozen state [153]. Surface-induced protein structural changes during freezing and freeze drying processes have also been well documented. Proteins undergo structural changes at the surface of ice crystals and/or other crystalline excipients, which can lead to aggregation or precipitation [154]. A substantial body of literature has demonstrated that the surface area of crystals is correlated to the degree of protein degradation (see Chang et al. and reference therein) [153].

During the drying stage of lyophilization, water is removed through the sublimation of crystalline (i.e., ice) and amorphous water by vacuum while the protein and amorphous phase of the formulation remain stable. At the end of the drying stage, the residual water content of the cake is reduced to approximately 0.5–1% [152]. As discussed earlier in the chapter, the native structure of the protein is a delicate balance between hydrophobic and hydrophilic forces. As both forces stem from the presence of bulk water, water is required for maintaining the protein native structure. Removal of the protein's hydration shell, which occurs during lyophilization, destabilizes the native protein conformation (see Carpenter et al. and references therein) [155]. In addition, studies on numerous proteins have shown that in the absence of appropriate stabilizers, proteins will be unfolded in the dried solid (reviewed in Carpenter et al.) [155]. Upon rehydration, the degree of damage, such as aggregation, in these samples directly correlates with the degree by which the protein is denatured in the solid state. In addition, structurally altered proteins are also more prone to chemical degradation [94]. Reducing posthydration damage thus depends on minimizing protein unfolding during freezing and drying. Thus, a stable lyophilized protein formulation needs to protect the proteins from the drying process by substituting the molecular interactions from water molecules. Formulation conditions that foster refolding during rehydration can also increase native protein recovery.

During the rapid processing steps, including filling, freezing, drying, reconstitution, and administration, the major degradation of proteins is usually physical damages, which typically lead to the formation of protein aggregates. The transition from a liquid state to a solid state slows the rate of physical changes sufficiently such that in the solid state, chemical degradation during storage is often a more prevalent and serious problem [155]. To protect the protein during these steps so that the dried dosage form immobilizes the native protein in an inert solid matrix, as well as to manipulate the lyophilized product properties so that the freeze-dried cake with appropriate stability and reconstitution properties is achieved, various excipients are used as stabilizers and bulking agents. While many of the mechanisms of excipient stabilization of proteins in liquid formulations are mediated through water, for example, hydrogen bonding, hydrophobic effect, and preferential hydration, in the dry state, excipient interaction with water is irrelevant. Rather, excipients confer their effects on proteins in the solid dosage form through direct binding with proteins and their effects on the physical properties of the powder [94]. In the following sections, the different classes of excipients used in lyophilized products are reviewed, including buffer agents, bulking agents, glass forming agents and stabilizers, surfactants, and preservatives (Table 4.3). Discussion of the rationale involved in excipient selection for solid dosage form is also provided.

TABLE 4.3 Common Excipients Used in Lyophilized Biopharmaceutical Products

Excipient Class	Representative Examples
Buffering agents	Acetate, citrate, histidine, glutamate, phosphate, Tris, glycine
Tonicity modifiers	Mannitol, glycine, dextrose
Bulking agents	Mannitol, glycine, sucrose, PEG, PVP
Stabilizers	Sucrose, trehalose
Surfactants	Polysorbates 20 and 80 (Tween 20 and Tween 80), poloxamer P188 (Pluronic F68)
Antioxidants	Methionine, sodium sulfite, ascorbic acid
Antimicrobials	Benzyl alcohol, *m*-cresol, phenol, methyl paraben, propyl paraben

Adapted with permission from Remmele et al. [152].

4.4.2 Buffer Agents

As previously discussed, choosing the right pH and buffer is a key starting point for stable formulation design. In the lyophilized state, pH has little meaning. However, pH needs to be maintained in liquid formulations prior to freeze drying, during freeze drying, and after reconstitution. Most lyophilized protein drug products on the market are formulated in the pH range between 4 and 8. It is generally recommended that pH should be maintained within 1 unit from the pK_a of a buffering agent for optimal buffering capacity. In addition, buffer agents should not be volatile or show pH changes upon freezing. Acetic acid, for example, has been shown to sublime during freeze drying. As a result, lower than desired concentration of the buffer agent is present after lyophilization, which consequently can cause pH variability and reduced buffering capacity of the reconstituted drug product. Buffers that do not sublime under vacuum, such as histidine and Tris, are therefore preferred over acetic acid. It has also been well documented that the disodium form of phosphate buffer crystallizes during freezing, which leads to the preferential concentration of the monosodium form of phosphate. As a result, the protein can be exposed to a pH of 4 or lower in the amorphous portion of the formulation during freezing and drying steps. In the presence of potassium ions as counterions, however, monobasic potassium phosphate preferentially crystallizes out of solution versus the dibasic potassium phosphate salt, resulting in a small pH change upon freezing of the buffer system. Alternative buffers to phosphate buffer include citrate, histidine, and Tris, which do not show large pH changes with freezing. In formulations with a pH targeted in the 4–6 range, glutamate (pK_a = 4.3) and histidine (pK_a = 6.1) are suitable buffer agents [152]. However, glutamate in some cases has caused pain upon injection [156].

4.4.3 Bulking Agents

Bulking agents are often required in lyophilized formulations to yield an elegant cake structure that is mechanically strong does not undergo collapse and/or eutectic melting. Residual moisture also needs to be relatively low (e.g., <1% residual moisture or 1 g of water per 100 g dried solid). If the cake mechanically collapses, it will not only be esthetically unacceptable but will also contain high residual moisture and become difficult to reconstitute (see Carpenter et al. and references therein) [151]. Bulking agents are generally used for low-dose (high-potency) drugs that do not support their own structure [97, 98]. Physically, the cake structure is formed by the bulking agent during the annealing step that follows the freezing step, during which the bulking agent crystallizes around the ice crystals. As a result, a highly porous cake structure is formed through which water can sublime during drying.

For formulations containing less than 50 mg/ml of the protein drug product, mannitol and glycine are commonly used as bulking agents. Mannitol has a long history of use, with a crystallization temperature that is sufficiently high such that it is crystallized out of solution after an annealing step that follows the freezing step. Crystalline mannitol also has a very high eutectic melting temperature (−1.4 °C) so that the cake can be easily processed and stored without collapsing or melting. Glycine is another commonly used bulking agent with a sufficiently high crystallization temperature. However, it is important to note that the rate of glycine crystallization is dependent on pH, where the rate is highest at pH ~6.5 and decreases when pH is adjusted below or above this value. In addition when adjusting the pH of a solution containing glycine hydrochloride using NaOH, NaCl is generated. Freeze concentration of this salt can potentially be destabilizing to the protein drug product.

For formulations that contain 50 mg/ml or more of the protein drug product (e.g., antibody formulations), the protein and a suitable stabilizer (i.e., sucrose) may be sufficient to form a proper cake structure as they provide adequate coverage of the ice crystal surface during lyophilization that provides a cake scaffold [152, 155]. In fact, increasing protein content to most formulations has been found to lead to a higher cake collapse temperature [155].

Polymers such as dextran and hydroxyethyl starch are attractive bulking agents as they have relatively high collapse temperatures. However, these polymers do not confer any stabilizing effects. As such, the usage of these polymers alone is not recommended, but they could prove useful in combination with protein stabilizers.

4.4.4 Stabilizers

With a properly selected pH and buffering agents and a crystallizing bulking agent that provides a mechanically strong and elegant cake structure, another major excipient component is stabilizers that form an amorphous phase with the protein in the dried solid and serve to protect the protein during drying, storage, and rehydration. Among numerous compounds tested, it appears that the most effective stabilizers are disaccharides. However, one group of compounds that should be avoided is the reducing sugars, including glucose, lactose, maltose, and maltodextrins [155]. These compounds may be effective at stabilizing proteins during lyophilization steps, but during storage in the solid state, they can degrade via the Maillard reaction between carbonyls of the sugar and free amino groups on the protein. As a result, the white cake containing the protein can become a brown syrup containing degraded protein.

The nonreducing disaccharides sucrose and trehalose are the most commonly used stabilizers in lyophilized formulations. Aside from their stabilizing effects in solution, sucrose and trehalose are relatively effective at protecting the proteins during freezing and usually excellent at inhibiting unfolding during drying and storage. During freezing, usually a stabilizer concentration of 5 wt/vol% or higher is needed for optimal protection. For protection during drying and storage, a weight ratio of the stabilizer to protein from 1:2 to 1:4 is required for adequate protein stability with optimal stability reported at around 5:1 [152, 155]. These ratios are suitable for low protein concentration formulations (<50 mg/ml). In practice, a range of stabilizer concentration should be tested to determine the optimal level of stabilizer that is needed to retain protein native structure in the dried solid as well as during rehydration.

The stabilization of biopharmaceuticals by disaccharides in lyophilized products can be attributed to two effects. Protection during freezing is attributed to the cosolute's ability to stabilize the protein native conformation via the preferential exclusion mechanism as free water is still present during freezing [94]. During drying, the preferential exclusion mechanism no longer applies as the bulk water, as well as the hydration shell of the protein, is removed. Disaccharides have been postulated to protect proteins during drying as well as in the dried state through water replacement. The presence of hydrogen bond between the carbohydrates and proteins in lyophilized samples has been confirmed by many studies (see Ohtake et al. and references therein) [94], where the level of hydrogen bonding of the protein in the dried state with disaccharides is similar to those with water in an aqueous solution.

An additional requirement for the stabilizer is that the final dried powder has a glass transition temperature T_g, defined as the temperature at which an amorphous (glass) material in a hard and brittle state transition into a molten state, well above the storage temperature. For example, if it is anticipated that the storage temperature may reach 30 °C, then a product containing the protein that has a T_g value >50 °C should be stable. Since residual moisture reduces the T_g of a material, the T_g criteria must apply to the maximum water content allowed by the product specification. Sucrose and trehalose are both glass forming compounds with high T_g values (52–70 °C and 77–79 °C for sucrose and trehalose, respectively), making them suitable as glass forming stabilizers in the dried state. In addition, it is desirable for the stabilizer to form an amorphous phase with the protein during freezing and drying steps that has a high apparent glass transition temperature T'_g, defined as the temperature at which a material in the glass state transitions into a viscous liquid. During annealing and drying steps, it is necessary for the freeze concentrated portion of the formulation, for example, proteins and excipients, to remain in the glass state to avoid collapse. Thus, annealing and freeze drying must be carried out at temperatures below the T'_g of the frozen drug product. The glass forming sucrose and trehalose have relatively high T'_g values, −32 °C and −30 to −27 °C, respectively and as such the lyophilization cycle or the shelf temperature can be maintained at relatively high temperatures during freeze-drying. Higher temperatures additionally reduce drying time due to higher rates of sublimation. As the drying steps are by far the longest steps in the lyophilization cycle, shorter times and higher temperatures can considerably reduce costs in scaled-up operations. In this regard, low-molecular-weight stabilizers with lower glass transition temperatures (e.g., sorbitol with a T_g of −3 to 8 °C and a T'_g of −45 °C) are not suitable stabilizers for lyophilized products. Some low-molecular-weight molecules such as glycerol can be advantageous if it improves protein stability when mixed with larger stabilizers such as sucrose in small amounts such that the glass transition temperatures are not significantly increased. Generally, the T'_g value has been found to increase almost linearly with increasing molecular weight, and protein molecules have high T'_g values. For example, albumin and ovalbumin have T'_g values −10 and −11 °C, respectively. Thus, it is desirable to formulate the protein at high concentrations to achieve relatively high T'_g values.

4.4.5 Surfactants

In addition to stabilizers, surfactants are frequently needed in a lyophilized formulation to achieve adequate process and storage stability of biopharmaceutical products. Surfactants at low levels have been shown to be effective at minimizing the extent of aggregate formation during lyophilization and upon reconstitution. By far the most commonly used nonionic surfactants are polysorbate 20 and polysorbate 80, and they are often used at concentrations slightly above their CMCs. Similar to their protective effects in solution, surfactants protect proteins in lyophilized formulation by suppressing interface-/surface-induced denaturation and thereby subsequent aggregation. Specifically, surfactants can prevent surface-induced denaturation and aggregation during mixing, filtration, and filling operations prior to lyophilization, prevent ice–water interface-induced protein unfolding and aggregation during the freezing step, protect protein against aggregation during the drying step (although they are not as effective as disaccharides), and prevent aggregation during rehydration. As in liquid formulations, it is generally advised to use the lowest amount of surfactants necessary to achieve desired stabilization. Autoxidation of polysorbates also need to be taken into consideration. Care must be given to the stability of the drug product during long-term storage, as trace amounts of organic peroxides or other impurities may be found from polysorbates that can increase oxidation risks for proteins.

4.4.6 Preservatives

In some protein products, oxidation can become an important stability concern. Antioxidants have been used to minimize chemical or covalent degradation of proteins. Commonly used antioxidants include sodium sulfite, ascorbic acid, and selenium sulfate [152]. L Methionine has also been used. In general, an antioxidant must be effective at the formulation pH. As with surfactants, the minimum amount of antioxidants should be used, with the antioxidant to protein ratio as an important consideration for determining an effective antioxidant concentration. In addition, photooxidation from light exposure, dissolved oxygen levels, and oxidative stress from other formulation excipients (e.g., polysorbates) also need to be taken into consideration. This may require that the antioxidant be present in sufficient excess.

In most cases, antimicrobial preservatives are not required in lyophilized drug products. However, if multiple injections from the same drug container will be made, antimicrobials will be needed. Among the range of antimicrobial preservatives approved for parenteral use, benzyl alcohol, *m*-cresol, methyl paraben, propyl paraben, phenol, and propyl gallate have been used [152]. The challenge in developing a formulation containing antimicrobial compounds is achieving effectiveness while maintaining protein stability during processing and storage. As discussed earlier, antimicrobial compounds are destabilizing to proteins at concentrations needed for maintaining sterility. It is thus recommended that antimicrobial preservatives be only considered if there is a need for multidose vial formulation, and other options are not available.

4.5 CONCLUSION AND FUTURE OUTLOOKS

Biologics-based therapeutics has become an important sector of the pharmaceutical industry and will continue to rapidly grow in the near future due to their numerous advantages, including efficacy and specificity. However, production costs remain extremely high, and achieving adequate stabilization during processing and storage remains a challenge. Research in the past 20 years from both academia and industry has vastly increased our knowledge regarding the degradation pathways of proteins. Combined with investigations of the mechanisms by which cosolutes or additives mediate protein chemical and physical stabilities, there is now a wealth of knowledge available to the formulation scientists to rationally select excipients and design formulations. Having an understanding of the dominant pathway(s) by which a particular drug product in a particular dosage form degrade chemically and physically is one of the first, and necessary, steps toward successful formulation development. Based on the degradation pathways, excipients can then be selected based on their mechanisms of action. As described in this review, a reasonable number of excipients are currently available to protect proteins from different degradation pathways. However, the need for new excipients to be developed and approved remains high, in particular, those with novel stabilization mechanisms. Such efforts are currently underway, as well as technological advances that reduce the costs of protein production and refolding.

REFERENCES

1. Frokjaer, S. and Otzen, D.E. (**2005**) Protein drug stability: A formulation challenge. *Nature Reviews Drug Discovery*, **4** (4), 298–306.
2. Lehninger, A.L., Nelson, D.L. *et al.* (**1993**) *Principles of Biochemistry*, Worth Publishers, New York, NY.
3. Dill, K.A. (**1990**) Dominant forces in protein folding. *Biochemistry*, **29** (31), 7133–7155.
4. Jaenicke, R. (**1988**) Is there a code for protein folding? in *Protein Structure and Protein Engineering*, vol. **39** (eds R. Huber and E.L. Winnacker), Springer-Verlag, Berlin, pp. 16–36.
5. Jaenicke, R. (**1991**) Protein folding – Local structures, domains, subunits, and assemblies. *Biochemistry*, **30** (13), 3147–3161.
6. Jaenicke, R. and Rudolph, R. (**1989**) in *Protein Structure: A Practical Approach* (ed. T.E. Creighton), IRL Press, Oxford, pp. 191–223.
7. Pace, C.N., Shirley, B.A. *et al.* (**1996**) Forces contributing to the conformational stability of proteins. *FASEB Journal*, **10** (1), 75–83.
8. Chi, E.Y., Krishnan, S. *et al.* (**2003a**) Roles of conformational stability and colloidal stability in the aggregation of recombinant human granulocyte colony-stimulating factor. *Protein Science*, **12** (5), 903–913.
9. Chi, E.Y., Krishnan, S. *et al.* (**2003b**) Physical stability of proteins in aqueous solution: mechanism and driving forces in nonnative protein aggregation. *Pharmaceutical Research*, **20** (9), 1325–1336.
10. Krishnan, S., Chi, E.Y. *et al.* (**2002**) Aggregation of granulocyte colony stimulating factor under physiological conditions: Characterization and thermodynamic inhibition. *Biochemistry*, **41** (20), 6422–6431.
11. Cromwell, M.E., Hilario, E. *et al.* (**2006**) Protein aggregation and bioprocessing. *AAPS*, **J8** (3), E572–E579.

12. Rathore, N. and Rajan, R.S. (**2008**) Current perspectives on stability of protein drug products during formulation, fill and finish operations. *Biotechnology Progress*, **24** (3), 504–514.

13. Jiskoot, W., Randolph, T.W. *et al.* (**2012**) Protein instability and immunogenicity: roadblocks to clinical application of injectable protein delivery systems for sustained release. *Journal of Pharmaceutical Sciences*, **101** (3), 946–954.

14. Porter, S. (**2001**) Human immune response to recombinant human proteins. *Journal of Pharmaceutical Sciences*, **90** (1), 1–11.

15. Schellekens, H. (**2002a**) Bioequivalence and the immunogenicity of biopharmaceuticals. *Nature Reviews Drug Discovery*, **1** (6), 457–462.

16. Schellekens, H. (**2002b**) Immunogenicity of therapeutic proteins: Clinical implications and future prospects. *Clinical Therapeutics*, **24** (11), 1720–1740.

17. Rosenberg, A.S. (**2006**) Effects of protein aggregates: An immunologic perspective. *AAPS*, **J8** (3), E501–E507.

18. Braun, A., Kwee, L. *et al.* (**1997**) Protein aggregates seem to play a key role among the parameters influencing the antigenicity of interferon alpha (IFN-alpha) in normal and transgenic mice. *Pharmaceutical Research*, **14** (10), 1472–1478.

19. Moore, W.V. and Leppert, P. (**1980**) Role of aggregated human growth-hormone (Hgh) in development of antibodies to Hgh. *Journal of Clinical Endocrinology and Metabolism*, **51** (4), 691–697.

20. Ratner, R.E., Phillips, T.M. *et al.* (**1990**) Persistent cutaneous insulin allergy resulting from high-molecular-weight insulin aggregates. *Diabetes*, **39** (6), 728–733.

21. Murano, G. (**1997**) FDA perspective on specifications for biotechnology products – from IND to PLA. *Development of Specifications for Biotechnology Pharmaceutical Products*, **91**, 3–13.

22. Zhu, C., Liu, L. *et al.* (**2012**) Water-soluble conjugated polymers for imaging, diagnosis, and therapy. *Chemical Reviews*, **112** (8), 4687–4735.

23. Chang, B.S. and Hershenson, S. (**2002**) Practical approaches to protein formulation development, in *Rationale Design of Stable Protein Formulations – Theory and Practice* (eds J.F. Carpente and M.C. Manning), Kluwer Academic/Plenum Publishers, New York, pp. 1–25.

24. Brange, J. (**2000**) Physical stability of proteins, in *Pharmaceutical Formulation and Development of Peptides and Proteins* (eds S. Frokjaer and L. Hovgaard), Taylor and Francis, London, pp. 89–112.

25. Cleland, J.L., Powell, M.F. *et al.* (**1993**) The development of stable protein formulations – A close look at protein aggregation, deamidation, and oxidation. *Critical Reviews in Therapeutic Drug Carrier Systems*, **10** (4), 307–377.

26. Goolcharran, C., Khossravi, M. *et al.* (**2000**) Chemical pathways of peptide and protein degradation, in *Pharmaceutical Formulation and Development of Peptides and Proteins* (eds S. Frokjaer and L. Hovgaard), Taylor and Francis, London, pp. 70–88.

27. Manning, M.C., Patel, K. *et al.* (**1989**) Stability of protein pharmaceuticals. *Pharmaceutical Research*, **6** (11), 903–918.

28. Volkin, D.B. and Middaugh, C.R. (**1992**) in *Stability of Protein Pharmaceuticals. Part A: Chemical and Physical Pathways of Protein Degradation* (eds T.J. Ahern and M.C. Manning), Plenum Press, New York, p. 215.

29. Manning, M.C., Chou, D.K. *et al.* (**2010**) Stability of protein pharmaceuticals: an update. *Pharmaceutical Research*, **27** (4), 544–575.
30. Clarke, S. (**1987**) Propensity for spontaneous succinimide formation from aspartyl and asparaginyl residues in cellular proteins. *International Journal of Peptide and Protein Research*, **30** (6), 808–821.
31. Miroliaei, M. and Nemat-Gorgani, M. (**2001**) Sugars protect native and apo yeast alcohol dehydrogenase against irreversible thermoinactivation. *Enzyme and Microbial Technology*, **29** (8–9), 554–559.
32. Stratton, L.P., Kelly, R.M. *et al.* (**2001**) Controlling deamidation rates in a model peptide: Effects of temperature, peptide concentration, and additives. *Journal of Pharmaceutical Sciences*, **90** (12), 2141–2148.
33. Li, S., Schoneich, C. *et al.* (**1995**) Chemical instability of protein pharmaceuticals: Mechanisms of oxidation and strategies for stabilization. *Biotechnology and Bioengineering*, **48** (5), 490–500.
34. Lam, X.M., Yang, J.Y. *et al.* (**1997**) Antioxidants for prevention of methionine oxidation in recombinant monoclonal antibody HER2. *Journal of Pharmaceutical Sciences*, **86** (11), 1250–1255.
35. Wasylaschuk, W.R., Harmon, P.A. *et al.* (**2007**) Evaluation of hydroperoxides in common pharmaceutical excipients. *Journal of Pharmaceutical Sciences*, **96** (1), 106–116.
36. Volkin, D.B., Mach, H. *et al.* (**1997**) Degradative covalent reactions important to protein stability. *Molecular Biotechnology*, **8** (2), 105–122.
37. Lai, M.C. and Topp, E.M. (**1999**) Solid-state chemical stability of proteins and peptides. *Journal of Pharmaceutical Sciences*, **88** (5), 489–500.
38. Fischer, S., Hoernschemeyer, J. *et al.* (**2008**) Glycation during storage and administration of monoclonal antibody formulations. *European Journal of Pharmaceutics and Biopharmaceutics*, **70** (1), 42–50.
39. Smales, C.M., Pepper, D.S. *et al.* (**2000**) Mechanisms of protein modification during model anti-viral heat-treatment bioprocessing of beta-lactoglobulin variant A in the presence of sucrose. *Biotechnology and Applied Biochemistry*, **32** (Pt 2), 109–119.
40. Dong, A.C., Prestrelski, S.J. *et al.* (**1995**) Infrared spectroscopic studies of lyophilization-induced and temperature-induced protein aggregation. *Journal of Pharmaceutical Sciences*, **84** (4), 415–424.
41. Fink, A.L. (**1998**) Protein aggregation: Folding aggregates, inclusion bodies and amyloid. *Folding and Design*, **3**, R9–R23.
42. Kendrick, B.S., Carpenter, J.F. *et al.* (**1998a**) A transient expansion of the native state precedes aggregation of recombinant human interferon-gamma. *Proceedings of the National Academy of Sciences of the United States of America*, **95** (24), 14142–14146.
43. Kendrick, B.S., Cleland, J.L. *et al.* (**1998b**) Aggregation of recombinant human interferon gamma: Kinetics and structural transitions. *Journal of Pharmaceutical Sciences*, **87** (9), 1069–1076.
44. Kim, Y.S., Cape, S.P. *et al.* (**2001**) Counteracting effects of renal solutes on amyloid fibril formation by immunoglobulin light chains. *Journal of Biological Chemistry*, **276** (2), 1626–1633.
45. Kim, Y.S., Wall, J.S. *et al.* (**2000**) Thermodynamic modulation of light chain amyloid fibril formation. *Journal of Biological Chemistry*, **275** (3), 1570–1574.

46. Lumry, R. and Eyring, H. (**1954**) Conformation changes of proteins. *Journal of Physical Chemistry*, **58**, 110–120.
47. Sanchez-Ruiz, J.M. (**1992**) Theoretical analysis of Lumry–Eyring models in differential scanning calorimetry. *Biophysical Journal*, **61**, 921–935.
48. Zale, S.E. and Klibanov, A.M. (**1983**) On the role of reversible denaturation (unfolding) in the irreversible thermal inactivation of enzymes. *Biotechnology and Bioengineering*, **25** (9), 2221–2230.
49. Webb, J.N., Webb, S.D. *et al.* (**2001**) Partial molar volume, surface area, and hydration changes for equilibrium unfolding and formation of aggregation transition state: High-pressure and cosolute studies on recombinant human IFN-gamma. *Proceedings of the National Academy of Sciences of the United States of America*, **98** (13), 7259–7264.
50. Pace, C.N. (**1986**) Determination and analysis of urea and guanidine hydrochloride denaturation curves. *Methods in Enzymology*, **131**, 266–280.
51. Pace, C.N. and Shaw, K.L. (**2000**) Linear extrapolation method of analyzing solvent denaturation curves. *Proteins – Structure Function and Genetics*, Suppl. 4, 1–7.
52. McQuarrie, D.A. (**1976**) *Statistical Mechanics*, Harper & Row, New York, NY.
53. George, A., Chiang, Y. *et al.* (**1997**) Second virial coefficient as predictor in protein crystal growth. *Macromolecular Crystallography*, **276** (Pt A), 100–110.
54. George, A. and Wilson, W.W. (**1994**) Predicting protein crystallization from a dilute-solution property. *Acta Crystallographica Section D: Biological Crystallography*, **50**, 361–365.
55. Guo, B., Kao, S. *et al.* (**1999**) Correlation of second virial coefficients and solubilities useful in protein crystal growth. *Journal of Crystal Growth*, **196** (2-4), 424–433.
56. Salinas, B.A., Sathish, H.A. *et al.* (**2010**) Understanding and modulating opalescence and viscosity in a monoclonal antibody formulation. *Journal of Pharmaceutical Sciences*, **99** (1), 82–93.
57. Ducruix, A., Guilloteau, J.P. *et al.* (**1996**) Protein interactions as seen by solution X-ray scattering prior to crystallogenesis. *Journal of Crystal Growth*, **168** (1–4), 28–39.
58. Porschel, H.V. and Damaschun, G. (**1977**) Determination of virial-coefficients of protein solutions by means of X-ray small-angle scattering and interpretations. *Studia Biophysica*, **62** (1), 69.
59. Receveur, V., Durand, D. *et al.* (**1998**) Repulsive interparticle interactions in a denatured protein solution revealed by small angle neutron scattering. *FEBS Letters*, **426** (1), 57–61.
60. Velev, O.D., Kaler, E.W. *et al.* (**1998**) Protein interactions in solution characterized by light and neutron scattering: Comparison of lysozyme and chymotrypsinogen. *Biophysical Journal*, **75** (6), 2682–2697.
61. Haynes, C.A., Tamura, K. *et al.* (**1992**) Thermodynamic properties of aqueous alpha-chymotrypsin solutions from membrane osmometry measurements. *Journal of Physical Chemistry*, **96** (2), 905–912.
62. Schaink, H.M. and Smit, J.A.M. (**2000**) Determination of the osmotic second virial coefficient and the dimerization of beta-lactoglobulin in aqueous solutions with added salt at the isoelectric point. *Physical Chemistry Chemical Physics*, **2** (7), 1537–1541.
63. Behlke, J. and Ristau, O. (**1999**) Analysis of the thermodynamic non-ideality of proteins by sedimentation equilibrium experiments. *Biophysical Chemistry*, **76** (1), 13–23.

64. Patro, S.Y. and Przybycien, T.M. (**1996**) Self-interaction chromatography: A tool for the study of protein–protein interactions in bioprocessing environments. *Biotechnology and Bioengineering*, **52** (2), 193–203.
65. Tessier, P.M., Lenhoff, A.M. *et al.* (**2002a**) Rapid measurement of protein osmotic second virial coefficients by self-interaction chromatography. *Biophysical Journal*, **82** (3), 1620–1631.
66. Tessier, P.M., Vandrey, S.D. *et al.* (**2002b**) Self-interaction chromatography: A novel screening method for rational protein crystallization. *Acta Crystallographica Section D: Biological Crystallography*, **58**, 1531–1535.
67. Tessier, P.M., Johnson, H.R. *et al.* (**2003**) Predictive crystallization of ribonuclease A via rapid screening of osmotic second virial coefficients. *Proteins*, **50** (2), 303–311.
68. Tessier, P.M., Sandler, S.I. *et al.* (**2004**) Direct measurement of protein osmotic second virial cross coefficients by cross-interaction chromatography. *Protein Sciences*, **13** (5), 1379–1390.
69. Winzor, D. J., D. J. Scott, *et al.* (**2007**). A simpler analysis for the measurement of second virial coefficients by self-interaction chromatography. *Analytical Biochemistry* **371** (1): 21–25.
70. Bengali, A.N. and Tessier, P.M. (**2009**) Biospecific protein immobilization for rapid analysis of weak protein interactions using self-interaction nanoparticle spectroscopy. *Biotechnology and Bioengineering*, **104** (2), 240–250.
71. Tessier, P.M., Jinkoji, J. *et al.* (**2008**) Self-interaction nanoparticle spectroscopy: A nanoparticle-based protein interaction assay. *Journal of the American Chemical Society*, **130** (10), 3106–3112.
72. Connolly, B.D., Petry, C. *et al.* (**2012**) Weak interactions govern the viscosity of concentrated antibody solutions: High-throughput analysis using the diffusion interaction parameter. *Biophysical Journal*, **103** (1), 69–78.
73. Shire, S.J., Shahrokh, Z. *et al.* (**2004**) Challenges in the development of high protein concentration formulations. *Journal of Pharmaceutical Sciences*, **93** (6), 1390–1402.
74. Liu, J., Nguyen, M.D. *et al.* (**2005**) Reversible self-association increases the viscosity of a concentrated monoclonal antibody in aqueous solution. *Journal of Pharmaceutical Sciences*, **94** (9), 1928–1940.
75. Cromwell, M.E.M., Carpenter, J.F., Scherer, T., Randolph, T.J. (**2008**) Opalescence in Antibody Formulations is a Solution Critical Phenomenon. In Proceseedings of the 236th ACS National Meeting, Philadelphia, PA, USA.
76. Mason, B.D., Zhang, L. *et al.* (**2011**) Opalescence of an IgG2 monoclonal antibody solution as it relates to liquid–liquid phase separation. *Journal of Pharmaceutical Sciences*, **100** (11), 4587–4596.
77. Sukumar, M., Doyle, B.L. *et al.* (**2004**) Opalescent appearance of an IgG1 antibody at high concentrations and its relationship to noncovalent association. *Pharmaceutical Research*, **21** (7), 1087–1093.
78. Nishi, H., Miyajima, M. *et al.* (**2010**) Phase separation of an IgG1 antibody solution under a low ionic strength condition. *Pharmaceutical Research*, **27** (7), 1348–1360.
79. Yadav, S., Liu, J. *et al.* (**2010**) Specific interactions in high concentration antibody solutions resulting in high viscosity. *Journal of Pharmaceutical Sciences*, **99** (3), 1152–1168.

80. Yadav, S., Shire, S.J. *et al.* (**2012**) Viscosity behavior of high-concentration monoclonal antibody solutions: correlation with interaction parameter and electroviscous effects. *Journal of Pharmaceutical Sciences*, **101** (3), 998–1011.

81. Chi, E.Y., Weickmann, J. *et al.* (**2005**) Heterogeneous nucleation-controlled particulate formation of recombinant human platelet-activating factor acetylhydrolase in pharmaceutical formulation. *Journal of Pharmaceutical Sciences*, **94** (2), 256–274.

82. Bee, J.S., Randolph, T.W. *et al.* (**2011**) Effects of surfaces and leachables on the stability of biopharmaceuticals. *Journal of Pharmaceutical Sciences*, **100** (1), 4158–4170. doi: 10.1002/jps.22597Epub 2011 Apr 26.

83. Pinholt, C., Hartvig, R.A. *et al.* (**2011**) The importance of interfaces in protein drug delivery – Why is protein adsorption of interest in pharmaceutical formulations? *Expert Opinion on Drug Delivery*, **8** (7), 949–964.

84. Chi, E.Y., Ege, C. *et al.* (**2008**) Lipid membrane templates the ordering and induces the fibrillogenesis of Alzheimer's disease amyloid-beta peptide. *Proteins*, **72** (1), 1–24.

85. Chi, E.Y., Frey, S.L. *et al.* (**2010**) Amyloid-beta fibrillogenesis seeded by interface-induced peptide misfolding and self-assembly. *Biophys*, **J98** (10), 2299–2308.

86. Jones, E.M., Dubey, M. *et al.* (**2012**) Interaction of tau protein with model lipid membranes induces tau structural compaction and membrane disruption. *Biochemistry*, **51** (12), 2539–2550.

87. Tyagi, A.K., Randolph, T.W. *et al.* (**2009**) IgG particle formation during filling pump operation: A case study of heterogeneous nucleation on stainless steel nanoparticles. *Journal of Pharmaceutical Sciences*, **98** (1), 94–104.

88. Liu, L., Randolph, T.W. *et al.* (**2012**) Particles shed from syringe filters and their effects on agitation-induced protein aggregation. *Journal of Pharmaceutical Sciences*, **101** (8), 2952–2959.

89. Bee, J.S., Nelson, S.A. *et al.* (**2009b**) Precipitation of a monoclonal antibody by soluble tungsten. *Journal of Pharmaceutical Sciences*, **98** (9), 3290–3301.

90. Basu, P., Krishnan, S. *et al.* (**2013**) IgG1 aggregation and particle formation induced by silicone–water interfaces on siliconized borosilicate glass beads: a model for siliconized primary containers. *Journal of Pharmaceutical Sciences*, **102** (3), 852–865.

91. Carpenter, J.F., Randolph, T.W. *et al.* (**2009**) Overlooking subvisible particles in therapeutic protein products: Gaps that may compromise product quality. *Journal of Pharmaceutical Sciences*, **98** (4), 1201–1205.

92. Bee, J.S., Davis, M. *et al.* (**2009a**) Aggregation of a monoclonal antibody induced by adsorption to stainless steel. *Biotechnology and Bioengineering*, **105** (1), 121–129.

93. Kamerzell, T.J., Esfandiary, R. *et al.* (**2011**) Protein–excipient interactions: mechanisms and biophysical characterization applied to protein formulation development. *Advanced Drug Delivery Reviews*, **63** (13), 1118–1159.

94. Ohtake, S., Kita, Y. *et al.* (**2011**) Interactions of formulation excipients with proteins in solution and in the dried state. *Advanced Drug Delivery Reviews*, **63** (13), 1053–1073.

95. Jorgensen, L., Hostrup, S. *et al.* (**2009**) Recent trends in stabilising peptides and proteins in pharmaceutical formulation – Considerations in the choice of excipients. *Expert Opinion on Drug Delivery*, **6** (11), 1219–1230.

96. Wang, W. (**1999**) Instability, stabilization, and formulation of liquid protein pharmaceuticals. *International Journal of Pharmaceutics*, **185**, 129–188.

97. Pramanick, S., Singodia, D. *et al.* (**2013a**) Excipient selection in parenteral formulation development. *Pharma Times*, **45** (3), 65–77.

98. Pramanick, S., Singodia, D. *et al.* (**2013b**) Excipient selection in parenteral formulation development. *Pharma Times*, **45** (3), 65–77.

99. Kumar, R. (**2009**) Role of naturally occurring osmolytes in protein folding and stability. *Archives of Biochemistry and Biophysics*, **491** (1–2), 1–6.

100. Lentzen, G. and Schwarz, T. (**2006**) Extremolytes: Natural compounds from extremophiles for versatile applications. *Applied Microbiology and Biotechnology*, **72** (4), 623–634.

101. Arakawa, T. and Timasheff, S.N. (**1982**) Stabilization of protein-structure by sugars. *Biochemistry*, **21** (25), 6536–6544.

102. Lee, J.C. and Timasheff, S.N. (**1981**) The stabilization of proteins by sucrose. *Journal of Biological Chemistry*, **256** (14), 7193–7201.

103. Timasheff, S.N. and Arakawa, T. (**1988**) Mechanism of protein precipitation and stabilization by co- solvents. *Journal of Crystal Growth*, **90** (1–3), 39–46.

104. Auton, M. and Bolen, D.W. (**2007**) Application of the transfer model to understand how naturally occurring osmolytes affect protein stability. *Methods in Enzymology*, **428**, 397–418.

105. Liu, Y. and Bolen, D.W. (**1995**) The peptide backbone plays a dominant role in protein stabilization by naturally occurring osmolytes. *Biochemistry*, **34** (39), 12884–12891.

106. Arakawa, T., Philo, J.S. *et al.* (**2004**) Elution of antibodies from a Protein-A column by aqueous arginine solutions. *Protein Expression and Purification*, **36** (2), 244–248.

107. Maity, H., Karkaria, C. *et al.* (**2009**) Effects of pH and arginine on the solubility and stability of a therapeutic protein (Fibroblast Growth Factor 20): relationship between solubility and stability. *Current Pharmaceutical Biotechnology*, **10** (6), 609–625.

108. Tischer, A., Lilie, H. *et al.* (**2010**) L-Arginine hydrochloride increases the solubility of folded and unfolded recombinant plasminogen activator rPA. *Protein Sciences*, **19** (9), 1783–1795.

109. Inoue, N., Takai, E. *et al.* (**2014**) Specific decrease in solution viscosity of antibodies by arginine for therapeutic formulations. *Molecular Pharmaceutics*, **11** (6), 1889–1896.

110. Shukla, D. and Trout, B.L. (**2010**) Interaction of arginine with proteins and the mechanism by which it inhibits aggregation. *Journal of Physical Chemistry*, **B114** (42), 13426–13438.

111. Arakawa, T., Ejima, D. *et al.* (**2007**) Suppression of protein interactions by arginine: A proposed mechanism of the arginine effects. *Biophysical Chemistry*, **127** (1–2), 1–8.

112. Baynes, B.M. and Trout, B.L. (**2004**) Rational design of solution additives for the prevention of protein aggregation. *Biophys*, **J87** (3), 1631–1639.

113. Shukla, D. and Trout, B.L. (**2011**) Preferential interaction coefficients of proteins in aqueous arginine solutions and their molecular origins. *Journal of Physical Chemistry*, **B115** (5), 1243–1253.

114. Hofmeister, F. (**1888**) Zur Lehre von der Wirkung der Saltz[About the science and effects of salts]. *Archiv für Experimentalle Pathologie und Pharmakologie*, **24**, 247–260.

115. Zhang, Y. and Cremer, P.S. (**2006**) Interactions between macromolecules and ions: The Hofmeister series. *Current Opinion in Chemical Biology*, **10** (6), 658–663.

116. Zhang, Y. and Cremer, P.S. (**2010**) Chemistry of Hofmeister anions and osmolytes. *Annual Review of Physical Chemistry*, **61**, 63–83.

117. Omta, A.W., Kropman, M.F. *et al.* (**2003**) Negligible effect of ions on the hydrogen-bond structure in liquid water. *Science*, **301** (5631), 347–349.
118. Lee, H.J., McAuley, A. *et al.* (**2011**) Molecular origins of surfactant-mediated stabilization of protein drugs. *Advanced Drug Delivery Reviews*, **63** (13), 1160–1171.
119. Chou, D.K., Krishnamurthy, R. *et al.* (**2005**) Effects of Tween 20 and Tween 80 on the stability of Albutropin during agitation. *Journal of Pharmaceutical Sciences*, **94** (6), 1368–1381.
120. Joshi, O., Chu, L. *et al.* (**2009**) Adsorption and function of recombinant Factor VIII at the air–water interface in the presence of Tween 80. *Journal of Pharmaceutical Sciences*, **98** (9), 3099–3107.
121. Randolph, T.W. and Jones, L.S. (**2002**) Surfactant–protein interactions, in *Rational Design of Stable Protein Formulations, Theory and Practice*, vol. **13** (eds J.F. Carpenter and M.C. Manning), Kluwer Academic/Plenum Publishers, New York, p. 198.
122. Kerwin, B.A. (**2008**) Polysorbates 20 and 80 used in the formulation of protein biotherapeutics: structure and degradation pathways. *Journal of Pharmaceutical Sciences*, **97** (8), 2924–2935.
123. Kiese, S., Papppenberger, A. *et al.* (**2008**) Shaken, not stirred: mechanical stress testing of an IgG1 antibody. *Journal of Pharmaceutical Sciences*, **97** (10), 4347–4366.
124. Kishore, R.S., Pappenberger, A. *et al.* (**2011**) Degradation of polysorbates 20 and 80: studies on thermal autoxidation and hydrolysis. *Journal of Pharmaceutical Sciences*, **100** (2), 721–731.
125. Wang, W., Wang, Y.J. *et al.* (**2008**) Dual effects of Tween 80 on protein stability. *International Journal of Pharmaceutics*, **347** (1-2), 31–38.
126. Singh, S.R., Zhang, J. *et al.* (**2012**) Effect of polysorbate 80 quality on photostability of a monoclonal antibody. *AAPS PharmSciTech*, **13** (2), 422–430.
127. Tomioka, H. (**1983**) Purification and characterization of the tween-hydrolyzing esterase of *Mycobacterium smegmatis*. *Journal of Bacteriology*, **155** (3), 1249–1259.
128. Tharmalingam, T., Ghebeh, H. *et al.* (**2008**) Pluronic enhances the robustness and reduces the cell attachment of mammalian cells. *Molecular Biotechnology*, **39** (2), 167–177.
129. Mustafi, D., Smith, C.M. *et al.* (**2008**) Multi-block poloxamer surfactants suppress aggregation of denatured proteins. *Biochimica et Biophysica Acta*, **1780** (1), 7–15.
130. al-Rubeai, M., Emery, A.N. *et al.* (**1992**) The effect of Pluronic F-68 on hybridoma cells in continuous culture. *Applied Microbiology and Biotechnology*, **37** (1), 44–45.
131. Vrkljan, M., Foster, T.M. *et al.* (**1994**) Thermal stability of low molecular weight urokinase during heat treatment. II. Effect of polymeric additives. *Pharmaceutical Research*, **11** (7), 1004–1008.
132. Won, C.M., Molnar, T.E. *et al.* (**1998**) Stabilizers against heat-induced aggregation of RPR 114849, an acidic fibroblast growth factor (aFGF). *International Journal of Pharmaceutics*, **167** (1–2), 25–36.
133. Andersson, M.M., Breccia, J.D. *et al.* (**2000**) Stabilizing effect of chemical additives against oxidation of lactate dehydrogenase. *Biotechnology and Applied Biochemistry*, **32** (Pt 3), 145–153.
134. Serno, T., Carpenter, J.F. *et al.* (**2010**) Inhibition of agitation-induced aggregation of an IgG-antibody by hydroxypropyl-beta-cyclodextrin. *Journal of Pharmaceutical Sciences*, **99** (3), 1193–1206.
135. Liu, W.R., Langer, R. *et al.* (**1991**) Moisture-induced aggregation of lyophilized proteins in the solid state. *Biotechnology and Bioengineering*, **37** (2), 177–184.

136. Costantino, H.R., Langer, R. *et al.* (**1994**) Solid-phase aggregation of proteins under pharmaceutically relevant conditions. *Journal of Pharmaceutical Sciences*, **83** (12), 1662–1669.
137. Ressing, M.E., Jiskoot, W. *et al.* (**1992**) The influence of sucrose, dextran, and hydroxypropyl-beta-cyclodextrin as lyoprotectants for a freeze-dried mouse IgG2a monoclonal antibody (MN12). *Pharmaceutical Research*, **9** (2), 266–270.
138. Hora, M.S., Rana, R.K. *et al.* (**1992**) Lyophilized formulations of recombinant tumor necrosis factor. *Pharmaceutical Research*, **9** (1), 33–36.
139. Tsai, P.K., Volkin, D.B. *et al.* (**1993**) Formulation design of acidic fibroblast growth-factor. *Pharmaceutical Research*, **10** (5), 649–659.
140. Zhou, S., Zhang, B. *et al.* (**2010**) Comparative evaluation of disodium edetate and diethylenetriaminepentaacetic acid as iron chelators to prevent metal-catalyzed destabilization of a therapeutic monoclonal antibody. *Journal of Pharmaceutical Sciences*, **99** (10), 4239–4250.
141. Evans, R.K., Xu, Z. *et al.* (**2000**) Evaluation of degradation pathways for plasmid DNA in pharmaceutical formulations via accelerated stability studies. *Journal of Pharmaceutical Sciences*, **89** (1), 76–87.
142. Evans, R.K., Nawrocki, D.K. *et al.* (**2004a**) Development of stable liquid formulations for adenovirus-based vaccines. *Journal of Pharmaceutical Sciences*, **93** (10), 2458–2475.
143. Evans, R.K., Zhu, D.M. *et al.* (**2004b**) Characterization and biological evaluation of a microparticle adjuvant formulation for plasmid DNA vaccines. *Journal of Pharmaceutical Sciences*, **93** (7), 1924–1939.
144. Zeng, Y., Ramsey, J.D. *et al.* (**2011**) Identifying stabilizers of plasmid DNA for pharmaceutical use. *Journal of Pharmaceutical Sciences*, **100** (3), 904–914.
145. Qi, P., Volkin, D.B. *et al.* (**2009**) Characterization of the photodegradation of a human IgG1 monoclonal antibody formulated as a high-concentration liquid dosage form. *Journal of Pharmaceutical Sciences*, **98** (9), 3117–3130.
146. Elder, D. P. and P. J. Crowley (**2012**) Antimicrobial Preservatives Part One: Choosing a Preservative System. Jan. 2012. Available: http://www.americanpharmaceuticalreview.com/Featured-Articles/38886-Antimicrobial-Preservatives-Part-One-Choosing-a-Preservative-System/.
147. Fransson, J., Hallen, D. *et al.* (**1997**) Solvent effects on the solubility and physical stability of human Insulin-like Growth Factor I. *Pharmaceutical Research*, **14** (5), 606–612.
148. Maa, Y.F. and Hsu, C.C. (**1996**) Aggregation of recombinant human growth hormone induced by phenolic compounds. *International Journal of Pharmaceutics*, **140** (2), 155–168.
149. Zhang, Y., Roy, S. *et al.* (**2004a**) Mechanism for benzyl alcohol-induced aggregation of recombinant human interleuki-1 receptor antagonist in aqueous solution. *Journal of Pharmaceutical Sciences*, **93** (12), 3076–3089.
150. Zhang, Y., Roy, S. *et al.* (**2004b**) Mechanism for benzyl alcohol-induced aggregation of recombinant human interleukin-1 receptor antagonist in aqueous solution. *Journal of Pharmaceutical Sciences*, **93** (12), 3076–3089.
151. Carpenter, J.F. and Chang, B.S. (**1996**) Lyophilization of Protein Pharmaceuticals, in *Biotechnology and Biopharmaceutical Manufacturing, Processing, and Preservation* (eds K.E. Davis and V.L. Wu), Interpharm Press, Buffalo Grove, pp. 199–264.
152. Remmele, R.L., Krishnan, S. *et al.* (**2012**) Development of stable lyophilized protein drug products. *Current Pharmaceutical Biotechnology*, **13** (3), 471–496.

153. Chang, B.S. and Yeung, B. (**2010**) Physical stability of protein pharmaceuticals, in *Formulation and Process Development Strategies for Manufacturing Biopharmaceuticals* (eds F. Jameel and S. Hershenson), John Wiley & Sons Inc., pp. 69–104.
154. Chang, B.S., Beauvais, R.M. *et al.* (**1996**) Physical factors affecting the storage stability of freeze-dried interleukin-1 receptor antagonist: Glass transition and protein conformation. *Archives of Biochemistry and Biophysics*, **331** (2), 249–258.
155. Carpenter, J.F., Pikal, M.J. *et al.* (**1997**) Rational design of stable lyophilized protein formulations: Some practical advice. *Pharmaceutical Research*, **14** (8), 969–975.
156. Gazerani, P., Wang, K. *et al.* (**2006**) Effects of subcutaneous administration of glutamate on pain, sensitization and vasomotor responses in healthy men and women. *Pain*, **124** (3), 338–348.

5

EXCIPIENT STANDARDS AND HARMONIZATION

R.C. Moreton
FinnBrit Consulting, Waltham, MA, USA

5.1 INTRODUCTION

Pharmaceutical excipients are materials, other than the active pharmaceutical ingredient (API), intentionally included in a drug product for ensuring that the drug product is acceptable to the regulatory authorities and patients in terms of manufacturability, performance, and appearance. Excipients do not treat or cure diseases, but they do allow the drug to be delivered to the patient in a convenient manner. Unformulated, most APIs are not particularly convenient for patients and may not even be absorbed properly. Excipients, when used correctly, change that. They are included in the formulation to make up for the deficiencies in the properties of the API, which would otherwise prevent the API from being converted into a medicinal product the patient can use.

Pharmaceutical excipients are a very diverse group of materials. They cover all the states of matter: gas, liquid, and solid (including semisolid), and they can be of natural (animal, vegetable, or mineral) or synthetic (including semisynthetic) origin. They can also be quite simple molecules such as sodium chloride or very complex high-molecular-weight polymers such as gelatin or cellulose. The manufacturing processes for excipients are also diverse, ranging from simple harvesting and extraction of natural products (e.g., starch and powdered cellulose) to total chemical synthesis of a polymer (e.g., povidone and polyacrylates) to recombinant technology (e.g., recombinant gelatin). Some pharmaceutical excipients are manufactured using batch

Pharmaceutical Excipients: Properties, Functionality, and Applications in Research and Industry, First Edition. Edited by Otilia M.Y. Koo.

processing, but many of the more common pharmaceutical excipients are manufactured using some form of continuous processing. The scale of manufacture is also different and may be measured in thousands of metric tons per annum. This is very much larger than the typical manufacture of APIs, and the scale of manufacturing operations in pharmaceutical excipient manufacture brings its own issues.

Very few excipients have been developed exclusively for the pharmaceutical industry. For most pharmaceutical excipients, the major use is in other industries, such as the food industry, or the oil industry. The pharmaceutical use of many excipients may be less than 10% of the total output. This can present problems when customers seek to tighten specifications because the manufacturer's main market is unlikely to be the pharmaceutical market, and the manufacturer will develop specifications and control strategies (e.g., in-process monitoring and finished excipient testing), which fit the requirements of their main market customers if there has to be a choice.

Excipient standardization began many years ago, and standards were available in many different compendia around the world. The current work to harmonize excipient monographs is a more recent project that has been worked on for about 25 years. Progress has been slow for a variety of different reasons, and the reasons are discussed in more detail in Section 5.7.

A dictionary definition for standardization is as follows:

> '*to make standard or uniform*'[1],

and for harmonization:

> '*to bring into consonance or accord*' [1].

However, in the context of excipients, we are discussing technical standards (i.e., specifications), and standardization refers to the setting of those specifications. In the pharmaceutical arena, harmonization has a very specific meaning relating to the work of the International Conference on Harmonisation (ICH) of Technical Requirements for Registration of Pharmaceutical Products for Human Use. In the context of pharmaceutical excipients, harmonization relates to the work of the Pharmacopoeial Discussion Group (PDG) in harmonizing excipient monographs and test methods (see Section 5.7.2.1).

5.2 THE EXCIPIENT LIFE CYCLE

As with any other product, a pharmaceutical excipient passes through what has become to be known as a "life cycle." The stages in the life cycle are similar to those for many other products:

- The "idea"
- Design
- Development

- Scale-up
- Commercial scale manufacture
- Launch
- Exclusive sales (if covered by a patent)
- Loss of exclusivity after patent expiration
- Discontinuation.

Unlike pharmaceutical finished products, there is no Hatch–Waxman extension to patents for pharmaceutical excipients, and this, along with the absence of a formal separate regulatory approval system for pharmaceutical excipients, has contributed to the reticence to introduce new chemical excipients in recent years.

For the introduction of a new chemical excipient, there has to be an identified unmet technical need which prompts the development of the novel excipient [2]. The stages of developing a new excipient are described in more detail in Chapter 7. Development of new excipients. Briefly, once the market need is assessed and the requirements for the performance of the excipient identified (analogous to the target product profile for drug products), the design phase can commence. This phase may be protracted in that many different approaches may be screened to identify the most appropriate. Once the requirements are identified, prototype materials can be made for initial testing. A decision will typically be made, on the basis of the results from such testing, as to whether or not the project should move forward, and the preferred design. An important point to remember is that the excipient design and development organization is beginning to acquire knowledge, experience, and understanding of the excipient, particularly as to how it performs during this phase of its life cycle.

The development phase is where the processing is defined. Processing can involve both chemical processing, for example, polymerization, addition, hydrolysis and oxidation reactions, and physical processing, for example, milling and grinding, agglomeration, and separation. In addition, the sequence of unit processes will be defined, together with the components of the equipment train. Again this will be an opportunity to acquire further knowledge, experience, and understanding of the excipient and its performance, but also on the manufacturing process, its limitations and capabilities.

It is this acquired knowledge, experience, and understanding of the excipient, its performance and the process of manufacture, that will be used to set the specification and control strategy for the new excipient. The final specifications (in-process, release, and sales specifications) will be a balance between the requirements for purity and specificity of the excipient, and the requirements for the performance of the excipient as it is intended to be used. In addition, there will be other parts to the specification which will reflect certain rules and regulations any pharmaceutical material is expected to comply with, for example, residual solvents and microbiology.

For an alternate source of an existing excipient (e.g., after patent expiration), the organization developing it will also go through a project with similar phases, except that the design phase will not be as protracted.

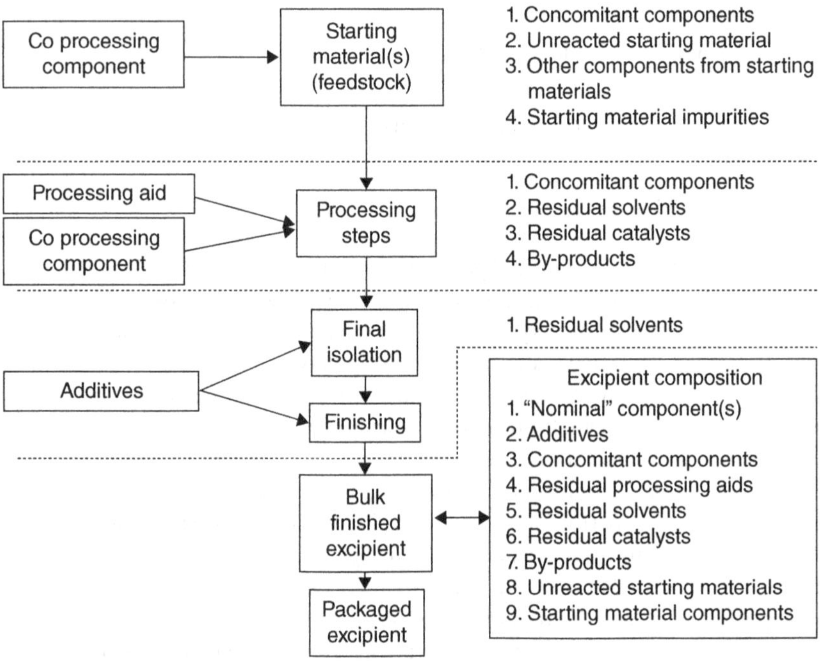

Figure 5.1 Sources of potential excipient components. From Ref. [3]; with permission.

5.3 EXCIPIENT COMPOSITION

Now that the developer of the new excipient has accumulated some understanding of the manufacture and performance of the new excipient, there is a need to understand the composition of the excipient. As discussed in Section 5.4, most excipients, in most applications, work (perform) because they contain other components. These are not additives, and they are not impurities. However, they are necessary for the proper performance of the excipient in the pharmaceutical formulation or product manufacturing process. In the USP–NF, they are referred to as concomitant components; others have referred to them as functional components, and this is probably a better term. In addition, there may be unreacted starting materials and reagents, residual processing aids, undesirable (read potentially detrimental) components, and so on. The origin of the different components in a pharmaceutical excipient is summarized in Figure 5.1.

The possible components of an excipient include the following:

- "Nominal" component(s)
- Additives
- Concomitant components
- Residual processing aids

- Residual solvents
- Residual catalysts
- By-products
- Unreacted starting materials
- Starting material components
- Degradants.

Note: Not all of these different components will necessarily be present in a particular excipient, but they could be. For a more complete discussion of excipient composition, the reader is referred to the IPEC Excipient Composition Guide [3].

The excipient developer will need to investigate the excipient composition to the extent that the available methodology will allow. They must be able to convince their potential customers that the new excipient (including a new alternate source) is well controlled and that they have sufficient understanding to be able support their customers in both delivery of excipient capable of providing the required performance and additional support when unexpected issues arise, as they inevitably will with a new excipient. It is also important to make a distinction between the types of investigations carried out during the development of a new excipient and the tests used to control the quality of the excipient in routine manufacture. The excipient developer will need to carry out more wide ranging tests at the development stage than they would use for routine quality control purposes. In addition, the excipient company should look to determine the best place to conduct such studies.

To give an example that is taken slightly out of context; when Penwest Pharmaceuticals (now JRS Pharma) launched their silicified microcrystalline cellulose product (ProSolv™), one of the questions raised by potential customers was how did they know there was no new covalent bonding in the coprocessed excipient? A study was carried out by an academic research group which showed, using an array of spectroscopic techniques, that there were no covalent bonds formed in the manufacture of silicified microcrystalline cellulose [4]. There are two points to be made; the research group was based in academia and therefore less likely to be influenced by "corporate," and the paper was published in a premier peer-reviewed journal, thus giving more credence to the results and conclusions compared to an in-house study in the excipient developer's own laboratories.

5.3.1 Processing Aids and Additives

One aspect of excipient composition that is very often overlooked is the presence of processing aids and/or additives. Processing aids are added to aid in the manufacture of the excipient itself. Processing aids have been defined as [5]:

> "*A material added to a manufacturing step for the purpose of facilitating the completion of that step or subsequent step.*"

The processing aid is added before the bulk excipient is finally available. Some of the processing aid(s) may be carried through to subsequent processing steps and may be present in the final excipient, all be it at very low levels. A recent example concerned the use of an antioxidant in the manufacture of a certain polyethylene glycol grade to suppress peroxide formation at an early stage in the process. Sufficient antioxidant was carried over to the final excipient to stabilize the pharmaceutical finished product. This only became apparent when the pharmaceutical manufacturer was looking to validate a second source of the excipient as part of a risk mitigation strategy. The critical process step for the manufacture of the alternative source excipient was carried out under nitrogen to avoid peroxide formation. Consequently, there was no added antioxidant and the finished product failed on stability.

Additives are somewhat different, and they may be defined as follows [5]:

> *"A substance added to the excipient to improve or maintain a characteristic such as a preservative, flow agent, antimicrobial, etc."*

Additives are materials added after the final processing step for the bulk excipient and are intended to improve the handling or storage of the excipient *per se*. Examples of additives include the use of silica as an anticaking agent in hydroxypropyl cellulose, and the use of propyl gallate as an antioxidant in ethyl cellulose.

As discussed in more detail in Section 5.6.5, additives are not permitted in pharmaceutical excipients stated to comply with a pharmacopeia monograph, unless their inclusion is specifically permitted in the monograph.

5.4 EXCIPIENT PERFORMANCE

As has been stated earlier, excipients are included in pharmaceutical formulations to help convert the API into a medicinal product which can conveniently be used by or administered to the patient. Excipients bring certain properties to the formulation which enable the formulation (and the drug) to function. These properties are collectively termed functionality or performance. It is important to remember that performance can relate to manufacture of the formulation, stability of the formulation, and/or *in vivo* performance of the formulation (i.e., after the formulation has been administered to the patient).

Excipient performance must arise from a combination of the chemical properties of the excipient, its physical properties, and the structure and surface morphology of the individual particles. Depending on the performance attribute, all four types of properties may contribute to the excipient performance. The key point is to understand which properties are important for the performance in a particular application (i.e., formulation).

Excipient performance can only be truly assessed in the context of the application. The performance attributes required for one formulation may be meaningless for another formulation, and *vice versa*. However, manufacturing a batch of product to confirm the acceptability of the performance of a particular delivery of an excipient is not economically viable for routine commercial manufacture. Other methods which can be used to predict excipient performance will therefore be required. These surrogate test methods may or not be included in the pharmacopoeia monograph.

In the future, more understanding will be required by potential users and the regulatory authorities, always recognizing that there may be limitations in the analytical methods with respect to the particular excipient. In the context of quality by design (QbD), it will be increasingly important to understand how and why excipients function in particular applications and how this relates to their composition and/or form. There is an implication that performance-critical components and/or properties should be controlled; however, attaining the requisite knowledge and understanding of what is critical for excipient performance in a particular application will continue to be a challenge. (QbD is discussed more fully in Section 5.8.2.)

The exact composition of many pharmaceutical excipients is not known, and there are several reasons for this. Although we may not know the precise details of the composition of a particular excipient, we do know that in certain cases excipient composition is linked to excipient performance. However, unlike with bulk active drugs where higher purity is preferred, many pharmaceutical excipients perform because they are not chemically "pure." Unfortunately, for many existing excipients, we do not know precisely which components of the excipient are crucial to its performance and the level at which they should be controlled.

For example, it is possible to prepare very high purity Dibasic Calcium Phosphate Dihydrate (DCP-D) USP. However, when used in direct compression, the very pure material did not perform as well as the regular material. DCP-D deforms during compaction by brittle fracture. The ability of the material to fracture is believed to be governed by the number of dislocations in the crystal lattice caused by foreign ions, and the very pure material did not fracture well because it had too few foreign ions in the crystal lattice.

In many cases, no suitable test methods are available due to the chemical nature of the molecule (e.g., poor solubility in common solvents). For pharmaceutical excipients, there has traditionally been an emphasis on physical parameters rather than chemical composition, and even where it is possible to better analyze the excipient, the link between excipient composition and performance is not well understood.

5.5 EXCIPIENT SPECIFICATIONS

Specifications, in the context of pharmaceutical excipients, may be defined as the combination of formal test methods and acceptable ranges for the results of such testing that define the excipient. In the United States (and some other countries), compliance with specification is one of the two prerequisites for use in pharmaceutical product manufacture (the other prerequisite being that the excipient is manufactured to acceptable standards of good manufacturing practice (GMP)). It is important to understand that there are two components to the specification: the method, and the range of acceptable results. Without a specified range anything could be acceptable, and without a formal test method how would we have confidence that the results obtained would be correct and that the excipient really did comply with its specification?

There can be several different specifications associated with a particular pharmaceutical excipient from a single supplier:

- Raw material specification: the ranges of tests and acceptable limits the raw materials and reagents used in the manufacture of the excipient are required to meet.
- In-process specification: the range of acceptable values for a particular test or tests that the excipient intermediate is expected to meet during processing and before the excipient is in its final form.
- Release specification: the tests and ranges of acceptable values the excipient is expected to meet before it can be released for sale as a pharmaceutical excipient.
- Sales specification: the tests and ranges of acceptable values the manufacturer warrants the excipient will meet on purchase. (In practice, the release specification will often have tighter acceptance limits than the sales specification.)
- Compendial (pharmacopoeial) monograph specification: the list of tests specified in the compendial monograph and the range of acceptable values the excipient is expected to meet, if the material is claimed to be in compliance with the monograph (not forgetting the requirement to be manufactured under acceptable GMP). (The manufacturer's release specification and sales specification may have tighter acceptance limits than the monograph limits, or they may have the same acceptance limits for particular tests.)
- Customer specification: the tests and ranges of acceptable values the customer has determined are necessary for the excipient to meet their particular requirements. The customer specification typically requires extra tests (outside the sales specification or monograph specification), or it may mean a tighter limits for one of the tests included in the sales specification or monograph specification, or both.

The reason for the tighter limits for the manufacturer's release specification, compared to the sales specification, is to ensure the excipient will always meet the sales specification allowing for the possibility that the test equipment at the excipient manufacturer's or user's testing laboratory may have a small but definite bias, and the sample tested at the user's testing laboratory will not be the same sample the original test results were generated on in the manufacturer's laboratory.

As we have discussed, the excipient specification will comprise a test or series of tests, together with limits to define the acceptable range of test results. The individual test or tests included in the specification (of any type) will fall into one of the following classifications:

- Chemical tests
- Physical tests
- Microbiological tests
- Other requirements.

There may be several different tests in any one of these groups for a given material.

5.5.1 Setting Excipient Specifications

For excipients having a monograph in a pharmacopoeia or other compendium, the specification and test methods are available in the official book, or on the official website. But how are these specifications arrived at and where do the methods come from? How are the specifications established for excipients without a compendial specification? In order to answer these questions, it is necessary to understand the life cycle of excipients and particularly the stages leading to commercial launch (see Section 5.2). For the purposes of this discussion, we will use the example of the development and subsequent launch of a new excipient, since this will give a greater insight than describing the development and launch of an alternative source of an existing excipient.

Excipient sales specifications should be considered from the start of any excipient development project, either for the introduction of a new source of an existing excipient, or the design and development of a new excipient. The eventual sales specification is analogous to the Quality Target Product Profile (QTPP) for pharmaceutical finished product development. For example, what constitutes success at each stage of the development project, and by what means should the success of the different stages in the development process be assessed? It should be noted that the assessment during the early stages of development may require more extensive testing than the eventual sales specification. Once the excipient development team is confident that they can scale up to the required scale of manufacture, they should begin to consider the eventual excipient sales specification in more detail. The eventual specification will need to assure the customer (excipient user) that the excipient is fit for purpose. It is highly likely that the eventual specification will embrace chemical, physical, and microbiological characteristics of the excipient. The objective will be to ensure that the manufacturing process and the final excipient are adequately controlled and that the excipient is fit for its intended purpose.

The initial testing of early development batches will focus on those excipient characteristics that the development team considers likely to relate to their assessment of the potential performance of the excipient. However, as the project moves through scale-up, a series of tests will need to be developed that address also the composition and safety aspects of the excipient, in addition to possible performance attributes.

Having gained knowledge and understanding of the excipient, its manufacture, its composition, its testing, and its performance during the development of the manufacturing process and scale-up to commercial manufacturing scale, the excipient developer/manufacturer will be in a position to finalize specifications for the new excipient. As stated, there will likely be several different specifications, including raw material, in-process, release, and sales specifications. Raw material specifications will not be discussed further. The in-process specifications will be set with the aim of ensuring that the excipient manufacturing process will consistently produce excipient that meets its release specification. The release specification will be set to ensure the excipient will consistently meet its sales specification. The sales specification informs the user of the excipient (or any other customer) of what they can expect and the degree of control they can expect.

One of the important aspects of setting a specification is the determination of the limits for a particular parameter. We tend to use the term "specification" to refer to both a group of tests and a single test. It is important to understand the context in which the term "specification" is used.

During the development of the excipient, the excipient developer will have gained knowledge, experience, and understanding of the excipient; how it is made, how well it addresses the project objectives, and so on. They will also have designed the manufacturing process and equipment train. It is possible they used an existing equipment train, or adapted an existing equipment train, but it is also possible they will have installed a new equipment train. Before commercial launch, they will have made several batches of the excipient and of each grade if more than one grade is to be offered for sale. These batches will be analyzed, and this data will be used to set limits (specifications) for different test parameters, provided there is enough data. The type of processing, batch versus continuous, will also influence the type of testing implemented and amount of data generated.

For an excipient manufactured using batch processing, the number of batches produced will govern the number of data points for the analysis. Process analytical methods (analogous to pharmaceutical process analytical technologies (PAT)) will likely be implemented in the equipment train, but the results reported for the batch will be the key data; the in-process data would only be supporting data. If an insufficient number of batches are available, the excipient developer/manufacturer could set tentative specifications for the initial launch and then finalize them when data from a sufficient number of batches were available.

With excipients manufactured by continuous processing, it is likely some form of continuous process monitoring (again, analogous to PAT) will be in place. In addition, the concept of batch size is different in continuous processing since there is no predefined quantity of material that is processed in one mass. In continuous processing, the batch size is usually taken as either a fixed time of output, for example, 1 day or 1 week, or as a fixed quantity of output, for example, 10 or 100 tons. Regardless of the way a batch is defined, there will likely be a lot more data. For example, an in-process control may provide an analytical result every minute; running continuously that sensor would generate 1440 data points per 24 hours, and that is a lot of data. Even a sample every 10 minutes would give 144 data points over 24 hours. In this context, more data give more confidence in the statistical analysis and the degree of variability and control.

The acceptance limits for a particular test will typically be set on the basis of a statistical analysis of the available data. Limits can be either two-sided with upper and lower limits, for example, for an assay, or they can be one-sided with an upper limit, for example, heavy metals, or a lower limit, for example, a minimum concentration of a particular component. In the process industries, it is common to use a six-sigma (6σ) approach to setting limits. (Note: this is not necessarily the same as implementing a six-sigma quality system; it simply uses the same statistics.) The reason for using $\pm 6\sigma$ is because the chance of failure is 3.4 per million; that is, there would be a chance that out of 1 million batches manufactured, or 1 million data points measured, three or four might fail. Using $\pm 6\sigma$, the excipient developer/manufacturer would set limits based on the mean result plus or minus six times the standard deviation. For example, for a mean of 120 and a standard deviation of 2, the $\pm 6\sigma$ limits would be 108–132.

There are certain presumptions underlying this $\pm 6\sigma$ approach, namely:

- The measurement method provides the necessary accuracy and precision and is robust.
- There is sufficient control of the process to allow fine adjustments to be made to the parameter under test.

Without these presumptions being met, the $\pm 6\sigma$ approach may not be appropriate (the limits would be too wide, or the degree of control too coarse).

One area where $\pm 6\sigma$ limits may not be appropriate is particle size. We can measure particle size down to the nanometer range if we need to, but we cannot necessarily control the particle size. For example, starch is a natural product. The grain size is governed by the plant species and the growing conditions. It does vary from species to species, and from 1 year to the next. Milling is a common unit process in excipient manufacture, and we can measure particle size to a precision of a few micrometers using a variety of different techniques. However, even with automatic feeders for the mills, and so on, in the author's experience, we do not have the control of the milling output that will allow us to use $\pm 6\sigma$ limits, and have sufficiently tight limits to satisfy the customer's needs. More typically, $\pm 3\sigma$ limits would be used in such circumstances. At first sight, $\pm 3\sigma$ limits would seem to be less robust, but they may represent a workable compromise. However, if the customer's limits would represent, for example, less than $\pm 2\sigma$, then there are likely to be supply issues and possibly product failures.

5.6 PHARMACOPEIAS AND OTHER COMPENDIA

There are many pharmacopoeias in the world. Many nations have their own pharmacopeias; however, many other nations use one of the three main pharmacopeias; that is, the European Pharmacopoeia, Japanese Pharmacopoeia, and the United States Pharmacopeia. These three pharmacopoeias are also the members of the Pharmacopeial Discussion Group (PDG) (see 5.7.2.1).

In general, a pharmacopoeia comprises at least three sections: General Notices, monographs (for excipients, APIs and in some cases products or product types), and General Chapters.

There are also other compendia which contain monographs for materials used as pharmaceutical excipients.

5.6.1 The European Pharmacopoeia, Japanese Pharmacopoeia, and United States Pharmacopeia

The three pharmacopoeias currently having the most influence in the global pharmaceutical arena are those from Europe, Japan, and the United States. These are also the main markets for drug products in the world. The United States Pharmacopeia-National Formulary is actually two separate books published in one combined set: the United States Pharmacopeia (USP) and the National Formulary (NF). (There are legal reasons for this, and the correct abbreviation is thus USP-NF.)

In the following discussion, the USP–NF will primarily be considered. Comparisons between the USP–NF, the European Pharmacopoeia (Ph.Eur), and Japanese Pharmacopoeia (JP) will be made where appropriate.

The JP and USP–NF are national pharmacopeias, whereas the Ph.Eur is a supranational or regional pharmacopeia. There are 36 national pharmacopoeia organizations that are signatories to the European Pharmacopoeia Convention, including the British Pharmacopoeia, the Deutsche Arzneibuch (German Pharmacopeia), and the Pharmacopée Française (French Pharmacopeia). The European Commission is also a signatory to the European Pharmacopoeia Convention, and the Ph.Eur is the pharmacopeia of the European Union. Those national pharmacopoeias that are members of the European Pharmacopoeia Convention accept the Ph.Eur monographs for materials that have such monographs. They also have their own individual monographs for materials that do not have Ph.Eur monographs. The formation of the European Pharmacopoeia was one of the early steps in harmonization, since it has harmonized many monographs and general chapters throughout the member pharmacopoeias.

The JP, Ph.Eur, USP–NF, and the other pharmacopeias have monographs for pharmaceutical excipients which are already included in marketed medicinal products. This is the general rule for proposing a new pharmacopoeial monograph; that the excipient be included in a medicinal product that is approved for commercial sale. There is an exception to this for the USP–NF whereby it is possible to propose and develop a pending monograph. The process for developing a pending monograph for the USP–NF was modified as of June 01, 2015. Under the revised process, the pending monograph would be subjected to the USP Notice and Comment process through publishing in the In-Process Revision section of Pharmacopeial Forum, and then approved by the appropriate expert committee. However, it would not be published in the USP–NF or supplement, until there is an FDA-approved commercial product on the market containing the excipient, at which time the monograph would be published in the next issue of Pharmacopeial Forum with a 6-month delay in implementation. The monograph would then be included in the next issue of the compendium or supplement. This will help to reduce the extended time in getting a new excipient monograph developed and approved if the monograph development process were initiated after FDA approval of the first commercial product. The Pending Monograph Guideline is available on the USP website.[1]

The Ph.Eur will also accept new excipients from marketed products; however, the criteria for final acceptance may be somewhat different. The process for adopting new monographs by the JP takes many years. There are fewer excipient monographs in the JP compared to the Ph.Eur or USP–NF.

[1] http://www.usp.org/sites/default/files/usp_pdf/EN/USPNF/pendingStandards/2015-06-01-pending-monograph-guideline.pdf. Note: The Guidance also describes two other possible uses of the pending monograph.

5.6.1.1 General Notices The USP General Notices [6] are mandatory and apply to all the contents of the USP–NF (both monographs and General Chapters). The requirements of the General Notices are mandatory for all USP monographs for Official Substances and Official Products. Originally, there were separate General Notices for the NF. In practice, the requirements of the General Notices of both the USP and the NF were very similar, and NF users were referred to the USP General Notices for many of the NF General Notices requirements. Today, there is one set of General Notices covering both the USP and NF parts of the compendium. (For simplicity and correctness, USP–NF will be used henceforth when referring to General Notices, unless circumstances dictate otherwise.)

The General Notices of the USP–NF explain, for example, the system of weights and measures, including the specification that all temperatures are defined as the Centigrade or Celsius system. However, the General Notices also explain the manufacture of Official Substances and Products, including manufacture according to recognized principles of GMP (3.10), compliance with the monograph and use of the USP or NF appellation (3.20), the USP–NF policy on such matters as additives (5.20.10), alternate test methods (6.30), and so on.

The JP and Ph.Eur also have mandatory General Notices.

5.6.1.2 General Chapters The General Chapters of the USP–NF are organized numerically into three sections:

- General Chapters 1–999: the provisions of these General Chapters are mandatory and thus apply theoretically to all monographs. However, it should be noted that the provisions will not apply if it is obvious that the provisions are not applicable to a particular Official Substance or Official Product (e.g., General Chapter <1> Injections does not apply if the product is not intended to be administered by injection).
- General Chapters 1000–1999: the provisions of these General Chapters are not mandatory and they are referred to as General Information Chapters. However, if a General Information Chapter is referred to in a test method in a monograph, it becomes mandatory for that monograph only.
- General Chapters 2000 and above: these General Chapters only apply to nutritional and dietary supplements. They do not apply to pharmaceutical substances or products. However, if a dietary or nutritional substance is used in a pharmaceutical product, the General Chapters below 1000 would apply.

The other pharmacopeias have different rules for the applicability of their General Chapters or their equivalent. The reader should familiarize themselves with the policies of the different pharmacopeias if they intend to use them.

The JP has a series of sections that may be considered analogous to the USP–NF General Chapters and General Information Chapters:

- General rules for crude drugs
- General rules for preparations

- General tests
- Processes and apparatus
- General information.

The Ph.Eur also contains the equivalent of the USP–NF General Chapters and General Information Chapters. However, they are presented in a format that is different again from both the JP and USP–NF.

5.6.1.3 Reference Substances Certain test methods are not absolute methods or the method is not sufficiently precise to allow the analyst to state unequivocally that the result can be assigned to a specific characteristic of the material with sufficient precision that the material is what it purports to be; thus, the result from the test has to be related to some standard material. Very often an official test method requires that the sample of the material being tested be compared to a standard lot or batch of the same material that has been determined to comply with the monograph for the particular test(s). The USP–NF refers to standards such as Reference Substances. Reference Substances are certified by the USP–NF Reference Standard Laboratory as being suitable for use in the testing associated with the particular monograph.

The requirement for a Reference Substance is determined during the initial development of a particular monograph. However, if there is a change in a test method for a particular monograph, the need for a Reference Substance may be revisited. Since the Reference Substance will be necessary for the testing of a substance for conformation that it complies with the monograph, a new monograph, or a revision to a monograph requiring a new Reference Substance will not be made official until supplies of the Reference Substance are available from USP–NF. This can delay implementation and is something monograph sponsors will need to factor into their plans, particularly if the Reference Substance is a by-product or a minor component that should be controlled.

The JP and Ph.Eur also use Reference Substances. However, the JP also uses Reference Spectra for both ultraviolet (UV) and infrared (IR) identity tests. The Reference Spectra are listed in separate sections of the book.

5.6.1.4 Use of Alternate Test Methods There is often confusion regarding the use of alternate test methods and/or procedures for Official Substances and Products. The USP–NF does allow the use of alternate test methods and/or procedures for the testing of Official Substances or Products. This is explicitly addressed in the USP–NF General Notices 6.30 Alternative and Harmonized Methods and Procedures. However, there are some obligations on the part of the laboratory using the alternate method or procedure. In summary, the alternate method must show the same or better sensitivity for the parameter being tested, and the reliability of the test should be the same or better than the monograph (official) method, that is, potential for false results, negative or positive should be no worse than the official method. The alternate method must also be appropriately validated. Finally, in the event of a dispute, the official method shall be the standard on which any decision to accept or reject the result or material shall be made.

TABLE 5.1 The "Tally" of Known Deaths Due to Ethylene Glycol/Diethylene Glycol Either Being Used in or Determined to Be an Adulterant of Medicines for Human Use

Country	Year	Incident
USA	1937	Sulfanilamide Elixir formulated with DEG – 107 deaths
South Africa	1969	Sedative formulated with DEG – 7 deaths
Italy	1985	DEG in wines from Austria – no known deaths
India	1986	Medicinal glycerin laced with DEG – 14 deaths
Nigeria	1990	Acetaminophen syrup containing DEG – 40 deaths (some sources estimate 200 deaths)
Bangladesh	1990–1992	Acetaminophen syrup containing DEG – 339 deaths
Haiti	1995/1996	Cough medicine containing DEG – 85 deaths
Panama	2006	Cough and antiallergy syrup containing DEG – 46 deaths (116 or 365 according to other sources)
USA	2006/7	Toothpaste containing DEG – no deaths
Panama	2007	Toothpaste containing DEG – no deaths
Nigeria	2008/9	Teething formula contaminated with DEG from propylene glycol – 84 deaths
Bangladesh	2009	Acetaminophen syrup for children adulterated with DEG – 24 deaths

Note: These are the reported cases that can be assigned to the use or adulteration; the actual number of deaths was likely higher.

The JP and Ph.Eur both also allow the use of alternate methods with similar restrictions and caveats as for the USP–NF.

If monograph users find alternate tests which provide advantages over the current official tests, the details should be communicated to the pharmacopeia. For example, the USP–NF will assess the suitability of such tests, and if the reliability and specificity are acceptable, may propose them as revisions.

5.6.1.5 USP–NF Modernization The USP–NF operates on 5-yearly revision cycles commencing with a meeting of the USP Convention (also held every 5 years). One of the objectives for the current revision cycle (2015–2020) is the modernization of USP–NF monographs and General Chapters where necessary.

Part of the impetus for this has been the rise in cases of economically motivated adulteration (EMA) seen in recent years, including glycerin and propylene glycol adulterated with ethylene glycol and diethylene glycol (several countries over the years – see Table 5.1), and heparin adulterated with oversulfated chondroitin sulfate (the United States and Germany) (see Section 5.8.1).

Another reason is that some monographs, particularly for excipients, still use "old" methodologies, such as some traditional "wet" chemical methods, and that there are better methods available which, if used, would make the monographs more effective in controlling excipients, and will also help exclude substandard materials from the pharmaceutical supply chain.

One of the approaches being adopted is to look at groups of similar monographs and to introduce methods that can distinguish between closely related substances.

A good example of such a group is vegetable oils. Some vegetable oils are relatively inexpensive (e.g., corn oil), and some are much more expensive. There is thus the potential for economically motivated adulteration. In order to detect such adulteration, monographs for the more expensive oils had “wet” chemistry-style limit tests for specific oils, often several individual oils. This made the testing very time consuming. The vegetable oils mainly comprise glycerides with small amounts of plant sterols. Using a combination of fatty acid composition, sterol composition, and triglyceride composition, it is possible to distinguish between the various oils and to detect the presence of other oils (adulterants).

The US FDA also has a Task Group focused on USP–NF modernization. At the time of writing, there were approximately 200 drug and product monographs and approximately 90 excipient monographs requiring updating to include more specific tests and/or to introduce better, more modern test methods.

5.6.2 Other National Pharmacopoeias

Nations having their own pharmacopoeia organizations outside of the signatories to the European Pharmacopoeia Convention, Japan, and the United States include Brazil, The People’s Republic of China, India, Russia, and Vietnam. The three major pharmacopeias are working with the pharmacopeia organizations in these different countries, and others, to help them; both to develop modern pharmacopoeias, and also to try to avoid a plethora of different, possibly conflicting, standards and specifications in the global market place, not only for excipients but also for APIs and General Chapters.

5.6.3 International Pharmacopoeia

The International Pharmacopoeia (Ph.Int.) is published by the World Health Organization. Its origins date back to 1874; however, the WHO only became involved in 1947. As such, it is the oldest example of a pharmacopoeia harmonization project. The current edition is the fifth edition, published in 2015 (at the time of writing, there had been no supplements published to the fifth edition). The aim of the International Pharmacopoeia is to attain global uniformity in the specifications for selected medicines. Since 1979, the drugs included in International Pharmacopoeia have been those from the WHO list of essential drugs. The WHO list of essential drugs is targeted at the less wealthy nations. The International Pharmacopoeia contains excipient monographs. The requirements of the International Pharmacopoeia only become official in a country when legislation is enacted or an order to that effect is promulgated within that country.

5.6.4 Other Compendia

Besides the pharmacopoeias, there are other compendia which may be relevant to pharmaceutical excipients. These compendia do not have the “official” status of a pharmacopeia; however, they may provide specifications that can be used for excipients in pharmaceutical or nonpharmaceutical applications.

5.6.4.1 Japanese Pharmaceutical Excipients As noted earlier, getting new monographs into the JP is a long process, and it was recognized early in the global harmonization effort that the JP did not contain monographs for some of the excipients included in the original list of harmonization candidates. To provide some form of specification for excipients already in use in Japan, including those on the harmonization list, but not having monographs in the JP, the Japanese Pharmaceutical Excipients Council (JPEC) has prepared a series of books, Japanese Pharmaceutical Excipients (JPE), which provide monograph specifications for excipients used in Japan but not included in the JP. There are several such books and they have been translated into English. However, each new edition does **not** automatically supersede the previous edition. For the most part, they contain different monographs, and all the individual monographs remain current unless superseded by an updated monograph published in a later edition.

Although the Japanese Pharmaceutical Excipient books are used in Japan, they do not have the status of a pharmacopoeia. Within Japan they are seen as a useful repository of excipient specifications which can be used as a basis for an excipient specification included in a drug product application.

5.6.4.2 Food Compendia Many pharmaceutical excipients have uses in the food industry. Indeed, their initial use may have been in food manufacture or preparation, but they were also found to be useful in the formulation and manufacture of pharmaceutical products.

Food Chemicals Codex The Food Chemicals Codex (FCC) was first published in 1966. It includes monographs for more than 1100 materials classified as food-grade chemicals, food additives, foods, vitamins, flavors, processing aids, and functional food ingredients. The FCC was published for many years under the auspices of the Institute of Medicine. However, in 2006, it was purchased by the USP Convention, Inc. and continues to be administered by the USP organization, although as a separate publication. The monographs in the FCC are agreed standards and they can aid in the specification of the materials and in the detection of substandard materials.

However, the FCC monographs, unlike USP–NF monographs, do not have regulatory status in that the FCC is not specifically mentioned in the US Federal Food Drug and Cosmetic Act. Nevertheless, the monographs may provide useful guidance on the specification of the materials. There is some overlap in monographs between the USP–NF and the FCC. Work is in hand to minimize any differences where possible.

Codex Alimentarius The Codex Alimentarius is a collection of internationally recognized standards for food and food ingredients. It is administered by the Codex Alimentarius Commission, which is jointly sponsored by two United Nations bodies; the WHO and the Food and Agriculture Organization (FAO). It was founded in 1963. Its publications include standards, codes of practice, procedures, and guidelines which cover areas such as food safety and food production. It does include GMPs for food and food chemicals. Note: Food GMPs differ in certain areas from pharmaceutical GMPs; in particular, with respect to the independence of the quality unit and change control.

5.6.5 Pharmacopeial Monographs

The specification in the pharmacopeia monograph (or other compendial monograph) is typically directed at the purity and safety of the excipient. It has often been stated that compliance with the pharmacopoeial monograph is the minimum standard for entry of a new source of an existing excipient into the global market place. Not all pharmaceutical excipients have monographs in a pharmacopeia. In such cases, the excipient manufacturer will establish their own specification, sometimes referred to as a noncompendial monograph. However, the general form of such a specification will very likely be similar to that of a pharmacopeia monograph. There are good reasons for this; the regulatory authorities and customers expect the material to be properly specified and controlled, and the style of the pharmacopeia monographs has become established over many years.

There are many similarities in the layout of monographs in the main pharmacopoeias; however, there are also some differences, and these are a consequence of the different legal and regulatory environments in which the different pharmacopoeias operate, together with the fact that they evolved independently of each other. For example, all the tests in a monograph in the USP–NF are considered mandatory; there are no nonmandatory tests in the monograph. This is also the case for monographs in the JP. The Ph.Eur has adopted a different approach. It has introduced a nonmandatory section into some of its excipient monographs concerning functionality-related characteristics (FRCs).

In the following discussion explaining the content of pharmacopoeia monographs, the emphasis will be on the revised monograph layout of the USP–NF. There are differences in the monographs between the pharmacopoeias. The reader should familiarize themselves with the pharmacopoeia applicable to their country or region.

A USP–NF excipient monograph in the revised format will typically contain the following sections:

- Definition and/or source of the excipient
- Identification
- Assay
- Impurities, for example:
  - Inorganic
  - Organic
  - Residual solvents
- Specific tests, for example:
  - pH
  - Water
  - Concomitant components (should they be identified/specified)
- Additional requirements, for example:
  - Packaging and storage
  - Labeling requirements.

The JP and Ph.Eur excipient monographs will contain similar tests. However, the JP and Ph.Eur monographs will also contain the following sections:

- Characteristics (Ph.Eur)/Description (JP)
  - Appearance
  - Solubility.

This information on appearance and solubility can also be found in the USP–NF, but not in the monograph. There is a separate subsection in the USP–NF entitled Description and Solubility, which is found in the Reference Tables section of the book. Since the description and solubility are not included in the monograph itself, or in the General Chapters with numbers <1000, they are not mandatory requirements in the USP–NF. As mentioned above, Ph.Eur excipient monographs may also contain a nonmandatory section on FRCs.

The monographs for certain excipients are being or have been harmonized via the work of the PDG, supported by the local International Pharmaceutical Excipients Councils, IPEC-Americas, IPEC Europe, and the JPEC (see Section 5.7.2.1).

Having given a list of the different sections of pharmaceutical excipient monographs in the pharmacopoeias, a discussion of the reasons for the inclusions of the different tests in the monograph is warranted. This discussion will be based on the layout of the modernized USP–NF monographs and will use mainly examples from the monographs for Dibasic Calcium Phosphate Dihydrate USP [7], Microcrystalline Cellulose NF [8], Anhydrous Lactose NF [9], and Povidone USP [10]. These materials are respectively an inorganic excipient, a polymeric excipient of natural origin (derived from wood), a disaccharide which is representative of a low-molecular-weight organic material, and a synthetic polymer. They were also included in the original list of excipient monographs to be considered for harmonization (see Section 5.7.2.1). The discussion will be based on the regulatory and legal considerations prevailing in the United States at the time of writing this chapter (Q1 2016). Other monographs will be referenced as necessary. Similar considerations will apply in Europe and Japan; however, the details will likely be different.

5.6.5.1 Definition and/or Source of the Excipient This section of the monograph provides a set of limitations that help to restrict the excipient to a given chemical structure, botanical source or manufacturing process. For example, Dibasic Calcium Phosphate Dihydrate USP is defined as containing [7]:

> *'not less than 98.0 percent and not more than 105.0 percent of dibasic calcium phosphate dihydrate* ($CaHPO_4{\cdot}2H_2O$)*.'*

Anhydrous Lactose NF is defined as follows [9]:

> *'O-β-d-galactopyranosyl-(1→4)-β-d-glucopyranose (β-lactose) or a mixture of O-β-d-galactopyranosyl-(1→4)-β-d-glucopyranose and O-β-d-galactopyranosyl-(1→4)-α-d-glucopyranose (α-lactose).'*

These two definitions refer to very specific chemical entities. By contrast, Microcrystalline Cellulose NF is defined as follows [8]:

> *'purified, partially depolymerized cellulose prepared by treating alpha cellulose, obtained as a pulp from fibrous plant material, with mineral acids.'*

In this instance, the material is defined as being prepared from a certain type of starting material (a pulp from fibrous plant material) by a specific type of process (hydrolysis by mineral acid). There is no chemical definition of microcrystalline cellulose, for the very good reason that we cannot define it properly in strict chemical terms.

The definition for Povidone USP is as follows [10]:

> *'Povidone is a synthetic polymer consisting essentially of linear 1-vinyl-2-pyrrolidinone groups, the degree of polymerization of which results in polymers of various molecular weights. The different types of Povidone are characterized by their viscosity in aqueous solution, relative to that of water, expressed as a K-value (see Specific Tests, K-value). The K-value of Povidone having a stated (nominal) K-value of 15 or less is NLT 85.0% and NMT 115.0% of the stated values. The K-value of Povidone having a stated K-value or a stated K-value range with an average of more than 15 is NLT 90.0% and NMT 108.0% of the stated value or of the average of the stated range. It contains NLT 11.5% and NMT 12.8% of nitrogen (N: 14.01), calculated on the anhydrous basis. It has a nominal K-value of NLT 10 and NMT 120. The nominal K-value is shown on the label.'*

The definition for Povidone gives both a chemical description and makes reference to different types of Povidone. The different types are typically referred to as different grades. Limits are given for the *K*-value and the chemical content. [Note: these are harmonized definitions, and the same definitions appear in the JP and the Ph.Eur.] Povidone is better characterized than, for example, microcrystalline cellulose; in part because it is soluble in water and certain organic solvents, whereas microcrystalline cellulose is not easily dissolved.

In some instances, it may be necessary to define the excipient in such a way so as to restrict the source. For example, Olive Oil NF is defined as follows [11]:

> *'the refined fixed oil obtained from the ripe fruit of Olea europaea Linné (Fam. Oleaceae). It may contain suitable antioxidants.'*

In this case, the excipient definition restricts the botanical source of the oil and specifies that it be the refined oil, and also allows the addition of antioxidants. Since olive oil is rich in unsaturated fatty acids, it is prone to oxidation (becomes rancid) and antioxidants will prevent this and extend the shelf life of the material.

The Olive Oil NF monograph also brings out another important point about monographed materials, or "Official Substances" as they are referred in the USP–NF. In order to claim compliance with the monograph, besides meeting specification, and being manufactured under appropriate standards of GMP, additives are not permitted unless expressly permitted in the monograph for the excipient. This point has not been well understood by either excipient manufacturers or excipient users in the past and has caused problems from time to time due to the inclusion of undeclared additives and subsequent incompatibilities. Where an additive is permitted, there is usually an accompanying "Labeling" requirement to declare the name and content of the additive (e.g., see Olive Oil NF [9]).

5.6.5.2 Identification The General Notices of the USP–NF [6] provide a very good explanation of why identification tests (ID tests) are included in the monographs. Section 5.40 Identity of the General Notices states as follows:

> "*A compendial test titled Identity or Identification is provided as an aid in verifying the identity of articles as they are purported to be, e.g., those taken from labeled containers, and to establish whether it is the article named in USP–NF. The Identity or Identification test for a particular article may consist of one or more procedures. When a compendial test for Identity or Identification is undertaken, all requirements of all specified procedures in the test must be met to satisfy the requirements of the test. Failure of an article to meet all the requirements of a prescribed Identity or Identification test (i.e., failure to meet the requirements of all of the specified procedures that are components of that test) indicates that the article is mislabeled and/or adulterated.*"

The importance of this statement is put into context when one considers the GMP requirements for pharmaceutical product manufacture. In the United States, the FDA has certain expectations that must be met in order for a manufactured pharmaceutical finished product to be in compliance with GMP, including the following [12]:

> "*(6)(d) Samples shall be examined and tested as follows:*
>
> (1) *At least one test shall be conducted to verify the identity of each component of a drug product. Specific identity tests, if they exist, shall be used.*
> (2) *Each component shall be tested for conformity with all appropriate written specifications for purity, strength, and quality. In lieu of such testing by the manufacturer, a report of analysis may be accepted from the supplier of a component, provided that at least one specific identity test is conducted on such component by the manufacturer, and provided that the manufacturer establishes the reliability of the supplier's analyses through appropriate validation of the supplier's test results at appropriate intervals.*"

In context of this statement, "manufacturer" means the manufacturer of the pharmaceutical finished product, that is, the user of the excipient. It is also clear from the above statement that the user of the pharmaceutical excipient must perform the test(s), regardless of whether or not the excipient supplier performs the test, and regardless of whether or not the ID test result is included on the certificate of analysis (CoA) supplied with the excipient delivery.

Another important point is the specificity of the ID test. Very often, the Identification section will list a number of ID tests. Sometimes they are very specific, but more often they will detect only one aspect of the material rather than the specific molecule, for example, the use of a sodium test for a sodium salt will only indicate that sodium is present, it will not indicate whether or not the correct anion is present. The specificity comes from carrying out all the ID tests listed in the monograph. The expectation is that all the different procedures listed under Identity in the monograph will be carried out. This is clearly stated in the USP General Notices 5.40 Identity as follows [6]:

> "*When a compendial test for Identity or Identification is undertaken, all requirements of all specified procedures in the test must be met to satisfy the requirements of the test. Failure of an article to meet all the requirements of a prescribed Identity or Identification test (i.e., failure to meet the requirements of all of the specified procedures that are components of that test) indicates that the article is mislabeled and/or adulterated.*"

In the USP–NF specific tests to detect potential adulterants are included in the Identification section of the monograph. This is because, under 21 CFR §211.84 (6)(d)(1) it is mandatory for the pharmaceutical manufacturer to carry out a specific identity test where it is available [12].

5.6.5.3 Assay Pharmacopoeial monographs for active substances are concerned with strength, purity, and efficacy. However, since pharmaceutical excipients are not intended to impart a pharmacologic effect, efficacy does not apply to them (although they may have an effect on human physiology, for example, the effect of high doses of sorbitol and other polyols on the motility of the gastrointestinal tract). However, we do have to consider strength and "purity." However, the concept of excipient "purity" is misleading, since many excipients function because they are not "pure" (see Sections 5.3 and 5.4). The assay relates to excipient composition, but it is not the only measure of excipient composition. For pharmaceutical excipients, there is the concept of the composition profile [3].

Of the four monograph materials selected for use as examples, Dibasic Calcium Phosphate Dihydrate USP has an assay; a back-titration using edetate disodium solution as the titrant and zinc sulfate solution as the back-titrant [7]. Neither Anhydrous Lactose NF [9] nor Microcrystalline Cellulose NF [8] has an assay. This is relatively common with pharmaceutical excipients in that many of them do not have assays and for those that do have assays, the assay is often nonspecific; for example, the assay for Povidone USP is a nonspecific nitrogen determination [10].

5.6.5.4 Impurities For APIs there is an expectation that the bulk active drug will be as pure as possible, and this is clear from the ICH Q3A (R2) document [13]. The goal is to reduce impurities below a maximum level commensurate with patient safety. The US FDA refers to all components of APIs and pharmaceutical excipients, other than the primary component and any additives permitted in the monograph, as impurities. There is thus a perception that "impurities" are bad. This is unfortunate as it sends the wrong message as far as pharmaceutical excipients are concerned.

"Impurities" is the wrong term to use with pharmaceutical excipients. The IPEC Excipient Composition Guide classifies excipient components into several categories and "impurities" is not one of those categories [3]. Instead, there is a category termed "undesirable components," and these are the components of an excipient that should be kept below a maximum level to ensure they remain at acceptable levels in the context of the patient's use of the pharmaceutical product. Other components may need to be controlled, but they are not associated with safety issues. This is an important concept for excipients because many excipients in many applications rely on the presence of other components (concomitant components) for their performance benefits. For pharmaceutical excipients, unlike APIs, the presence of concomitant components may be necessary to achieve the requisite performance in certain applications (formulations) (e.g., see the discussion on dibasic calcium phosphate dihydrate in Section 5.4).

The USP–NF requirements relating to impurities are explained in Sections 5.60 and 5.60.10 of the USP General Notices. Section 5.60 is entitled Impurities and Foreign Substances, and reads as follows [6]:

> "*Tests for the presence of impurities and foreign substances are provided to limit such substances to amounts that are unobjectionable under conditions in which the article is customarily employed.*"

The important point is that the impurities and foreign substances are controlled:

> "*to amounts that are unobjectionable under conditions in which the article is customarily employed.*"

Most excipients have been used in pharmaceutical products for many decades. Their composition has probably not changed beyond the typical variation seen in any manufactured product. Thus, it can be argued that any amounts of any concomitant materials in the excipients will also have been around for decades, and therefore should be:

> "... *unobjectionable under conditions in which the article is customarily employed.*"

However, where improved or novel analytical methods show that hitherto unknown, but objectionable components are present in the excipient, then there is an obligation to control them, notwithstanding that the excipient (and objectionable component) may have been used commercially for a number of years.

Section 5.60.10 of the USP General Notices is entitled Other Impurities in *USP* and *NF* Articles, and reads as follows:

> "*The presence of any unlabeled other impurity in an official substance is a variance from the standard if the content is 0.1% or greater. The sum of all* Other Impurities *combined with the monograph-detected impurities may not exceed 2.0% (see* Ordinary Impurities *<466>), unless otherwise stated in the monograph.*"

[Note: ICH Q3A, as originally drafted, included a similar limit of 0.1% for unidentified impurities. However, this limit was modified in the (R1) version to read 0.10%.]

Certain categories of API are exempted from the provisions of General Notices Section 5.60.10. However, Section 5.60.10 also makes it clear that known toxins must be declared separately:

"*Any substance known to be toxic shall not be listed under* Other Impurities."

USP General Chapter <466> Ordinary Impurities defines ordinary impurities as follows:

"*Ordinary impurities are defined as those species in drug substances and/or drug products that have no significant, undesirable biological activity in the amounts present. These impurities may arise out of the synthesis, preparation, or degradation of compendial articles.*"

It is clear from reading General Chapter <466> that it is intended to be applied to APIs. The General Chapter also states quite clearly that it does not apply to known impurities or to concomitant components. It also defines concomitant components:

"*Concomitant components are defined as species characteristic of many drug substances that are not considered to be impurities in the Pharmacopeial sense.*"

It thus seems logical that unobjectionable (i.e., not undesirable) concomitant components of pharmaceutical excipients

"*... are not considered to be impurities in the Pharmacopeial sense.*"

The term "impurity" as used in the USP–NF really relates to the "undesirable components" of the IPEC Composition Guide [3]. There are three types of impurities that can be listed in a USP–NF monograph for a pharmaceutical excipient:

- Inorganic impurities
- Organic impurities
- Residual solvents.

Inorganic Impurities Inorganic impurities are inorganic salts that may be present as a consequence of the raw materials used in the manufacture of the excipient, or they may be by-products of the processing. They have been traditionally controlled using two tests: Residue on Ignition (also known as Sulfated Ash in the Ph.Eur) and Heavy Metals. The test for residue on ignition gives an estimate of the total nonvolatile inorganic materials present in the excipient. Note: The Residue on Ignition test will not detect volatile inorganic ions such as ammonium.

Heavy metals are generally toxic, and there is a need to keep them below an acceptable limit. The basic principle of heavy metals test is to release the heavy metal from the excipient, typically by some form of digestion of the sample, traditionally using high temperature oxidation of the excipient matrix for organic materials, and then to precipitate the heavy metal ions as sulfides. Neither the sample preparation nor the quantitation is ideal. It is well known that certain more volatile heavy metals (e.g., mercury and arsenic) can be lost during the heating of the sample.

There are thus two issues: the extraction of the inorganic ions from the excipient matrix and the determination of the ions present. In reality, it is often simpler to determine small levels of metal cations than the associated anions. It has long been recognized that the USP–NF Heavy Metals test is not adequate. The decision has been made to replace the heavy metals test with more specific tests for elemental impurities: USP General Chapter <232> Elemental Impurities – Limits, and USP General Chapter <233> Elemental Impurities – Procedures. These two General Chapters were first published in Supplement 2 to USP 35-NF 30. They have since been harmonized as far as is possible with the ICH Q3D – Guideline for Elemental Impurities Step 4 document. [14] General Chapters <232> and <233> are now official and General Chapter <231> Heavy Metals will be deleted from the USP–NF as of January 01, 2018 with concomitant changes to the monographs that reference General Chapter <231>. Once implemented, these General Chapters will be mandatory for all monographs since they have numbers below 1000. The implementation will be governed by an amendment to the General Notices Section 5.60.30.

Organic Impurities As the name indicates, organic impurities are organic chemical compounds and may be present in the pharmaceutical excipient as a consequence of the raw materials and reagents, for example, unreacted starting materials, or as by-products of the manufacturing process, for example, oxidation products. For APIs, ICH Q3A (R2) applies [13], and there are specified reporting, identification, and qualification thresholds. However, ICH Q3A does not apply to pharmaceutical excipients, and the USP General Notices, General Chapters, and General Information Chapters that deal with impurities are directed to drug substances and drug products.

If we consider the four example monographs, the monograph for Dibasic Calcium Phosphate Dihydrate USP [7] does not contain any limit tests for organic impurities. This is not surprising since it is an inorganic material prepared from calcined limestone (lime) and phosphoric acid, and organic impurities would likely not survive the high-temperature calcining step for the conversion of limestone to lime.

The monograph for Microcrystalline Cellulose NF [8] does not contain any tests which are direct tests for specific organic impurities. However, this monograph does contain two limit tests, Water-Soluble Substances (not more than 0.25%) and Ether-Soluble Substances (not more than 0.05%), that will pick up residues from the acid hydrolysis of the wood pulp, and resin residues that have passed through the pulping and hydrolysis processes, respectively. There is also a test for Conductivity which relates to water-soluble ionic components. The tests for Water-Soluble Substances and Conductivity also relate to the efficiency of the washing process after hydrolysis.

The monograph for Anhydrous Lactose NF [9] contains one limit test, Protein and Light-Absorbing Impurities, which is intended to detect levels of, for example, milk protein, and is based on limits for UV absorption at certain wavelengths.

The most recent monograph for Povidone USP [10] contains limit tests for several organic impurities: Aldehydes (not more than 0.05%), Hydrazine (not more than 1 ppm), Vinylpyrrolidone (not more than 0.001%), 2-Pyrrolidone (not more than 3.0%), Peroxides (not more than 400 ppm as H_2O_2), and Formic Acid (not more than 0.5%). These are all starting materials, process intermediates, or reaction by-products. The differences in the limits for the different materials reflect both the differences in their toxicity and the capability of the process to reduce them to acceptable levels (and the ability of the analytical method to adequately quantify them).

Residual Solvents Residual solvents are the residues of volatile organic solvents that may be present in APIs, pharmaceutical excipients, and pharmaceutical products. The testing requirements in the USP–NF are given in General Chapter <467> Residual Solvents [15]. This is a mandatory USP General Chapter and

> "*... applies to all drug substances, excipients and products. All substances and products are subject to relevant control of solvents likely to be present in a substance or product.*"

The General Chapter contains lists of solvents divided into three categories: Class 1 (solvents to be avoided), Class 2 (should be limited), and Class 3 (low toxic potential). Class 1 solvents are toxic, carcinogenic, and/or environmental hazards. Methods are provided for the determination of residual solvent levels, and for calculating the daily intake based on the use of an excipient in a particular pharmaceutical formulation.

General Chapter <467> also states that

> "*It is only necessary to test for residual solvents that are used or produced in the manufacture or purification of drug substances, excipients or drug products.*"

In addition, it is important to note that when the degradation of a drug substance or excipient results in the formation of volatile organic solvents, these breakdown products will be treated as residual solvents and General Chapter <467> applies.

5.6.5.5 Specific Tests Specific Tests comprise those tests applicable to the substance in question that are not included in any of the other Sections of the USP monograph. Typical tests included in this section are pH, Water Determination, Viscosity, and other physical tests.

For example, there is one Specific Test for Dibasic Calcium Phosphate Dihydrate USP; Loss on Ignition [7]. The Specific Tests for Microcrystalline Cellulose NF [8] include Microbiological Enumeration, Tests for Specified Microorganisms, Conductivity, pH, Loss on Drying, Bulk Density, Particle Size Distribution, Water-Soluble Substances, and Ether-Soluble Substances. The Specific Tests for Anhydrous Lactose NF [9] include Clarity and Color of Solution, Microbiological Enumeration, Tests for Specified Microorganisms, Specific Rotation, Acidity or Alkalinity, Loss on Drying, Water Determination and Protein, and Light-Absorbing Impurities. The Specific Tests for Povidone USP [10] include pH, Water Determination, and *K*-Value (related to viscosity and thus molecular weight).

*5.6.5.6 **Additional Requirements*** This section typically does not contain tests but sets out further mandatory requirements such as for Packaging and Storage, Labeling, and a list of the USP Reference Standards required for testing to the complete monograph.

5.7 HARMONIZATION

The supply chains and markets for pharmaceutical products are now global. The supply chains encompass both APIs and excipients, and they can both be sourced globally. It is important to ensure that excipients and APIs of the requisite quality are used in the manufacture of pharmaceutical products. It would therefore be advantageous to avoid a plethora of different tests and specifications that would be wasteful of resources if the pharmaceutical manufacturer were required to test to the excipient specifications in all the different markets in which a particular product was sold. Having one specification that will serve all the main markets will help reduce the testing burden.

In addition, the registration of pharmaceutical products in the different markets round the world requires the submission of considerable amounts of data concerning the safety, efficacy, and manufacture of the API and the product. The latter is often referred to as the Chemistry, Manufacturing, and Controls (CMC) section of the application. There was a gradual realization that it would be beneficial if the marketing applications (dossiers) for pharmaceutical products could be similar in format, style, and content as this would avoid unnecessary duplication of effort.

These were the motivations behind the formation of the PDG and the ICH.

5.7.1 International Conference on Harmonisation

ICH was created in 1990, and is a collaboration between the regulatory authorities and pharmaceutical industry organizations from Europe, Japan, and the United States. ICH was created to find ways to compile the data for marketing applications into a common format and style, to look at best practices in how studies to support marketing applications should be carried out, and to avoid unnecessary repetition of work in submitting applications in the three main markets covered by ICH.

The work of ICH is carried out by Expert Working Groups and the work has been divided into four broad categories, each having a number of Expert Working Groups to address a series of key questions. The four broad categories are designated Q, S, E, and M as follows:

- Quality (CMC issues)
- Safety (safety and toxicology issues)
- Efficacy (clinical issues)
- Multidisciplinary (issues that cut across the boundaries of the categories listed above).

At first glance, ICH is not concerned with excipients. However, since excipients are part of most drug products, several ICH Guidelines will impact excipients, either directly or indirectly. A list of ICH Guidelines and their impact on excipients is presented in Table 5.2.

5.7.2 Excipient Harmonization

As stated earlier, one of the first stages in harmonization of excipient monographs was the founding/creation of the European Pharmacopoeia in 1964. The more recent efforts started with the formation of the PDG in 1989. The current work on the harmonization of excipient monographs parallels the ICH harmonization efforts in some ways, and formerly there were regular meetings between the ICH Q4B Expert Working Group and the PDG. However, the ICH Q4B Expert Working Group has now been disbanded.

5.7.2.1 Pharmacopoeial Discussion Group (PDG) The current work on the harmonization of excipient monographs is coordinated through the PDG, which, as stated earlier, comprises the three major pharmacopeias: JP, Ph.Eur, and USP–NF. PDG convenes twice yearly. There are also organizations with observer status at PDG, including WHO.

The motivation for the creation of PDG was that the three pharmacopeias had monographs for the same excipients, but the testing was different, and there was no mechanism whereby the regulatory authorities could recognize testing to the monograph from another pharmacopeia, and accept it. Pharmaceutical manufacturing companies were forced to undertake extra testing to ensure the excipient complied with all the relevant pharmacopeias, and to maintain the necessary specifications. This was burdensome to industry, particularly with the globalization of the pharmaceutical excipient market, pharmaceutical clinical development, and pharmaceutical product manufacture. With the active collaboration of the regulatory agencies and industrial organizations, an initial list of approximately 53 excipients was selected for the initial harmonization effort based on their frequency of use in pharmaceutical products. However, four from this initial list were subsequently removed from the harmonization process. The initial list was later augmented with a further 10 excipients. In addition, several further monographs were identified as being suitable for harmonization, including Sterile Water for Injection, Lactose for Inhalation, and Isomalt. At the time of writing, the total number of excipients in the PDG system was 65; these are listed in Table 5.3.

In addition to excipient monographs, the PDG also selected a number of General Chapters to be harmonized including six relating to biotechnological and biological drug products. This is logical since harmonized monograph specifications require harmonized test methods and interpretation. The list of General Chapters included in the harmonization effort at the time of writing is presented in Table 5.4.

The selected excipients and General Chapters were divided among the three pharmacopeias, with the designated pharmacopeia being the lead or coordinating pharmacopeia for the particular project. The lead pharmacopeia carries out the initial research

TABLE 5.2 ICH Guidelines Relevant to Pharmaceutical Excipients

	Topic	Comments
Q	Quality guidelines	These guidelines relate to chemistry, manufacturing, and controls (CMC) issues, and are likely to impact pharmaceutical excipient
Q1	Stability	This group of guidelines set out the stability requirements for bulk actives and pharmaceutical products. Pharmaceutical excipients are also expected to have appropriate stability. However, there may be alternative means to establish the stability for excipients [16]
Q2	Analytical validation	This guideline details the types of studies and the data required to validate four types of analytical test: • Identification tests • Quantitative tests for impurities' content • Limit tests for the control of impurities • Quantitative tests of the active moiety in samples of drug substance or drug product or other selected component(s) in the drug product While not directly dealing with pharmaceutical excipients, the guideline does provide useful information on how such excipient analytical methods might be validated, and the expectations relating to supporting data
Q3	Impurities	Q3A relates to impurities in bulk active drugs. As such it is not directly relevant to pharmaceutical excipients Q3B relates to impurities and degradants in drug products, and thus will impact pharmaceutical excipients, since any "impurities" in the excipients will appear in the drug products Q3C relates to residual solvents and does apply to pharmaceutical excipients Q3D relates to metal impurities, and will impact pharmaceutical excipient when it is finally issued (only at Step 1 in the ICH harmonization process at the time of writing)
Q4	Pharmacopoeias	The Q4B expert working group has been disbanded
Q5	Quality of biotechnological products	These guidelines, Q5A–Q5E, will likely be relevant to pharmaceutical excipients using biotechnological methods, including human or animal cell cultures and recombinant technologies
Q6	Specifications	Q6A applies to small molecule drug product and bulk actives. Q6B applies to biotechnological and biological products. While not directly applicable to pharmaceutical excipients, both guidelines will provide useful information on the setting of specifications
Q7	Good manufacturing practice	This guideline refers to GMP for bulk active pharmaceutical materials. As such it is not relevant to pharmaceutical excipients. It does not adequately address continuous processing

(continued)

TABLE 5.2 ***(Continued)***

	Topic	Comments
Q8	Pharmaceutical development	This guideline introduces quality by design concepts in the development of drug products. Since excipients will be a component of most drug products, it follows that this guideline will be relevant to excipients
Q9	Quality risk management	"This Guideline provides principles and examples of tools of quality risk management that can be applied to all aspects of pharmaceutical quality including development, manufacturing, distribution, and the inspection and submission/review processes throughout the life cycle of drug substances and drug (medicinal) products, biological and biotechnological products, including the use of raw materials, solvents, excipients, packaging and labeling materials" [17]
Q10	Pharmaceutical quality systems	This guideline relates to pharmaceutical products and bulk active drugs. It contains many of the same quality system elements as in, for example, ISO 9000
Q11	Development and manufacture of drug substances	This guideline applies to bulk active drug substances of either chemical or biotechnological origin. It introduces some of the QbD concepts for drug products discussed in Q8 adapted to bulk actives
S	Safety guidelines	These guidelines relate to active drugs and finished products. For pharmaceutical excipients there are other guidelines available [18]
E	Efficacy guidelines	Not relevant to pharmaceutical excipients
M	Multidisciplinary guidelines	These guidelines involve more than one general group of guidelines. Those potentially relevant to excipients are listed as follows
M4	Common technical document (CTD)	The CTD is a harmonized template for the assembly of relevant data for submission to the relevant regulatory authority to support a clinical trial application or for registration of a drug product. It is divided into five modules. Module 2 includes the Quality Overall Summary, and the CMC data are in Module 3. Excipient details are part of the information required to be submitted
M7	Genotoxic impurities	This is a new project at ICH and the concept paper was endorsed in June 2010. When finalized, it is anticipated that this guideline will impact pharmaceutical excipients
M8	Electronic common technical document (eCTD)	This is a new project at ICH, having started in 2011, and taking on the work of the CTD Quality Implementation Working Group operating under ICH M2 Electronic Standards. It is anticipated that the final guideline/template will impact excipients since excipients are an important part of the majority of pharmaceutical products

TABLE 5.3 Excipients Included in the PDG Harmonization Process (as of June 06, 2012)

	Excipient	Lead		Excipient	Lead
1	Alcohol	Ph.Eur	34	Saccharin, sodium	USP
2	Dehydrated alcohol	Ph.Eur	35	Saccharin, calcium	USP
3	Benzyl alcohol	Ph.Eur	36	Silicon dioxide	JP
4	Calcium disodium edentate	JP	37	Silicon dioxide, colloidal	JP
5	Calcium phosphate dibasic	JP	38	Sodium chloride	Ph.Eur
6	Calcium phosphate dibasic, anhydrous	JP	39	Sodium starch glycolate	USP
7	Carmellose calcium	USP	40	Starch, corn	USP
8	Carmellose sodium	USP	41	Starch, potato	Ph.Eur
9	Croscarmellose sodium	USP	42	Starch, rice	JP
10	Microcrystalline cellulose	USP	43	Starch, wheat	Ph.Eur
11	Cellulose, powdered	USP	44	Stearic acid	Ph.Eur
12	Cellulose acetate	USP	45	Sucrose	Ph.Eur
13	Cellulose acetate phthalate	USP	46	Talc	Ph.Eur
14	Citric acid, anhydrous	Ph.Eur	47	Titanium dioxide	USP
15	Citric acid, monohydrate	Ph.Eur	48	Ethyl paraben	Ph.Eur
16	Crospovidone	Ph.Eur	49	Propyl paraben	Ph.Eur
17	Ethylcellulose	Ph.Eur	50	Butyl paraben	Ph.Eur
18	Hydroxyethylcellulose	Ph.Eur	51	Glycerin	USP
19	Hydroxypropylcellulose	USP	52	Carmellose	JP
20	Hydroxypropylcellulose, low substituted	USP	53	Calcium carbonate	USP
21	Hypromellose	JP	54	Copovidone	JP
22	Hypromellose phthalate	USP	55	Gelatin, gelling type	Ph.Eur
23	Lactose, anhydrous	USP	56	Gelatin, nongelling grade	Ph.Eur
24	Lactose, monohydrate	USP	57	Glucose monohy-drate/anhydrous	Ph.Eur
25	Magnesium stearate	USP	58	Glyceryl monostearate	USP
26	Methylcellulose	JP	59	Mannitol	Ph.Eur
27	Methyl paraben	Ph.Eur	60	Propylene glycol	Ph.Eur
28	Petrolatum	USP	61	Sodium lauryl sulfate	USP
29	Petrolatum, white	USP	62	Starch, pregelatinized	JP
30	Polyethylene glycol	USP	63	Isomalt	Ph.Eur
31	Polysorbate 80	JP	64	Lactose for inhalation	USP
32	Povidone	JP	65	Sterile water for injections in containers	USP
33	Saccharin	USP			

TABLE 5.4 General Chapters to Be Harmonized in PDG

	General Chapter	Lead		General Chapter	Lead
1	Dissolution	USP	19	Inhalation	Ph.Eur
2	Disintegration	USP	20	Optical microscopy	USP
3/4	Uniformity of content/mass	USP	21	Powder fineness	USP
5a	Tests for specified microorganisms	Ph.Eur	22	Specific surface area	Ph.Eur
5b	Microbial enumeration	Ph.Eur	23	Porosimetry by mercury intrusion	Ph.Eur
5c	Limits for nonsterile products	Ph.Eur	24	Laser diffraction measurement of particle size	Ph.Eur
6	Bacterial endotoxins	JP	25	X-ray powder diffraction	Ph.Eur
7	Color (instrumental method)	Ph.Eur	26	Water–solid interaction	Ph.Eur
8	Extractable volume of parenterals	Ph.Eur	27	Thermal analysis	Ph.Eur
9	Particulate contamination	USP	28	Uniformity of delivered dose of inhalations	Ph.Eur
10	Residue on ignition	JP	29	Microcalorimetry	Ph.Eur
11	Sterility test	Ph.Eur	30	Density of solids	Ph.Eur
12	Analytical sieving	USP	31	Chromatography	PhEur
13	Bulk density and tapped density	Ph.Eur	32	Amino acid determination	USP
14	Conductivity	USP	33	Capillary electrophoresis	Ph.Eur
15	Gas pycnometric density of solids	Ph.Eur	34	Isoelectric focusing	Ph.Eur
16	Powder flow	USP	35	Protein determination	USP
17	Tablet friability	USP	36	Peptide mapping	USP
18	Metal impurities	USP	37	Polyacrylamide del electrophoresis	Ph.Eur

to identify the initial draft monograph for review, and collates the responses from the other pharmacopeias, and any public comments from the Official Inquiry Stage (Stage 4). Very early in its history, the PDG enlisted the help of the International Pharmaceutical Excipients Councils: IPEC-Americas, IPEC Europe, and JPEC (Japan); the three have been collectively referred to as IPEC. IPEC has provided a lot of feedback to PDG and there is a liaison meeting between PDG and IPEC (formerly just the original three PECs, but now the IPEC Federation – which at the time of writing also included IPEC China).

TABLE 5.5 The PDG Harmonization Process [19]

	Activity
Stage 1	Identification
Stage 2	Investigation
Stage 3	Proposal for expert committee review
Stage 4	Official inquiry
Stage 5	Consensus
Stage 5A	Provisional
Stage 5B	Draft sign-off
Stage 6	Regional adoption and implementation
Stage 6A	Adoption and publication
Stage 6B	Implementation
Stage 6C	Indication of harmonization
Stage 7	Interregional acceptance

5.7.2.2 The Harmonization Process The PDG harmonization process starts with the identification of the excipient (or General Chapter) as being a suitable candidate for harmonization. Thereafter, the work proceeds through a series of stages until the harmonized monograph (or General Chapter) is fully implemented (Stage 7). In reality, the work on a particular monograph or General Chapter is essentially complete with the completion of Stage 6 – Regional adoption and implementation; completion of Stage 6 includes final sign-off by the three pharmacopeias. The harmonization process used by the PDG is presented in Table 5.5.

Harmonization by Attribute Initial progress on the harmonization of the excipient monographs was slow, and at the end of the first 10 years no excipient monographs had been harmonized. On investigation, it was found that most of the monographs were about 80–90% harmonized, but there were one or two tests in the proposed monograph where consensus could not be reached. Very often, there were factors outside of the pharmaceutical arena that were preventing consensus. By way of an example, there was a problem with a particular test for one excipient; however, the test reagent that gave the best results was not permitted to be used in Japan for safety reasons, and the alternate method did not give the same results, and was not acceptable to the other two pharmacopeias.

In order to allow progress to be made, the concept of "Harmonization by Attribute" was introduced. This process effectively allows the pharmacopoeias to agree to disagree on certain tests listed in the harmonized monograph, but to proceed with allowing the harmonized tests to be recognized. The introduction of "Harmonization by Attribute" has allowed excipient monograph harmonization to move forward. Progress has been made. However, it should be noted that all the excipients that have achieved Stage 6, or better, thus far have been Harmonized by Attribute; none have been completely harmonized. However, under Harmonization by Attribute, the Definition and assay must be harmonized.

As listed in USP 37-NF 32 General Chapter <1196> [19] (the latest figures available at the time of writing), 35 of 61 excipient monographs had been harmonized by attribute (i.e., had reached Stage 6 in the PDG harmonization process, at least once) although three were undergoing further revision and none had been revised once and five were on their second revision. In addition, 22 of 36 general chapters had been harmonized, including 3 of the 6 general chapters relating to biotechnology drugs. One of the harmonized general chapters was undergoing revision, five had been revised once and one had been revised a second time.

For those monographs that are harmonized by attribute, copies of the tables showing which tests are harmonized are available, for example, on the USP website. At the time of writing, there were also discussions within PDG to develop a common approach to show which attributes are harmonized, and to explain the differences where they exist.

5.8 THE FUTURE

As with all things, the field of pharmaceutical excipients continues to evolve. This applies to the pharmacopoeias and excipient monographs, as well as other aspects of excipients. Changes in the regulatory field for both pharmaceutical products and excipients will also impact excipients and excipient monographs. At the time of writing, there was increased interest in excipients for several reasons, including QbD, potential for economically motivated adulteration, and supply chain integrity. It is generally recognized that issues related to supply chain integrity for pharmaceutical excipients extend beyond the remit of the pharmacopoeias and that the solution to such issues is within the realm of GMP (see USP–NF General Information Chapter <1078> Good Manufacturing Practices for Bulk Pharmaceutical Excipients [20]) and Good Distribution Practice (GDP – see USP–NF Draft General Information Chapter <1197> Good Distribution Practices for Bulk Pharmaceutical Excipients [21]).

5.8.1 Economically Motivated Adulteration (EMA)

EMA has been defined as follows [22]:

> "*Fraudulent, intentional substitution or addition of a substance in a product for the purpose of increasing the apparent value of the product or reducing the cost of its production, i.e., for economic gain.*"

In Haiti, in 1995, deaths were reported due to the mislabeling of industrial grade glycerin as pharmaceutical USP grade. The industrial grade glycerin contained ethylene glycol and diethylene glycol. The toxicity of ethylene glycol first came to the public's attention in the United States in 1937 when Sulfanilamide Elixir formulated with ethylene glycol caused several deaths. Several incidents of economically motivated adulteration have occurred over the years concerning glycerin, propylene glycol, and heparin. Most of these adulteration incidents have resulted in some deaths.

There have also been incidents with melamine in pet food and also melamine in milk (in China). While these last two examples did not concern pharmaceutical products, the underlying motivation for the adulteration was similar; these may be summarized as follows (although not all of them applied in all cases):

- The test methods for identification and assay were nonspecific, and not capable of differentiating between unadulterated and adulterated material.
- There was a significant price differential between the normal trade/industrial grade material and the pharmaceutical grade.
- Materials were available that, if blended with the pharmacopeia material or food material, could boost the nonspecific specification parameter.
- There was poor control of the source of the material and the supply chain.

The key point for the pharmacopeias concerned the test methods. Many traditional excipient monographs do not contain test methods that are specific to the material in question; instead, they will test only a part of the molecule. It was recognized that some of the test methods in many of these older excipient monographs are not capable of properly controlling the excipient and would not detect the type of adulteration seen in the EMA incidents. There has been a concerted effort within the pharmacopeias and regulatory agencies to identify those excipient monographs requiring updating to include more specific test methods. For example, the USP–NF monographs for glycerin and propylene glycol now include a test method and limits for the detection of ethylene glycol and diethylene glycol. The testing is also mandated through an FDA Guidance document. The method is also being introduced for certain other liquid excipients rich in hydroxyl groups. There is also a test method included in the monograph for Heparin Sodium USP, based on ^{1}H NMR, which can detect the presence of oversulfated chondroitin sulfate (used to adulterate the crude heparin sodium obtained for the preparation of Heparin Sodium USP).

In the light of these incidents, a review was undertaken to identify other USP–NF monographs that did not have specific tests for identity and used a nonspecific assay. Several monographs were identified, and work undertaken to add specific tests capable of identifying potential adulterants. In addition, for any new monographs, the specificity of the identification and assay methods will be assessed, and specific methods will likely be required, where technically feasible, before the new monograph can be accepted for inclusion in the USP–NF. Excipient monograph sponsors should take account of this in their proposals for monographs to be included in the USP–NF.

5.8.2 Excipients and Quality by Design

QbD was only recently introduced to the pharmaceutical industry. However, it was first proposed by Duran in about 1986 as part of his concept of Quality Planning. He introduced the QbD concept formally in 1992 [23]. The underlying concept is that "You cannot inspect quality into a product, it is already there." (W.E Deming, 1900–1993). If we want to improve the quality of a product, then we have to build that improved quality into the product, that is, by design. QbD concepts are now included in ICH Q8. There were several reasons for the introduction of QbD; however, a major reason was that the FDA was concerned that the traditional three-batch validation paradigm was not delivering the required product quality. The Agency was concerned at the number of product recalls and batch failures attributable to the lack of a robust product formulation and manufacturing process, among other things.

The basic premise of pharmaceutical QbD is that, by demonstrating enhanced understanding through a scientifically justifiable Design of Experiments (DoE), and thereby establishing a scientifically sound Design Space and Control Strategy during development and scale-up, there may be possibilities for regulatory relief for certain changes postcommercial launch. Initially, whether to adopt QbD or not was up to the individual applicants. However, more recently, it has become clear from the nature of some of the questions back to the applicants that the US FDA is requiring elements of QbD to be included in all new marketing applications, both for New Drug Applications (NDAs) and Abbreviated New Drug Applications (ANDAs).

The introduction of QbD has led to increased scrutiny of pharmaceutical excipients and a wider understanding that compliance with the pharmacopeia monograph may not provide sufficient control of our excipients for every application. In particular, in the absence of appropriate test methods, we may not have sufficient understanding of the variability of the excipient in a given application to be able to develop a robust formulation.

Excipient performance can only properly be assessed in the context of the particular finished product, and since it is uneconomic to manufacture a batch of drug product from every delivery of an excipient, it follows that surrogate methods will be required to allow the excipient user (the pharmaceutical product manufacturer) to be able to predict whether or not a particular delivery of an excipient will be suitable for the manufacture of their product(s). Some of these tests are likely to be extra to those included in the pharmacopeia monograph.

There have been two main responses from the pharmacopeias to the need for performance-related tests. The Ph.Eur introduced a nonmandatory section of the monograph for certain excipients that is termed FRCs. These tests are intended to relate to typical uses of the particular excipient. However, nonmandatory sections of monographs are not possible in either the JP or USP–NF, and the introduction of FRCs by the Ph.Eur has been a problem for the PDG and harmonization.

The USP–NF has adopted a different approach. The USP–NF now has General Information Chapter <1059> Excipient Performance [24]. In this General Information Chapter, the different functions of the excipients are addressed. The intent is to match the excipient functions to the category listing of USP and NF Excipients in the Table Excipients Listed by Functional Category [25]. Thus, the emphasis in the USP–NF approach is to evaluate the potential performance based on intended use, rather than to link such tests only to the material monograph without any reference to the intended use. Since it was first published, the USP–NF General Information Chapter <1059> has undergone a major revision, together with the list of excipients by categories.

More understanding will likely be required by users, and the regulatory authorities, always recognizing that there may be limitations in the analytical methods with respect to the particular excipient. In the context of QbD, it will likely be increasingly important to understand how and why excipients function in particular applications, and how this relates to their composition. Attaining the requisite knowledge and understanding of what is critical for excipient performance will continue to be a challenge.

It is also very likely that the surrogate performance tests that will be necessary to assure that a particular lot of an excipient is suitable for use in a particular application will not be monograph tests, that is, such tests will be in addition to the tests listed in the monograph. The pharmacopeia monograph will still represent the minimum standards that must be met for the excipient to be acceptable for use in the manufacture of pharmaceutical finished dosage forms. However, it would be impossible for the pharmacopoeias to develop a monograph for an excipient that covers every potential use of that excipient, and if they did there would be a considerable amount of unnecessary testing listed with no added benefit for patient safety. The surrogate performance tests should be agreed solely between the excipient supplier and the excipient user. This should not be taken to mean that pharmacopoeias and the monographs are not important; they remain just as important as they ever were. It is simply a comment on the fact that pharmaceutical formulation and manufacturing has moved on beyond the traditional requirements of the pharmacopeias for the safety and "purity" (i.e., composition) of excipients.

5.8.3 Excipient Composition

Going forward, we can expect more interest in the details of excipient composition as has already been discussed above. This is partly due to the need for increased understanding as required by QbD, but also because of the growing realization that the more recent drug candidates are typically prodrugs, and thus more labile than the parent drugs. There are examples in the public domain where the presence or absence of trace components in an excipient has caused the drug product to fail on stability.

In addition, as newer test methods and equipment become available, it seems logical that such methods and equipment should be used to investigate pharmaceutical excipients. We can thus expect additional further understanding to be developed as these newer methods and technologies are introduced and used.

5.8.4 Excipient Variability

Variability is inherent in everything, and excipients are no exception. It is not possible to get rid of excipient variability; we have to come to terms with it and find ways to minimize the effects of that variability. This will be particularly important for QbD formulation development projects [26].

Currently, there is much interest in excipient variability and its significance for pharmaceutical product performance. There is a need to address the issue of the consistency of formulation performance, particularly since the APIs and the means of

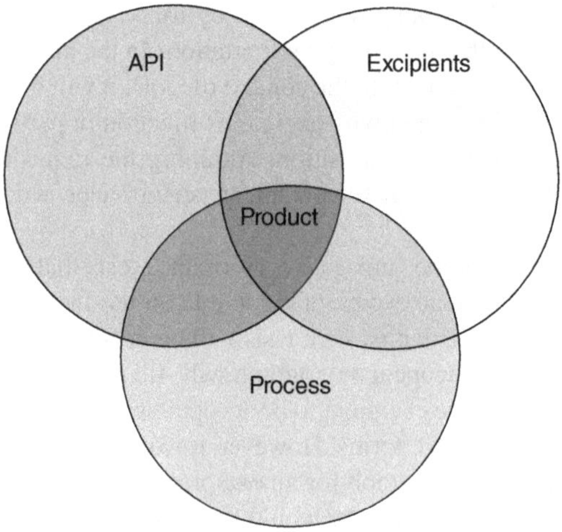

Figure 5.2 The components of a pharmaceutical formulation.

delivery (drug delivery systems, i.e., formulations) have become more sophisticated. There is a concern among the regulatory agencies, including the FDA, that there are drug products on the market that are not reliable and, should the delivery mechanism fail, have the potential to harm the patient.

Variability in formulation performance can arise from several different sources. The components of a formulation comprise the API and excipients, but there is a third component that cannot be ignored – the processing. This is summarized in Figure 5.2. On this basis, it seems logical to suggest that excipients, and excipient variability, along with API variability and process variability should have an influence on formulation performance variability.

However, this is not the complete story. The manufacture of pharmaceutical finished products (i.e., pharmaceutical formulations) is a complex process, and process dynamics come into play together with the interactions between the formulation components. Even this is not the complete story. Most pharmaceutical products are manufactured using batch processing, and there is also the influence of the operator. A more complete description of excipient variability is presented in Figure 5.3, where

$$\sigma^2_{\text{Interactions}} = \sigma^2_{\text{Interaction (1)}} + \sigma^2_{\text{Interaction (2)}} + \cdots + \sigma^2_{\text{Interaction }(n)}.$$

[Note 1: Variability is represented by σ^2 (variance) in the above equations. Often we refer to the standard deviation, that is, σ, in relation to the variability of a set of data. However, variances are additive, whereas standard deviations are not.

Note 2: The variance referred to is the variance that contributes to product variability. This will likely differ from the variance measured outside the particular application or formulation.]

The interactions can be powder–powder, powder–liquid, powder–process, and operator–process, for each component and for each operation. They may be binary,

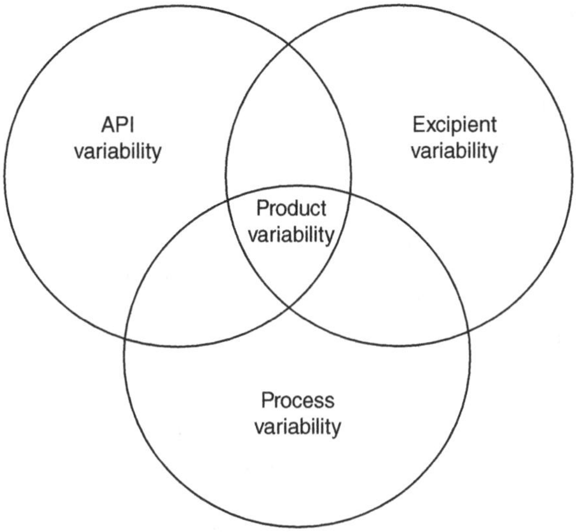

$$\sigma^2_{\text{Product}} = \sigma^2_{\text{API}} + \sigma^2_{\text{Excipients}} + \sigma^2_{\text{Process}} + \sigma^2_{\text{Interactions}}$$

Figure 5.3 Components of pharmaceutical product variability.

tertiary, or possibly even more complex. Not all possible interactions will have a significant influence on the final product, but some likely will.

The incorporation of excipient variability (and other variability) into QbD programs is a recurring issue, and will continue to be an issue. Obtaining excipient lots at the extremes of specification is generally not possible for technical and economic reasons. However, QbD does offer better options such as investigating other grades and preparing, for example, fractionation and dilution [27, 28].

5.8.5 Harmonization

Harmonization of excipient monographs will continue. There are still a number of excipients from the current list of candidates that require work. It is likely that new candidate excipients will also be sought. For example, USP 39-NF 34 contains more than 350 monographs, and the Handbook of Pharmaceutical Excipients [28] also contains monographs for a similar number of pharmaceutical excipients. The current PDG list represents less than 20% of the monographs contained in USP 39-NF 34 and the Handbook of Pharmaceutical Excipients, and there are other excipients not included in either publication.

The issues surrounding the Ph.Eur and FRCs have now been resolved. Where a test is included in the harmonized monograph which is considered a functionality-related characteristic, the Ph.Eur will include it in the main body of the monograph and reference the test in the FRC section of the monograph. However, the harmonization of several monographs was delayed for approximately 2 years until the issues were satisfactorily resolved.

5.9 CONCLUSION

Excipients are an important component of any pharmaceutical formulation. They will remain so for the foreseeable future. QbD will require that we continue to investigate our excipients using all methods available to obtain/derive the requisite understanding that will allow the development of robust formulations to the benefit of the patient.

The pharmacopeias have an important role to play in the future of excipients. Harmonization of excipient monographs and relevant General Chapters will continue. The elaboration of new monographs for excipients used in commercial pharmaceutical products, but not yet included in the pharmacopeias, will also continue, as will updating of test methods and procedures to take advantage of newer developments in analytical methodologies as they become available.

For many years, excipients were regarded simply as "inert carriers." QbD along with other initiatives and concepts has helped dispel that notion. For the future, we will need to be more focused on excipients, particularly excipient variability; coming to terms with it and developing experimental approaches to address it that are not technically impossible and/or economically prohibitive. However, it is also likely that the need for better understanding and performance surrogates will require that the pharmacopoeia monograph is increasingly seen as the absolute minimum; a place to start from in specifying excipients, rather that the complete specification. This will require both excipient suppliers and excipient users to collaborate in ways they have traditionally not done.

No one individual knows it all in any field. That is certainly the case for excipients. The way forward is through partnerships and collaboration, and those partnerships and collaborations should include the pharmacopoeias, and the dialog will necessarily include the regulatory agencies.

REFERENCES

1. The Merriam-Webster Pocket Dictionary (**1974**) , 3rd edn, Pocket Books, New York, NY, p. 668.
2. Moreton, R.C. (**1996**) Tablet excipients to the year 2001: A look into the crystal ball. *Drug Development and Industrial Pharmacy*, **22** (1), 11–23.
3. **2009**) *The IPEC Excipient Composition Guide*, IPEC-Americas/IPEC Europe, Arlington, VA, USA/Brussels, Belgium.
4. Tobyn, M.J., McCarthy, G.P., Staniforth, J.N., and Edge, S. (**1998**) Physicochemical comparison between microcrystalline cellulose and silicified microcrystalline cellulose. *International Journal of Pharmaceutics*, **169**, 183–194.
5. IPEC Glossary, November **2014**. http://ipecamericas.org/content/download-ipec-guidance-documents-here.
6. General Notices and Requirements (**2014**) *United States Pharmacopeia 38–National Formulary 32*, United States Pharmacopeia Convention, Inc., Rockville, MD, pp. 3–12.
7. Dibasic Calcium Phosphate Dihydrate USP (**2014**) *United States Pharmacopeia 37–National Formulary 32*, United States Pharmacopeia Convention, Inc., Rockville, MD, pp. 2088–2089.

8. Microcrystalline Cellulose NF (**2014**) *United States Pharmacopeia 37–National Formulary 32*, United States Pharmacopeia Convention, Inc., Rockville, MD, pp. 7231–7232.
9. Anhydrous Lactose NF (**2014**) *United States Pharmacopeia 37–National Formulary 32*, United States Pharmacopeia Convention, Inc., Rockville, MD, pp. 7364–7366.
10. Povidone USP (**2014**) *United States Pharmacopeia 38–National Formulary 32*, United States Pharmacopeia Convention, Inc., Rockville, MD, pp. 5480–5483.
11. Olive Oil NF (**2014**) *United States Pharmacopeia 38–National Formulary 32*, United States Pharmacopeia Convention, Inc., Rockville, MD, pp. 7426–7428.
12. Code of Federal Regulations, Title 21 – Food and Drugs, Part 211 – Current Good Manufacturing Practice for Finished Pharmaceuticals, §211.84 Testing and approval or rejection of components, drug product containers and closures. US Food and Drug Administration, 43 FR 45077, Sept. 29, 1978, as amended at 63 FR 14356, Mar. 26, 1998.
13. ICH Harmonised Tripartite Document: Impurities in New Drug Substances, Q3A (R2). Current *step 4* version dated 25 October 2006.
14. International Conference on Harmonisation of Technical Requirements for the Registration of Pharmaceuticals for Human Use, ICH Harmonised Guideline, Guideline for Elemental Impurities, Q3D, Current Step 4 version dated 16 December 2014.
15. General Chapter <467> Residual Solvents (**2014**) *United States Pharmacopeia 38–National Formulary 32*, United States Pharmacopeia Convention, Inc., Rockville, MD, pp. 339–354.
16. The IPEC Stability Program Guide **2010**, *IPEC-Americas, Arlington VA and IPEC Europe*, Brussels, Belgium, http://ipecamericas.org.
17. ICH Harmonised Tripartite Guideline Quality Risk Management Q9, Current *Step 4* version dated 9 November 2005, International Conference on Harmonisation of Technical Requirements for Registration of Pharmaceuticals for Human Use, Geneva, Switzerland, http://www.ich.org.
18. Guidance for Industry: Nonclinical Studies for the Safety Evaluation of Pharmaceutical Excipients. U.S. Department of Health and Human Services, Food and Drug Administration, Center for Drug Evaluation and Research (CDER) and Center for Biologics Evaluation and Research (CBER), **May 2005**, http://www.fda.gov/Drugs/GuidanceComplianceRegulatoryInformation/Guidances
19. General Information Chapter <1196> Pharmacopeial Harmonization (**2014**) *United States Pharmacopeia 37–National Formulary 32*, United States Pharmacopeia Convention, Inc., Rockville, MD, pp. 1107–1112.
20. General Information Chapter <1078> Good Manufacturing Practices for Bulk Pharmaceutical Excipients (**2014**) *United States Pharmacopeia 38–National Formulary 32*, United States Pharmacopeia Convention, Inc., Rockville, MD, pp. 1128–1148.
21. Draft General Information Chapter <1197> Good Distribution Practices for Bulk Pharmaceutical Excipients, (**2012**) Pharmacopeial Forum, 37(6): In-Process Revision.
22. Lutter, R., Addressing the Challenges of Economically-Motivated Adulteration in Public Meeting on Economically Motivated Adulteration, U.S. Food and Drug Administration, MD, May 01, 2009. http://www.fda.gov/NewsEvents/MeetingsConferencesWorkshops/ucm163619.htm (*Accessed July 13, 2016*).
23. Juran, J.M. (**1992**) *Juran on Quality by Design: The New Steps for Planning Quality into Goods and Services*, The Free Press, New York, NY.

24. General Information Chapter <1059> (**2014**) *Excipient Performance, United States Pharmacopeia 38–National Formulary 32*, United States Pharmacopeia Convention, Inc., Rockville, MD, pp. 1062–1091.

25. **2014**) *Excipients: USP and NF Excipients Listed by Category, United States Pharmacopeia 38–National Formulary 32*, United States Pharmacopeia Convention, Inc., Rockville, MD, pp. 7127–7136.

26. Moreton, R.C. (**2009**) Functionality and performance of excipients in a quality by design world. Part 2: Excipient variability, QbD and robust formulations. *American Pharmaceutical Review*, **12** (2), 24–27.

27. Moreton, R.C. (**2009**) Functionality and performance of excipients in a quality-by-design world. Part 4: Obtaining information on excipient variability for formulation design space. *American Pharmaceutical Review*, **12** (5), 28–33.

28. *The Handbook of Pharmaceutical Excipients* 7th Edition, Rowe RC, Sheskey PJ, Cook WG, Fenton ME eds, The Pharmaceutical Press, London and the American Pharmacists Association, Washington, DC (**2012**).

6

REGULATORY INFORMATION FOR EXCIPIENTS

CHRISTOPHER C. DEMERLIS, ALEXA SMITH,
AND DAVID R. SCHONEKER

Global Regulatory Affairs, Colorcon, Inc., West Point, PA, USA

6.1 INTRODUCTION

Very few drug products can be manufactured without the use of at least one excipient. Excipients are generally necessary for either the manufacture and/or the performance of the drug product and often make up the majority of the finished dosage form. The excipients frequently comprise a larger percentage of the finished drug product than the active ingredient. Inactive in a therapeutic sense, excipients are often regarded as the "nonfunctional" components of drug products; however, excipients can have significant impact on a dosage form's stability, identity, delivery, and processability.

There are various definitions for excipients. In 21 CFR 210.3(b) (8), Food and Drug Administration (FDA) defines an excipient as an inactive ingredient that is any component other than an active ingredient. The International Pharmaceutical Excipients Council (IPEC), a nonprofit trade association focused on excipient quality and safety, defines excipients as substances other than the API that have been appropriately evaluated for safety and are intentionally included in a drug delivery system [1]. The United States Pharmacopeia 35/National Formulary 30 (USP/NF) lists over 40 different functional categories for excipients [2]. Common functions include binders, disintegrants, fillers (diluents), lubricants, glidants (flow enhancers), colors, preservatives, coatings, flavors, and printing inks.

Pharmaceutical Excipients: Properties, Functionality, and Applications in Research and Industry,
First Edition. Edited by Otilia M.Y. Koo.

6.2 REGULATION OF EXCIPIENTS IN THE UNITED STATES

The regulation of excipients dates back to the Pure Food and Drug Act of 1906, which included legislation for seven synthetic colors. Interestingly, it was an excipient, diethylene glycol, in a sulfanilamide elixir that ended a 5-year legislative debate resulting in the Food, Drug and Cosmetic Act of 1938. The modern FD&C Act, Section 201 (g) (1), includes all components of drug products under the definition of a drug legislatively bringing excipients under the regulatory umbrella of drugs. However, there is no independent regulatory approval process for excipients. Excipients are only reviewed and approved within the context of a drug application. Approval of excipients is specific to the drug product, route of administration, and use level. Approval of an excipient in a New Drug Application (NDA) or an Abbreviated New Drug Application (ANDA) establishes a precedence of use in the United States for that route of administration and level of use. New excipients that are not fully qualified by existing data with respect to the proposed level of exposure, duration of exposure, or route of administration require additional supporting safety data. New excipient evaluation is discussed in Safety section.

The FDA maintains a database, the Inactive Ingredient Database (IID), which lists excipients that have been previously approved in an NDA or ANDA. The IID is intended as an aid to industry to assist in development of drug products. The IID also includes the route of administration, dosage form, and the maximum potency per dosage unit previously approved for each excipient. Generally, once an inactive ingredient has been previously approved for a particular route of administration, the excipient is not considered new and may require a less extensive review the next time it is included in a drug application [3].

Recent discussions between industry and FDA have focused on the possible use of a "family approach" to assess the safety of related excipients such as different viscosity or molecular weight grades of a polymer excipient. This approach is currently being discussed and is expected to provide some flexibility to use safety information that brackets a range of related excipients to support the safety of a particular grade in the family when no specific safety information may be available for that particular grade. This approach would provide significant benefits to both industry and the FDA and would provide for appropriate assurances of patient safety. Other than being included and approved in an NDA/ANDA, there is no other mechanism to have an excipient listed in the IID at this time.

Excipients for use in OTC monograph products must be suitable and safe for their intended use as described in 21 CFR §330.1(e). They must not interfere with the effectiveness or quality of the drug product.

6.3 COLOR ADDITIVES AND FLAVORS

Unlike other excipients, color additives and flavors do have a regulatory scheme independent of drug applications. These substances are evaluated for safety in processes outside of the drug review process.

All color additives in the United States are subject to premarket approval by the FDA. The FDA has a well-defined process for obtaining approval of a color additive and the petition process is described in 21 CFR Part 71. Information related to safety, specifications, manufacture, application, and estimated exposure must be provided to the FDA. Color additives to be incorporated into drug products must be preapproved by the FDA. The FDA provides, by regulation, for the use of a variety of color additives that are either subject to certification (21 CFR Part 74) or exempt from certification (21 CFR Part 73). Only colors that are specifically listed for the intended drug application can be used. Applicable 21 CFR references should be provided in the drug application.

In the case of a new flavoring substance, such substances can be evaluated by the Flavor and Extract Manufacturers Association (FEMA) of the United States Expert Panel to determine if they are Generally Recognized As Safe (GRAS). Flavoring substances are determined to be GRAS by the FEMA Expert Panel pursuant to the authority granted in Section 201(s) of the FD&C Act. References to the FEMA GRAS evaluations can be included in the drug application to support the safe use of a particular flavor.

6.4 INTRODUCTION TO IPEC

The IPEC Federation, established in 2010, is a global organization created to promote quality and safety in pharmaceutical excipients. The IPEC Federation, based in Belgium, includes the regional IPEC organizations:

- IPEC-Americas
- IPEC Europe
- IPEC Japan
- IPEC China
- IPEC India.

IPEC-Americas has also formed three partnerships recently in Brazil, Argentina, and Mexico that will help to expand IPEC's reach into Latin America so that the growing pharmaceutical and excipient industry can participate in IPEC's activities and utilize the IPEC Guidelines to improve quality and protect patients.

IPEC has been developing guidelines, programs, and proposals on various aspects of excipient control over the last 20 years and has developed more than 12 industry guidance documents that pursue harmonization across IPEC Federation organizations. The association has taken leadership positions on important topics such as USP's chapter on residual solvents and FDA's Guidance on melamine. IPEC also sponsors various regulatory and educational conferences and webinars.

6.5 EXCIPIENT INFORMATION FOR DRUG PRODUCT APPLICATIONS

Excipient information must be included in Investigational New Drug (IND) Applications as described in 21 CFR §312.23(a)(7) under the chemistry, manufacturing, and controls information. Excipient information required for NDA is also described in

21 CFR §314.50(d)(1)(ii)(a) and for ANDA in 21 CFR §314.94(a)(9). The CFR also lists specific additional excipient requirements for parenteral, ophthalmic, and topical drug products. Applicable FDA Guidance should also be consulted for excipient information needed for drug applications.

The US regulations for INDs, NDAs, and ANDAs all require that information be provided regarding the excipients used in the manufacture of the drug product regardless of whether they appear in the final dosage form. Drug product compositional information appears in Section P.1 of the ICH Common Technical Document (CTD). Also required in this section is a description of the quality of the excipients used. Compendial references, 21 CFR references, or supplier specifications can be used to describe the quality standards for each excipient. A discussion of the excipient characteristics that may influence drug product performance should also be included in Section P.2.1.2 of the CTD [4]. United States Pharmacopeia General Chapter <1059> provides a useful overview of the key functional categories of excipients, tests that may assess excipient performance, and test procedures that may not be presented in the compendial monographs [2].

Section P.4 of the CTD, Control of Excipients, requires specific information on the control of excipients [4]. See Table 6.1. The specifications and analytical procedures for testing the excipients should be provided. Analytical method validation information, including experimental data, for the analytical procedures used for testing the excipients should be provided when compendial methods are not used. In addition, a justification for the excipient specifications proposed by the Applicant should be provided in this section. Additional information is required for excipients of human or animal origin as well as for new or novel excipients. Excipients used for the first time in a drug product or by a new route of administration, full details of manufacture, characterization, and controls, with cross-references to supporting safety data (nonclinical and/or clinical) should be provided according to the drug substance format [4].

For mixed excipients, information on the individual components of the mixture is required. The type of information expected includes the following:

- List of excipient components (chemical/compendial names including viscosity and MW information)

TABLE 6.1 Section P.4 of the CTD, Control of Excipients

ICH CTD Section	ICH Guideline/FDA Guidance Reference
P.4.1 Specifications	Q6B
P.4.2 Analytical Procedures	Q2A; Q6B
P.4.3 Validation of Analytical Procedures	Q2A; Q2B; Q6B
P.4.4 Justification of Specifications	Q3C; Q6B
P.4.5 Excipients of Human or Animal Origin	Q5A; Q5D; Q6B
P.4.6 Novel Excipients	FDA Guidance for Industry, Nonclinical Studies for the Safety Evaluation of Pharmaceutical Excipients

- Quantitative percentages for each component
- Specification (USP/NF, PhEur, JP/JPE, 21 CFR, etc.)
- Evidence of prior use in the same route of administration at similar levels (especially important for ANDAs, Japanese drug applications).

6.6 DRUG MASTER FILES

Currently, Drug Master File (DMF) systems for excipients exist in the United States, Canada, and Japan to support drug product applications. A DMF, in the case of excipients, is a voluntary mechanism to allow manufacturers of excipients to convey confidential details about their products to the regulatory agency without providing them directly to the Applicant. The International Pharmaceutical Excipients Council of the Americas (IPEC-Americas) Master File Guide is an industry guide that can be used to format uniform excipient information for DMF submissions [5].

In the United States, a Type IV DMF is used for excipients, colorants, flavor, essence, or material used in their production. DMFs can be used to provide information to the FDA to support an IND Application, NDA, ANDA, Biological License Application (BLA), Veterinary Drug Application or another DMF [6, 7].

A Type V DMF is used for FDA-accepted reference information. Generally, the FDA requires that all toxicology information for an excipient be provided in a Type V DMF. However, many existing Type IV DMF's contain toxicology and safety information. Toxicology information for a new excipient should be submitted in a Type V DMF. Before submitting a Type V DMF, the holder must first submit a letter of intent to the FDA [6].

An excipient DMF is not required by US regulation. It is submitted solely at the discretion of the holder. It is not approved or disapproved, and the FDA maintains the DMF as a confidential document. In order for an Applicant to reference information in a DMF in their application, the DMF holder must issue a letter of authorization on behalf of the Applicant allowing the FDA to reference the DMF during their review of the application. The DMF may contain confidential manufacturing and controls information, technical data, and/or safety information to support the safety and quality of excipients.

For formulated excipient mixtures such as film coating systems, color additive mixtures, printing inks, and flavors, proprietary names can be used in a drug application if the quantitative composition of the mixture is provided. The quantitative information is frequently provided in a DMF with an authorization reference letter from the DMF holder.

Generally, the FDA does not review DMFs for compendial excipients and discourages their submission unless there is very specific reason. CMC and safety information for a compendial excipient used for a new route of administration or where total dosing may affect the safety and efficacy of a drug product is required and this type of information, if confidential, may be submitted in a Type IV and/or a Type V DMF.

The regulatory responsibilities of a DMF holder are cited in 21CFR §314.420. A DMF is required to contain a listing of persons authorized to incorporate information in the DMF by reference that must be updated annually by the holder. If the DMF holder adds, changes, or deletes significant information in the file, the holder is required to notify in writing each person authorized to reference that information. The DMF holder is also required to provide an annual report to the DMF on the anniversary of the original submission that includes the updated list of authorized persons and a list of all changes and additional information submitted during the previous year [6]. Currently, there are no fees associated with Type IV or Type V DMFs in the United States.

A Product Master File (PR-MF) can be used in Canada for the submission of confidential excipient information. Although current guidance refers to Product Master Files, a 2008 draft guidance, DMFs, adopts the DMF nomenclature [8]. The Canadian DMF system operates similarly to the US system except submission of a DMF and authorized references to a DMF require payment of fees to Health Canada. In the Canadian DMF system, excipients, including colorants, flavors, and other additives, may be submitted in Type III DMF. Similarly to the US DMF system, submission of a DMF is voluntary but requires the holder to issue a letter authorizing the Applicant to incorporate by reference information in the DMF in their drug application. Canadian DMFs must be updated by way of report every 5 years.

A drug master file system (MF) that can be utilized for excipients was established in Japan as a result of the revision of the Pharmaceutical Affairs Law in April 2005 [9, 10]. The MF system is used by the MF registrant for the protection of intellectual property (confidential information, trade secrets, etc.) so that the review of pharmaceutical products can be carried out more efficiently. The MF is submitted to the Pharmaceutical and Medical Device Agency (PMDA).

This represented a significant change in Japan. Suppliers of excipients in Japan can provide confidential product manufacturing and characterization information directly to the regulators while maintaining the confidentiality of the information.

Information for raw materials, active pharmaceutical ingredients (APIs), excipients, and other additives may be filed separately by their manufacturers in an MF. A guideline has been issued by the Ministry of Health, Labour, and Welfare. The guideline describes circumstances where the Master File can be used such as in the application for registration, application of changes to the registered items, and minor changes for registered items. In addition, the guideline discusses circumstances where the MF registrant should inform the Applicant regarding changes in the items in the MF and when amendments for the MF should be completed [10]. The IPEC-Americas Excipient Master File Guide was used by MHLW as a reference during the development of the MF Guideline for excipients for Japan.

IPEC developed the Excipient Master File Guide to provide a standard format for providing confidential excipient information to DMF systems [5]. The format is coordinated and harmonized with the electronic ICH CTD for presenting chemistry, manufacturing, and controls and safety information (Figure 6.1). IPEC reviewed various existing guidance during the preparation of the Excipient Master File Guide. The initial focus of the guide is to assist in the improvement of the DMF system

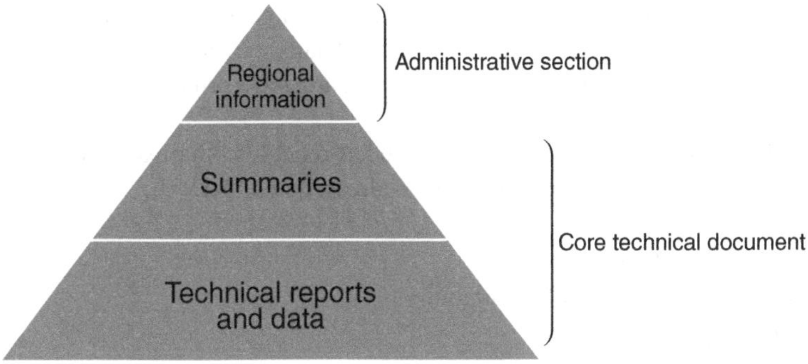

Figure 6.1 IPEC excipient master file guide format.

in the United States; however, the intent is to eventually develop it as a global guide.

According to the IPEC guide, information for the DMF includes description and characterization, method of manufacture, process controls, specifications, and nonclinical safety assessment. The scope of the IPEC guide is for the submission of technical, regulatory, and safety information for the following situations:

- Existing excipients not fully described by monographs (i.e., mixtures of excipients)
- New (novel) excipients
- New route of administration/application for existing excipients
- Biopharmaceutical excipients.

6.7 SUPPORTING REGULATORY INFORMATION NECESSARY FOR EXCIPIENTS

In order to utilize an excipient in a drug product, users need to obtain a significant amount of data regarding various regulatory issues about the excipient manufacturer, distributor, and the excipient itself. IPEC developed the Excipient Information Package (EIP) Guide to provide a uniform format for excipient suppliers to present this information [11].

The EIP is composed of three documents that are designed to work together as a package but can also be useful as stand-alone documents. The EIP documents are as follows:

- Product Regulatory Datasheet (PRD)
- Site Quality Overview
- Site and Supply Chain Security Overview.

These documents are designed to provide valuable information regarding the excipient and its manufacture. The EIP Guide details topics that should be covered in each document and a recommended standardized format for presenting the information.

The PRD is focused on information needed related to the excipient itself. The following key data is recommended to be addressed in a PRD [9]:

- Regulatory status
- Drug Master File (DMF) availability
- BSE/TSE Information
- Allergens/Hypersensitivities Information
- GMO Information
- Residual Solvents Information
- Metal catalyst and metal reagent residues
- Kosher/Halal status.

The Site Quality Overview and the Site and Supply Chain Security Overview are both focused on the facility providing the excipient. These documents are useful in obtaining basic information on the quality systems employed by the supplier as well as information on how the supplier ensures the protection of the product and the continuity of supply. Following are the key topics covered in these two documents:

- Site Quality Overview
  - Compliance evidence such as ISO registration or other external certifications or audit programs
  - Basic details of compliance with the elements of the Joint IPEC-PQG Good Manufacturing Practice Guide for Pharmaceutical Excipients
- Site and Supply Chain Security Overview
  - Supply chain security during storage and distribution
  - Security information including facility, computer, and personnel
  - Safety and Environmental information.

6.8 NEW DEVELOPMENTS IN THE UNITED STATES AFFECTING EXCIPIENTS

The US Congress passed the Food and Drug Administration Safety and Innovation Act (FDASIA) in July 2012. This act is the largest change to drug and medical device legislation in many years. FDASIA contains an entire section on amendments related to drug supply chain and contains several changes specific to the control of excipient safety and quality. The new legislation includes an updated definition of cGMP to include specifically the responsibility for ensuring the quality safety of materials used in the manufacture of drug products. Section 501 (21 I.S.C. 351) is amended by adding at the end the following flush text: "For purposes of paragraph (a)(2)(B), the term 'current good manufacturing practice' includes the implementation of oversight and controls over the manufacture of drugs to ensure quality, including managing the risk of and establishing the safety of raw materials, materials used in the manufacturing of drugs, and finished drug products."

The FDA will now be required to issue guidance or regulations to implement various changes contained in the legislation that address several different areas of drug and medical device regulation. The legislation requires a significant increase in supplier controls throughout the life cycle of the drug product and the FDA is already working on drafting regulations to implement the drug supply chain requirements of FDASIA.

As a result of FDASIA, all domestic and foreign excipient manufacturing facilities will now be required to be identified in drug applications. The FDA is required by the act to create a Unique Facility Identifier (UFI) system to maintain an electronic database containing the registration and listing information of all drug facilities including excipient manufacturing facilities. Drug manufacturers will be required under the new legislation to list in their drug application additional information about suppliers of excipients used in the manufacture of the drug product. This will include the names and addresses of the supplier's manufacturing facilities, the UFI, and a point-of-contact e-mail address.

Also significant to excipients, the FDA must report on their website the number of domestic and foreign excipient suppliers audited beginning in 2014. In the past, excipients were only inspected by the FDA for cause or as a result of a special situation [12]. Under FDASIA, all registered facilities, including excipient manufacturers, will be inspected using a risk-based approach to inspection interval.

6.9 SAFETY EVALUATION OF EXCIPIENTS

6.9.1 Introduction

New excipients are needed for various challenges facing formulation scientists. For example, in the small molecule area, new excipients are needed to overcome problems with poorly soluble drugs and less stable drug substances. The commonly used excipients will not always be appropriate for these types of drug substances. Complex drug delivery systems also require new excipients to improve drug product performance.

There have been increasing numbers of macromolecular drugs that have specific stability requirements and newly designed excipients are needed to develop the stable oral delivery of macromolecules. There is a growing need for excipients that are inert and safe for stabilizing proteins in a liquid product at refrigerated and higher temperatures.

Toxicological studies for excipients should be conducted under Good Laboratory Practice (GLP) regulations and guidelines. The test article should be well characterized. The FDA is seeking comments on whether to amend the GLP regulations [13]. The FDA decided to require a GLP quality system for all facilities/laboratories, as well as to more completely address nonclinical studies as they are presently conducted. The FDA will evaluate the need to modify the existing regulations.

6.9.2 The Evaluation of the Safety of Excipients

According to the FDA and ICH Guidance, an excipient is considered new or novel if it is used for the first time in a human drug product. The current regulatory environment discourages the development of new excipients. The use of new excipients is discouraged since there is no independent regulatory approval process. New excipients are only reviewed in the context of an NDA. Pharmaceutical companies generally use excipients that have been used in previously approved drug products and are reluctant to use new excipients since they do not want to experience delays in the approval of their drug application. Supporting toxicological data must be provided for new excipients. In the long term, a new global regulatory system is needed to independently evaluate excipient safety.

The development of a regulatory process for new excipient review and evaluation has been an IPEC-Americas priority since the organization's inception in the early 1990s. The IPEC-Americas Safety Committee had a series of meetings that resulted in a publication by the committee with recommendations for excipient safety testing based on the route of administration [14]. The FDA new excipient guidance is largely based on the IPEC recommendations and the guidance was finalized in 2005. The FDA Guidance discusses safety testing generally required to establish safety of a new excipient, which is very similar to those required for a new drug [15]. However, unlike drugs, excipients are designed to be pharmacologically inactive. Testing should be evaluated for a new excipient on a case-by-case basis. The United States Pharmacopeia published the IPEC Safety Guidelines as General Chapter <1074> Excipient Biological Safety Evaluation Guidelines.

The FDA Guidance references the International Conference on Harmonization (ICH) safety testing guidelines for the conduct of the testing for new excipients. The FDA Guidance also discusses testing strategies proposed for short-, intermediate-, and long-term use. The toxicological tests are outlined in Table 6.2 with the relevant ICH Guideline reference. Specific FDA CDER Guidance should also be used in the evaluation of new excipients. The objective is to describe the toxicokinetics of the new excipient and to evaluate the preclinical safety regarding the intended use. The goal is to use an integrated risk assessment and identify the potential of the new excipient to produce any adverse effects.

If toxicity and any pharmacologic effect were absent in subchronic studies, a 6-month study may be sufficient. When toxicity is detected in shorter duration studies or in rodents, a chronic study in nonrodents of 9–12 months may be appropriate. The FDA Guidance discusses approaches to be used to evaluate carcinogenic potential of an excipient.

Excipient Safety and Toxicity is an excellent reference for excipient safety evaluation [16]. Additional articles provide supporting information to evaluate the safety of excipients [17, 18].

TABLE 6.2 ICH Testing Guidance

Guideline	ICH Reference
Safety pharmacology	ICH S7A
Short-term use 14 days or less	
Acute	ICH M3 (R2)
ADME	ICH S3A, S3B
Genotoxicity	ICH S2 (R1)
1-month repeated dose study	ICH M3 (R2)
Reproductive toxicology	ICH S5 (R2)
Intermediate-term use 2 weeks to 3 months	
Subchronic 3-month study	ICH M3 (R2)
Long-term use more than 3 months	
6-month rodent study	ICH S4
Chronic study in mammalian nonrodent	ICH S4
2-year bioassay in 2 species or a bioassay in a rodent species plus an alternative study	ICHS1A, S1B, S1C (R2)
Other routes of exposure: pulmonary, injectable, topical	
Photosafety data	ICH S10

A tiered approach to testing can be considered to evaluate the safety of new excipients [19]. *In vitro* assays could be used before conducting the more expensive *in vivo* studies. A potential new excipient could be eliminated if an *in vitro* test yields a positive result. The new excipient can then be subjected to a tiered approach for testing, which could consist of three tiers. Tier 1 testing uses the minimal data set applicable to all compounds while Tier 2 testing generates more extensive data. Tier 3 studies should be performed on a case-by-case basis with consideration of all available data and based on the FDA excipient testing guidance.

Many pharmaceutical excipients are also used as food additives and GRAS substances. Studies will frequently be designed to meet both the FDA Redbook and the ICH Guidelines for food and pharmaceutical applications. The testing strategy for a new excipient should be evaluated on a case-by-case basis and the testing program should be discussed with the appropriate FDA review division for guidance.

The ICH M3 (R2) Guidance on Nonclinical Safety Studies for the Conduct of Human Clinical Trials and Marketing Authorization for Pharmaceuticals provides harmonized guidance for the nonclinical safety studies to support the various stages of clinical development among the regions of European Union (EU), Japan, and the United States [20]. The guidance discusses the type and duration of nonclinical safety studies and their timing to support the Conduct of Human Clinical Trials and Marketing Authorization for Pharmaceuticals. The guideline discusses dose selection for general toxicity studies. Limit doses for acute, subchronic, and chronic toxicity studies of 1000 mg/kg/day for rodents and nonrodents are considered appropriate in all cases except where a dose of 1000 mg/kg/day does not result in a mean exposure margin of 10-fold to the clinical exposure and the clinical dose exceeds 1 g/day, then the doses in the toxicity studies should be limited by a 10-fold exposure margin or a dose of 2000 mg/kg/day or the maximum feasible dose (MFD), whichever is lower. The ICH S3A Toxicokinetics Guideline discusses the general principles that should be considered in the design of a GLP study such as sampling time points and dose levels [21].

6.9.2.1 Subchronic Studies Repeated dose studies are discussed in ICH M3 (R2) Guideline and the study duration is related to the duration of the clinical trial for the drug product. Studies can be conducted in two mammalian species (one nonrodent). The guideline discusses the recommended duration for the repeated dose toxicity studies.

The FDA Excipient Guidance discusses three clinical use periods in products for 14 or fewer days per treatment episode (short-term use), more than 2 weeks (intermediate use) but less than or equal to 3 months or more than 3 months (long-term use). Limit doses that are appropriate are also discussed in the guidance.

Excipients are frequently developed and used in food and dietary supplement products and the FDA Redbook should be considered when designing studies and selecting dose levels. Range finding preliminary studies are conducted for 90-day studies to select dose ranges.

6.9.2.2 Chronic and Carcinogenicity Studies The FDA Excipient Guideline discusses several approaches that may be used to evaluate the carcinogenic potential of a new excipient using the ICH Guidelines if the excipient is intended for long-term use. The ICH S1A Guideline on the Need for Carcinogenicity Studies of Pharmaceuticals discusses factors to consider for carcinogenicity testing. [22–24].

The ICH S1C Guideline on the Dose Selection for Carcinogenicity Studies of Pharmaceuticals discusses the criteria for selection of the high dose for carcinogenicity studies. All relevant information should be considered for dose and species/strain selection for these types of studies.

6.9.2.3 Genotoxicity Testing The ICH S2 (R1) Guideline combined the ICH S2A and the S2B Guidelines.

A battery of tests is a reasonable approach because no single test is capable of detecting all genotoxic mechanisms for tumorigenesis [25, 26].

The guideline discusses two testing options for the standard battery. Option one uses a test for gene mutation (Ames Test), a cytogenetic test for chromosomal damage, or an *in vitro* mouse lymphoma Tk gene mutation assay and an *in vivo* test for genotoxicity. Option two utilizes a test for gene mutation (Ames Test) and an *in vivo* assessment of genotoxicity with two different tissues, usually an assay for micronuclei using rodent hematopoietic cells and a second *in vivo* assay.

The guideline discusses recommendations for *in vitro* and *in vivo* tests such as dose selection, study design, and protocols. Guidance on the evaluation of test results and follow-up test strategies are also discussed.

6.9.2.4 Developmental Toxicity ICH S5 (R2) The ICH S5 (R2) Guidance discusses the types of studies that can be conducted to evaluate potential effects of medicinal products on reproduction and development [27]. The guideline discusses studies such as the following:

1. Fertility study
2. Embryo and fetal development study
3. Pre- and postnatal development study.

An embryo and fetal development study is conducted in two mammalian species, the preferred species being rats and rabbits. The guideline discusses the study design for the detection of effects on reproduction.

To allow detection of immediate and latent effects of exposure, observations should be continued through one complete life cycle from conception on one generation through conception in the following generation. The sequence can be subdivided into the following stages:

A. Premating to conception
B. Conception to implantation
C. Implantation to closure of the hard palate
D. Closure of the hard palate to the end of pregnancy
E. Birth to weaning
F. Weaning to sexual maturity.

6.9.2.5 Biotechnology-Derived Pharmaceuticals ICH S6 (R1) The ICH S6 (R1) Guideline discusses the appropriate timing and framework for conducting preclinical safety studies for protein therapeutics derived using recombinant DNA technology [28]. The guideline was updated including species selection, study design, immunogenicity assessments, developmental and reproductive toxicity testing, and carcinogenicity testing. This guidance can be used to evaluate the toxicity of biotechnology-derived excipients.

6.9.2.6 Immunotoxicology Studies ICH S8 The ICH S8 Immunotoxicology Guidance [29] provides recommendations on nonclinical testing approaches to identify compounds that have the potential to be immunotoxic and provides guidance on a decision-making approach for immunotoxicity testing. The guidance can be consulted for potential immunotoxicity questions regarding excipients.

6.9.2.7 Safety Pharmacology Studies ICH S7A and S7B The ICH S7A and S7B Safety Pharmacology Guidelines can be used to identify potential undesirable pharmacodynamic properties of an excipient that may have relevance to its human safety [30]. Vital systems are considered such as cardiovascular, respiratory, and central nervous systems.

6.9.2.8 OECD Guidelines for the Testing of Chemicals: Section 4 Health Effects The OECD *Guidelines for the Testing of Chemicals* is a collection of approximately 100 internationally agreed testing methods used by government, industry, and independent laboratories to identify and characterize potential hazards of new and existing chemical substances, chemical preparations, and chemical mixtures. They are used primarily in regulatory safety testing, chemical notification, and chemical registration. The OECD Guidelines can also be used in the evaluation of the safety of new and existing excipients [31].

6.9.2.9 Impurities: Safety Considerations Impurities are defined in various ICH Guidance including ICH Q3A (R2) [32], Q3B (R2) [33], and Q3C (R5) [34], and the ICH Guidance can be used to evaluate impurities for excipients. ICH Q7 defines an impurity as any component of the intermediate or API that is not the desired entity. The ICH M7 Guidance titled "Guidance for the Assessment and Control of DNA Reactive (Mutagenic) Impurities for Pharmaceuticals to Limit Potential Carcinogenic Risk" was finalized in June 2014. The guideline states that excipients used in existing marketed products and flavoring agents are excluded from this guideline. Application of this guidance to new excipients is not intended but the risk assessment principles of this guidance for limiting potential carcinogenic risk can be used.

The ICH Q3D Expert Working Group (EWG) has developed a guideline on elemental impurities that will be used to control patient exposure to these materials from drug products. The United States Pharmacopeia (USP) has published two new General Chapters <232> and <233> on Elemental Impurities that contain a subset of the elemental impurities that are listed in the ICH Q3D draft guideline. USP intends to harmonize their list of elemental impurities and the Permitted Daily Exposure (PDE) limits in <232> with those listed in ICH Q3D in the near future.

The PDE limits listed in ICH Q3D only apply to the finished drug products, not the excipients. In fact, there is no requirement for the excipients to meet any specific limit for each element unless there is a specific requirement in the excipient compendial monograph. This is completely appropriate since excipients get used at different levels in drug products and drug products are dosed differently. Therefore, the patient only gets exposed to what ends up in the drug product at the listed dosing regimen. Excipient manufacturers are encouraged to investigate the potential for their products to contain elemental impurities so that they can provide adequate information to their customers as the ICH Q3D requirements are implemented. That said, it is important to understand that there is no regulatory requirement for an excipient manufacturer to meet related to elemental impurities other than to explain to their customers what they may know or not know about their products. The ultimate responsibility for demonstrating that a drug product complies with the PDEs listed in ICH Q3D or USP General Chapter <232> falls on the drug product manufacturer. They can utilize a combination of risk assessment, information they may get from their suppliers and testing that they may perform to make this determination.

The situation for the evaluation of impurities for excipients is more complex since excipients are multicomponent and may be less well defined. Their functionality may be dependent on the presence of components other than the labeled entity. In order to distinguish these components from true impurities the appropriate term when discussing excipients is "minor component" or "concomitant component," for example, the water of crystallization in magnesium stearate required for optimum lubricant efficacy. The IPEC Composition Guide [35] provides an explanation of components (impurities) in excipients, the establishment of an excipient composition profile, and other useful information to qualify components (impurities) in excipients.

There have been suggestions that new and existing excipients should be subjected to the same testing procedures for genotoxic impurities as proposed for APIs in the ICH M7 guideline. A report was published that put such recommendations for existing excipients into proper perspective [36] This report clarified that the levels of genotoxic impurities that may be present in existing excipients that have been commonly used for many years do not need further controls on the levels present. However, additional considerations may be needed for new excipients to assess appropriate levels.

6.9.2.10 Future Trends The Environmental Protection Agency asked the National Research Council to report on the state of toxicity testing and a report was published called Toxicity Testing in the 21st Century: A Vision and a Strategy [37]. The goal is to identify new mechanisms of chemical activity in cells, to prioritize the backlog of untested chemicals for more extensive evaluations and develop better predictive models of human response to toxicants. The focus is to change from a system based on whole-animal testing to one based on *in vitro* methods that use cells, cell lines, or cellular components. The Tox21 program relies on unbiased screening methods that do not assume any prior knowledge about what a chemical might do in the cell. The program also considers animal welfare by adopting animal testing strategies in line with the 3-Rs (replacement, refinement, and reduction). The Tox21 program will also influence the safety testing of excipients in the future.

There is considerable activity in the development of new and innovative excipients. Recent excipient innovations are discussed in several publications [38] and include excipients for orally disintegrating tablets (ODT) and controlled-release formulations. In the future, the application of nanotechnology may be evaluated for developing novel excipients for new therapeutic solutions.

Emerging excipients may become important in developing new APIs for new or more specific therapeutic targets. New and novel excipients include some of the following examples:

- Excipients for ODTs
- Coprocessed mixtures
- Synthetic polymers
- Natural products (or natural polymers) modified with synthetic polymers, small molecules, or other combinations
- Synthetic polymers modified with small molecules or other combinations
- Controlled-release formulations
- Nanotechnology.

Challenges in formulation development represent a need for new excipients. Some of these challenges include the following:

- Demand for ideal filler-binder for direct compression manufacturing process for tablets
- Increased speed of tablet machines and manufacturing efficiency drives demand for excipients with good compressibility and low weight variation at shorter dwell times
- Overcome loss of compaction with wet granulation and high moisture sensitivity
- Need to modulate solubility, permeability, and stability of drug substances.

In the United States, the FDA maintains the IID that lists excipients used in approved drug products, their route of administration, and the maximum dosage (maximum potency per dosage unit) [3]. However, neither the US nor ICH Guidance distinguishes between new chemical entities and modifications of approved excipients, coprocessed mixtures of existing excipients, or approved excipients proposed for a new route of administration as new excipients. Some of these excipients may not require the full battery of tests listed in the FDA Guidance. In these cases, excipient and pharmaceutical manufacturers must evaluate the appropriate safety testing needed. FDA may request additional testing during the review of the drug product application containing the new excipient.

The safety data needed for new or novel excipients can be viewed as a continuum based on the type of new excipient. A simple graph presented in Figure 6.2 demonstrates the level of supporting data needed based on the type of new excipient.

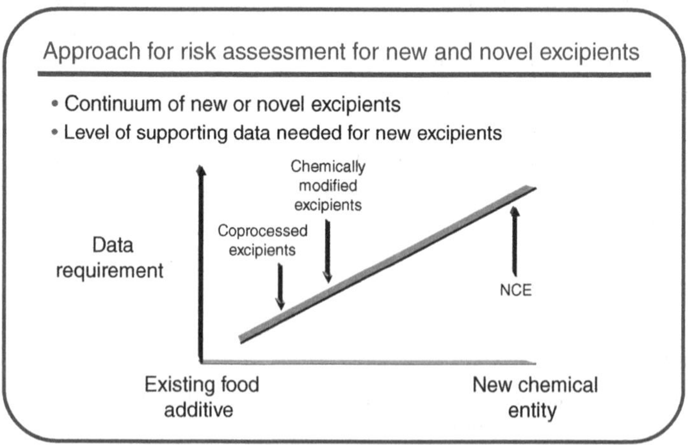

Figure 6.2 Level of supporting data needed based on the type of new excipient.

6.10 THE IPEC NEW EXCIPIENT SAFETY EVALUATION PROCEDURE

In 2007, the IPEC-Americas Safety Committee proposed and developed the IPEC Novel Excipient Safety Evaluation Procedure, which is an independent excipient safety review procedure. This process was developed to reduce the cost and uncertainty related to use of novel excipients in pharmaceutical formulations, thereby encouraging their use in drug development programs and assisting drug formulation innovation [19, 39, 40].

6.10.1 The IPEC-Americas Procedure

In August 2005, IPEC-Americas presented a proposal for an independent excipient safety evaluation procedure to the FDA staff. A supportive letter was received from the FDA on February 7, 2006. The FDA agreed to review the first excipient safety evaluation expert panel submission for consistency with FDA procedures. The FDA lauded IPEC's efforts to expedite the development of new excipients and to form the expert committee process. The FDA stated that it is appropriate to include the opinion of the committee in a DMF. The FDA reviewer must still form an independent opinion of the use of the new excipient based on review of the safety data. The decision must be made in association with a specific drug application. The FDA recommended a pilot program for the evaluation of the first excipient.

These efforts resulted in the evaluation of the BASF novel excipient Polyoxyl (Macrogol) 15 hydroxystearate (Solutol® HS 15) [41]. In September 2007, NEEC reviewed the first safety package for BASF's Solutol HS 15 and the first submission was completed in 2008 for BASF's Solutol HS 15.

In September 2007, the panel began reviewing the safety information for Solutol HS 15 and their conclusions were subsequently submitted to FDA staff. In May 2008, in a letter to IPEC, the FDA concluded that "The issues considered by the expert panel reviewers in the weight-of-evidence determination on the safety of Solutol HS 15 are the same as would be considered by a reviewing division," indicating that the IPEC process provides a reasonable proxy for FDA review.

The IPEC Procedure includes the evaluation of new excipients by the New Excipient Evaluation Committee (NEEC), which is composed of toxicologists selected by the Chair of the NEEC. The NEEC is comprised of three expert general toxicologists who are rotated off for the next excipient evaluation.

The NEEC's primary function is to evaluate compliance of the excipient data with the FDA Guidance and make recommendations to the excipient manufacturer if data gaps exist in the excipient dossier. The expert committee acts independently of the IPEC-Americas Safety Committee and its members must have confidentiality agreements in place. The NEEC is comprised of three experts in general toxicology and ideally members have experience in industrial, academic, and/or regulatory toxicology including experience in toxicology laboratories. If the committee decides that an expert in one area of toxicology is needed to help in the evaluation, a request to the excipient manufacturer will be made for permission to include the expert on the committee.

An excipient safety dossier in CTD format (to facilitate subsequent FDA review) is submitted to the NEEC Chair who in turn distributes it to other committee members. It is recommended that excipient dossiers be prepared according to IPEC's Master File Guide [5]. Once agreement is reached, the final draft report is sent to the excipient sponsor for review and comment.

The committee report will contain the following at a minimum:

1. A discussion of chemical and toxicological data and human safety concerns based upon intended use of the excipient
2. Opinions on conformance with data needs according to the CDER Guidance
3. Identification of data gaps
4. Points of reviewer disagreement if not resolved with the reasons identified in the final draft report.

For More Information on the IPEC New Excipient Safety Evaluation Procedure contact:

IPEC-Americas
3138 N. 10th Street
Suite 500
Arlington, VA 22201
571-814-3449
e-mail: ipecamer@ipecamericas.org

In conclusion, the IPEC Procedure provides an independent evaluation of the safety of a new excipient. The procedure can also be used to obtain support for higher levels of use of an existing excipient and for new grades not currently listed in the IID. The procedure provides the benefit of expanded uses of existing excipients. A positive panel conclusion increases likelihood of use of a new excipient by pharmaceutical companies.

The long-term goal is to convince regulatory authorities of the utility of an independent excipient review.

Various excipients have been evaluated using IPEC's New Excipient Safety Evaluation Procedure in recent years. The following section describes an example of the use of the IPEC Procedure for evaluating new grades and higher levels of use for an existing excipient.

6.10.2 The Novel Excipient Evaluation Committee Review of Surelease®

Surelease® Ethylcellulose Dispersion Type B NF is manufactured by Colorcon. It is a formulated plasticized dispersion containing the rate-controlling polymer ethylcellulose in water. Surelease is used in approved drug products in the United States, Europe, and other countries; it is also used in commercial dietary supplements in the

United States and Europe. Surelease has been used globally for more than 20 years to modify drug release and taste masking. Its ingredients are commonly found in foods, dietary supplements, and drug products.

The differences in Surelease grades are based on plasticizers, stabilizers, and additives used. The original grade, Surelease® E-7-19010, is listed in the IID. But two new grades – Surelease E-7-19040 and Surelease NG E-7-19050 – are not listed yet in the IID. Their absence could raise concerns about using them in drug product applications, especially generic applications. Colorcon used the IPEC Procedure to assure potential customers about the safety of the new grades and to gain support for higher use levels.

Colorcon entered into an agreement to develop an independent safety evaluation of the new Surelease grades. The use of the IPEC procedure for a complex excipient such as Surelease was valuable because Surelease is not a simple mixture. A package containing all supporting information was submitted to the NEEC including the following:

- Several safety and analytical studies that were conducted to bridge to existing Surelease E-7-19010 toxicological data
- A report by a GRAS panel for the dietary supplement use of the new Surelease grades
- A request for an official USP–NF monograph submitted to the USP (coprocessed excipient by the proposed USP–NF monograph acceptance criteria)
- Dossiers prepared by Colorcon in CTD format

Based on the NEEC safety review, a human equivalent dose (HED) was established for Surelease E-7-19040 and Surelease NG E-7-19050. The expert panel issued a report to be used to support and market the new grades and higher use levels of Surelease. The New Excipient Safety Evaluation Procedure helped Colorcon and their customers gain regulatory acceptance of the new Surelease grades and supports higher levels of use of the excipient. The report is now included in Colorcon's DMF for Surelease.

As new excipients emerge, it is important to recognize their potential use in various complex delivery systems. The IPEC procedure offers a path forward because regulatory agencies can use the NEEC report to assist their evaluations of new pharmaceutical excipients.

The IPEC New Excipient Safety Evaluation Procedure provides a useful method for independently evaluating the safety of new excipients including coprocessed mixtures of existing excipients, physical, and chemical modification of existing excipients, higher use levels of existing excipients, and NCEs. The excipient sponsor can use the NEEC's report to support the use of a new excipient in a drug development approval process.

6.11 TOTAL EXCIPIENT CONTROL SYSTEM

Excipients are included in the Federal Food, Drug, and Cosmetic Act as components of drugs and drug products and are considered as drugs as defined in the act [section 201(g)(1)(d)]. In addition, the act includes excipients as drugs (components and drug products) that must be manufactured in conformance with current good manufacturing practice [501(a)(2)(B)] and as a drug whose name appears in an official compendium that must meet the standards set forth in the official compendium [501(b)].

The IPEC-Pharmaceutical Quality Group (PQG) Excipient GMP Guide is an industry guide that is used for the manufacture of excipients. The PQG is an organization based in the United Kingdom who collaborated with IPEC to develop this guide. The guide discusses excipient GMP and supply chain issues. The IPEC-PQG Excipient GMP Guide has been published in the US Pharmacopeia as General Chapter <1078>.

Patient Safety is paramount and recent events have demonstrated how pharmaceutical ingredients can cause harm when they are not designed and manufactured for these intended uses and not sourced through secure supply chains. It is important that excipients are controlled throughout their life cycle. It is necessary to build a system for Total Excipient Control (TEC) that can contribute significantly to improving patient safety [42].

Excipients are used in almost all approved drug products and are essential to the performance of the product. Most excipients used in pharmaceutical products are manufactured to comply with pharmacopoeial standards. Excipients are very different from APIs since they are used in many different drug products and with very different functional characteristics depending on the formulation type.

IPEC Guidelines fill a specific need related to the area of excipient control. A list of IPEC Guidelines and White Papers is included in Table 6.3. These guidelines and white papers have been used globally by many companies and regulatory authorities to establish appropriate standards for excipient control.

IPEC's efforts to assist the Pharmacopeial Discussion Group (PDG) with monograph harmonization have also helped to establish scientifically justified test methods

TABLE 6.3 List of IPEC Guidelines and White Papers

IPEC Excipient Guidelines and White Papers
IPEC-PQG Excipient Good Manufacturing Practices Guide
IPEC Excipient Good Distribution Practices Guide
IPEC Excipient Good Manufacturing Practices Audit Guide
IPEC Excipient Good Distribution Practices Audit Guide
IPEC White Paper on Excipient Pedigree
IPEC Excipient Qualification Guide
IPEC Excipient Information Protocol Guide
IPEC Excipient Quality Agreement Guide
IPEC Excipient Certificate of Analysis Guide
IPEC Excipient Stability Guide
IPEC Excipient Composition Guide
IPEC-Americas Excipient Significant Change Guide
IPEC-Americas Excipient Master File Guide

and specifications for excipients that can be used globally. Many proposals from IPEC have resulted in providing the excipient information and expertise needed to modernize and harmonize existing monographs in the United States Pharmacopeia/National Formulary (USP–NF), European Pharmacopeia (PhEur), and the Japanese Pharmacopeia (JP) to meet today's standards of quality.

6.11.1 Third-Party Auditing and Certification

IPEC established the International Pharmaceutical Excipients Auditing (IPEA) program in 2001 to perform qualified third-party audits of excipient manufacturers whereby the audit reports can be shared with user companies in industry. This was the first attempt at developing a qualified third-party certification program.

IPEC has more recently worked with several other trade associations to develop the EXCiPACT Excipient GMP standard and certification scheme that has been used by a number of excipient companies around the world to provide credible excipient GMP audit information to their customers. The EXCiPACT Excipient GMP standard is essentially an annex to ISO-9001 that outlines all the additional GMP considerations beyond the normal ISO-9001 quality management system requirements that are needed for a quality excipient. IPEC has also worked with the American National Standards Institute (ANSI) and NSF International to develop the ANSI/NSF/IPEC 363 Excipient GMP standard, which is one document that contains all the elements of GMP and quality management systems that exist in the EXCiPACT standard and ISO-9001. Third-party certification programs can utilize the ANSI/NSF/IPEC 363 standard to provide certification of an excipient manufacturing facility when the facility may not be already certified to meet ISO-9001.

The formation of these qualified third-party certification schemes is a major step forward in providing the industry with an alternative for obtaining GMP audit information from their suppliers. Some users find that their suppliers are not willing to permit audits since they purchase a very small amount of the excipient. That same supplier may, however, be willing to become certified by a qualified third-party certifier, which is a means to provide qualified GMP audit information to multiple customers at a reduced cost.

6.11.2 Utilizing IPEC's Excipient Control Resources for Total Excipient Control (TEC)

IPEC has created many guides that users, makers, and distributors can utilize to develop their excipient control programs. The Excipient Qualification Guideline was developed to provide overall guidance on the use of a number of the related IPEC Guidelines to build a credible excipient qualification program from both the maker's and the user's perspectives.

The IPEC Significant Change Guide for Bulk Pharmaceutical Excipients is an industry guide that can be used as a general reference to evaluate the significance of changes for excipients. The IPEC Guide establishes uniform considerations for evaluating the significance of changes involving the manufacture of excipients [43].

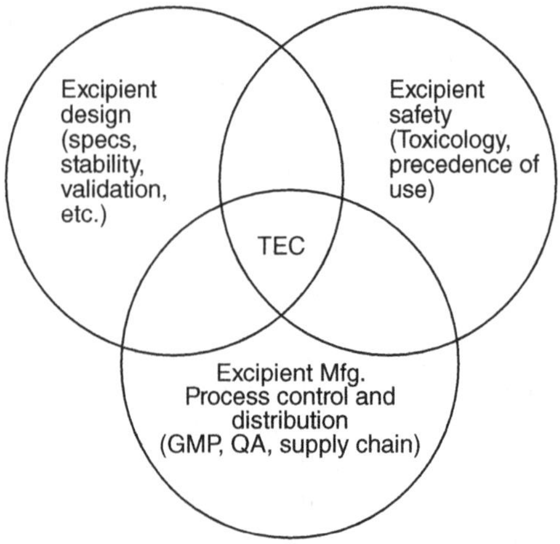

Figure 6.3 Three main areas of control within the total excipient control system.

The IPEC guide has been published in the *United States Pharmacopeia* as general chapter <1195> Significant Change Guide for Bulk Pharmaceutical Excipients [2].

It is important to work toward the development of a system of TEC that will utilize all of the existing IPEC guidelines, programs, and proposals to build an overall control system from the marketing a chemical as an excipient to the pharmaceutical industry to the time the patient takes the drug product containing the excipient.

The TEC system involves three main areas of control (Figure 6.3). Excipient design controls include design criteria set to meet the requirements for the intended use incorporating quality by design (QbD). Excipient safety involves the information that has been developed to support the safe use of the excipient in the intended application at the proposed levels of use by the patient. Excipient manufacturing process control and distribution is the area of control traditionally covered by GMPs, auditing, QC testing, information sharing, and supply chain security.

IPEC will determine where there may be gaps in the current IPEC Guidance as it relates to these concepts and develop additional guidelines or programs to help establish TEC. Some areas that IPEC has been working on are as follows.

6.11.3 Validation versus Process Capability

Validation is one of the most common issues discussed during pharmaceutical company audits of their supplier's excipient manufacturing facilities. Pharmaceutical company auditors request validation data for the excipient manufacturing or cleaning processes. Excipient manufacturers do not always have formal validation studies compiled in the manner typical in pharmaceutical industry. This is many times a terminology problem rather than an actual issue of control.

Excipient manufacturers will typically have an extensive amount of process capability data that is essentially the same type of data that a pharmaceutical company identifies as "validation data." This data is stored in sophisticated large-scale computerized process control systems and is not in the same format usually used within the pharmaceutical industry. This process capability data can provide appropriate excipient control and can provide information typical to pharmaceutical type validation studies.

IPEC is currently drafting an excipient validation guide that will clarify how validation studies should be handled for excipients. This guide will help the excipient manufacturers convert their process capability data into usable information that will provide pharmaceutical users with an understanding for how the excipient processes are controlled.

6.12 EXCIPIENT COMPOSITION: ADDITIVES AND PROCESSING AIDS

IPEC published an Excipient Composition Guide that provides details about the types of components that are typically part of an excipient's composition and how to develop an appropriate excipient composition profile. However, there is still confusion in the industry and regulatory community related to the presence of additives and residual processing aids that have always been part of various excipient's composition.

The use of quality by design principles by pharmaceutical companies would require information concerning the additives and residual processing aids present in an excipient so that the user can develop an adequate understanding of any interactions between the additive or processing aid and the other components of a drug product. There have been many examples of where the presence of a small level of an antioxidant or other additive in an excipient has affected drug product stability or other performance properties.

Companies are concerned about compendia and regulatory complications if additives or residual processing aids are disclosed. Therefore, companies are currently struggling to have appropriate discussions concerning the presence of additives and residual processing aids.

To address this situation, IPEC-Americas is compiling a list of additives and processing aids that are commonly used in excipients. The actual excipients that contain these additives and processing aids are not being identified due to confidentiality concerns. IPEC-Americas will submit this list of well-known additives and processing aids (typically GRAS or excipient materials on their own right) to the US FDA and request a meeting to discuss how IPEC-Americas can work with the FDA to find a mechanism to provide specific additive and processing aid information to the FDA. This information could also be placed into the USP–NF monographs to identify these materials when they are present if there are no confidentiality concerns. In some cases, however, due to confidentiality, disclosure of this information to the FDA may need to occur through the use of a DMF.

IPEC will pursue similar discussions with other regulatory agencies globally to obtain similar flexibility if possible. It is critical that these discussions occur because additives and residual processing aids exist in numerous excipients and this information is generally not well known by drug product manufacturers and regulators.

6.12.1 Visible Particles in Excipients

Visible particles are another topic related to excipient composition. The presence of these particles in excipients is typically not a contamination issue. These particles may be off-color charred particles from heat in the manufacturing process, small amounts of metal particles that typically occur from normal manufacturing process equipment wear or other types of particles that are visibly different than the main excipient particles. Typically, these particles have no safety implications and are technically unavoidable.

IPEC has established a guideline on Technically Unavoidable Particles (TUPs) that discusses how to determine when the presence of these particles is acceptable depending on the type, number, and size of the particles. This guideline also discusses appropriate testing methodologies for assessing these visible particles and what may be necessary to characterize the particle's identity.

6.12.2 Elemental Impurities

Another topic related to the excipient composition profile is elemental impurities as the ICH Q3D Guideline and the USP General Chapters on Elemental Impurities begin to be implemented. In the past, the heavy metals limit test was performed that did not really provide useful information about the actual levels in excipients due to limitations in the test methodology.

Many excipients will contain some level of elemental impurities, but these levels may not really impact the potential for the drug product to meet the PDE limits due to low use levels of the excipient in the drug formulation. Some excipients (especially mined excipients and excipients derived from natural sources) contain amounts of elemental impurities, however, which may cause some drug products to exceed the proposed limits. Therefore, since some drug products may need to be reformulated it is critical drug product manufacturers investigate whether their drug products have a potential to exceed the PDEs as early as possible during the implementation phase for the ICH Q3D Guideline.

IPEC has been working with the excipient industry to encourage them to conduct at least a limited set of analytical studies now that the ICH Q3D limits are known so that this information can be shared with the pharmaceutical users in the future along with any information they may have about expected excursions in the typical levels. The pharmaceutical users will then need to utilize the information they may get from their suppliers along with any of their own testing or other related literature information they may be able to obtain to perform appropriate calculations and risk assessments related to the levels of elemental impurities that may exist in the finished dosage form.

6.12.3 Other Areas of Interest for Total Excipient Control

A number of other areas have been identified by IPEC to develop additional guidance that can be useful for establishing a system of TEC. The following topics are currently being evaluated by various IPEC committees to establish guidelines:

- QbD – Excipient variability in chemical and physical properties and the effect of variability on drug product manufacture and performance. The guidelines will address QbD sampling processes and concepts for developing robust formulations that are resistant to excipient variability.
- Coprocessed excipients – Supporting analytical data to build safety bridging arguments to component safety data. Coprocessed excipients will be defined and guidance will be given to encourage the use of these materials that are typically designed for purpose in the pharmaceutical industry.
- Atypical actives – Excipients being used as APIs that are not manufactured using ICH Q7 GMPs. Atypical actives are needed for many OTC and generic drug applications but may not be available in the future unless appropriate regulatory flexibility can be defined by regulators to minimize liability issues for suppliers who will never be able to implement ICH Q7 GMPs. The guideline will discuss risk management techniques that can be used to properly assess the quality of these materials. Once developed, this guideline will be discussed with the FDA to determine a reasonable action plan for the use of a flexible approach for assessing GMPs used for these materials.

IPEC is dedicated to continue to develop appropriate guides and white papers that can be used to fill these gaps and ultimately finds ways of combining all the IPEC tools into a workable system of TECs that begin with excipient design and end with patient consumption.

REFERENCES

1. IPEC-Americas Glossary. *International Pharmaceutical Excipients Council of the Americas* [Online] (Cited 6 November 2012). http://ipecamericas.org/glossary.
2. United States Pharmacopeial Convention (**2011**) *United States Pharmacopeia 35/National Formulary 30*, United States Pharmacopeial Convention, Inc., Rockville, MD.
3. Inactive Ingredient Search for Approved Drug Products: Frequently Asked Questions. *Food and Drug Administration* [Online] (Cited 6 November 2012). http://www.fda.gov/Drugs/InformationOnDrugs/ucm080123.htm.
4. International Conference on Harmonisation of Technical Requirements for Registration of Pharmaceuticals for Human Use. The Common Technical Document for the Registration of Pharmaceuticals for Human Use: Quality – M4Q (R1), **2002**.
5. The International Pharmaceutical Excipient Council. *IPEC-Americas Excipient Master File Guide*, **2004**.

6. *Guideline for Drug Master Files.* Food and Drug Administration, Center for Drug Evaluation and Research, **1989**.
7. Food and Drug Administration, Center for Biologics Evaluation and Research. Submitting Type V Drug Master Files to the Center for Biologics Evaluation and Research, **2001**.
8. Health Canada, Health Products and Food Branch. Draft Guidance Document, Drug Master Files (DMFs), **2008**.
9. Kosaku, H. and Wheelright, S.M. (**2005**) *Changes to the Japanese Pharmaceutical Affairs Law. BioProcess International*, **3** (1), 14–16.
10. Ministry of Health, Labour and Welfare, Pharmaceutical and Food Safety Bureau, Evaluation and Licensing Division. Guideline for the Utilization of Master Files for Bulk Drugs, etc. **2005**. Food and Drug Inspection No. 0210004.
11. The International Pharmaceutical Excipient Council. IPEC Excipient Information Guide: Template and User Guide, **2009**.
12. Food and Drug Administration, Office of Regional Operations and the Center for Drug Evaluation and Research. Guide to Inspections of Bulk Pharmaceutical Chemicals, **1991**, Reformatted in 1994.
13. Federal Register (**2010**). Proposed Rule, **75** (244), 80011–80013.
14. *A New Approach to the Safety Assessment of Pharmaceutical Excipients.* Borzelleca, J F, et al. **24**, **1996**, *Regulatory Toxicology Pharmacology* 149-154.
15. Food and Drug Administration. Guidance for Industry: Nonclinical Studies for the Safety and Evaluation of Pharmaceutical Excipients, **2005**.
16. Kotkoskie, L.A. and Weiner, M.L. (**2000**) *Excipient Toxicity and Safety*, Marcel Dekker, Inc., New York.
17. *Trends in Excipient Safety Evaluation.* DeMerlis, C C, et al. 6, **2011**, *International Journal of Toxicology*, Vol. **30**, pp. 600–610.
18. Osterberg, R. and See, N. (**2003**) *Toxicity of Excipients – A Food and Drug Administration Perspective. International Journal of Toxicology*, **22** (5), 377–380.
19. *The IPEC New Excipient Safety Evaluation Procedure.* Brock, W, et al. **2009**, *Pharmaceutical Technology*, Vol. **33**, pp. 72–82.
20. International Conference on Harmonisation of Technical Requirements for Registration of Pharmaceuticals for Human Use. Guidance on Nonclinical Safety Studies for the Conduct of Human Clinical Trials and Marketing Authorizations for Pharmaceuticals – M3 (R2), **2009**.
21. International Conference on Harmonisation of Technical Requirements for Registration of Pharmaceuticals for Human Use. Note for Guidance on Toxicokinetics: The Assessment of Systemic Exposure in Toxicity Studies – S3A, **1994**.
22. International Conference on Harmonisation of Technical Requirements for Registration of Pharmaceuticals for Human Use. Need for Carcinogenicity Studies Pharmaceuticals – S1A, **1995**.
23. International Conference on Harmonisation of Technical Requirements for Registration of Pharmaceuticals for Human Use. Testing for Carcinogenicity Studies of Pharmaceuticals – S1B, **1997**.
24. International Conference on Harmonisation of Technical Requirements for Registration of Pharmaceuticals for Human Use. Dose Selection for Carcinogenicity Studies of Pharmaceuticals – S1C (R2), **2008**.

25. International Conference on Harmonisation of Technical Requirements for Registration of Pharmaceuticals for Human Use. Guidance on Genotoxicity Testing and Data Interpretation for Pharmaceuticals Intended for Human Use – S2 (R1), **2011**.
26. *ICH Guidelines: Inception, Revision, and Implications for Drug Development.* Brown, P, et al. **2010**, *Toxicological Sciences*, Vol. **118**, pp. 356–367.
27. International Conference on Harmonisation of Technical Requirements for Registration of Pharmaceuticals for Human Use. Detection of Toxicity to Reproduction for Medicinal Products and Toxicity to Male Fertility – S5 (R2), **1993**.
28. International Conference on Harmonisation of Technical Requirements for Registration of Pharmaceuticals for Human Use. Preclinical Safety Evaluation of Biotechnology-Derived Pharmaceuticals – S6 (R1), **1997**.
29. International Conference on Harmonisation of Technical Requirements for Registration of Pharmaceuticals for Human Use. Immunotoxicity Studies For Human Pharmaceuticals – S8, **2005**.
30. International Conference on Harmonisation of Technical Requirements for Registration of Pharmaceuticals for Human Use. Safety Pharmacology Studies for Human Pharmaceuticals – S7A, **2000**.
31. OECD Guidelines for the Testing of Chemicals. *OECDiLibrary* [Online] (Cited 7 November 2012). http://www.oecd-ilibrary.org/environment/oecd-guidelines-for-the-testing-of-chemicals-section-4-health-effects_20745788.
32. International Conference on Harmonisation of Technical Requirements for Registration of Pharmaceuticals for Human Use. Impurities in New Drug Substances – Q3A (R2), **2006**.
33. International Conference on Harmonisation of Technical Requirements for Registration of Pharmaceuticals for Human Use. Impurities in New Drug Products – Q3B (R2), **2006**.
34. International Conference on Harmonisation of Technical Requirements for Registration of Pharmaceuticals for Human Use. Impurities: Guideline for Residual Solvents – Q3C (R5), **2011**.
35. The International Pharmaceutical Excipients Council. The IPEC Excipient Composition Guide, **2009**.
36. *A Perspective on Testing of Existing Pharmaceutical Excipients for Genotoxic Impurities.* Brusick, D J. **2009**, *Regulatory Toxicology and Pharmacology*, Vol. **55**, pp. 200–204, Elsevier Inc.
37. Committee on Toxicity Testing and Assessment of Environmental Agents. Toxicity Testing in the 21st Century: A Vision and a Strategy. s.l.: The National Academy of Sciences, **2007**.
38. *Applications of complementary polymers in HPMC hydrophilic extended release matrices.* Rajabi-Siahboomi, A R and Tiwari, S B. 7, **2009**, Drug Delivery Technology, *Vol.* **9**, pp. 20–27.
39. *Eye on excipients.* DeMerlis, C C, et al. 3, **2010**, *Tablets and Capsules*, Vol. **9**, pp. 33–41.
40. *Marketing new excipients: Clearing regulatory hurdles.* DeMerlis, C C and Goldring, J. 11, **2003**, *Pharmaceutical Technology*, Vol. **27**, pp. 102–108.
41. *Solutol HS15 as a new excipient.* Ku, S and Velagaleti, R. 11, **2010**, Pharmaceutical Technology, *Vol.* **34**, pp. 108–110.
42. *Total excipient control: A pathway to increased patient safety.* Schoneker, D R. 4, **2011**, *Pharmaceutical Technology*, Vol. **35**, pp. S34–S38.
43. The International Pharmaceutical Excipient Council of the Americas. IPEC-Americas Significant Change Guide for Bulk Pharmaceutical Excipients, **2005**.

7

DEVELOPMENT OF NEW EXCIPIENTS

KARL KOLTER AND FELICITAS GUTH

BASF SE, Global Research & Formulation Nutrition & Health, Ludwigshafen, Germany

7.1 INTRODUCTION

Most currently used excipients are quite old and have been in use for decades. The majority of them were not developed specifically for pharmaceutical applications but initially for cosmetics, food, and even technical applications; only later were they used in pharmaceutical applications. Because of their long history of use and experience gained with them, they are considered to be safe. These materials are usually characterized by a rather simple structure or a straightforward manufacturing method and are very often of natural origin or partially modified by chemical synthesis (e.g., sugars, sugar alcohols, salts, minerals, starches, modified starches, microcrystalline cellulose, and cellulose derivatives).

Over the last two decades, drug delivery systems (DDS) have gained importance and high growth rates are expected in the future. Very often, these DDS require particular, highly functional excipients in order to achieve the targeted product properties [1]. At the same time, the number of new actives launched per year has significantly decreased [2, 3], and this has consequently led to increased efforts to develop special dosage forms in order to extend the life cycle of a drug. Furthermore, highly functional excipients are used in common dosage forms where they markedly improve drug performance and quality as well as lower manufacturing costs. The availability of new excipients provides more opportunities for formulators to achieve *in vitro* and *in vivo* correlation in the performance of the drug. Also, the use of new and special manufacturing technologies (e.g., melt extrusion) might well require novel materials in order to achieve the best results [4].

Pharmaceutical Excipients: Properties, Functionality, and Applications in Research and Industry, First Edition. Edited by Otilia M.Y. Koo.

7.1.1 Types of New Excipients

In principle, there are three types of excipients and the development processes for these vary significantly:

1. Modified excipients (physical or purity changes)
2. Coprocessed excipients (excipient formulations)
3. Novel excipients (new chemical entities).

7.1.1.1 Modified Excipients Generally, the characteristics of an excipient are defined by its chemical nature, physical appearance, and properties. The chemical structure and composition determine the toxicological characteristics and safety. Thus, they cannot be changed easily. However, the physical properties can be varied and optimized for specific applications. Typical physical properties are, for instance, particle size, morphology, and structure. These are varied quite often. Thus, numerous fillers such as lactose, mannitol, and microcrystalline cellulose are available in different grades, each having benefits in certain applications and dosage forms [5, 6]. The requirements for excipients based on monographs in pharmacopoeias and directives from the authorities have become more stringent, resulting in improved quality. For certain drugs and dosage forms, special grades of well-established excipients are now being demanded by customers. In this regard, purity might well be significant since undesired reactive components can cause instability of the drug. A change of the particle structure of an excipient may allow easier handling and improve its application-related properties. Examples of grades with special purity properties are Tween™ 80 HP from Croda (low aldehydes) and Kollidon® 30 LP from BASF (no peroxides). Examples of excipients with special physical features are Kollidon® VA 64 Fine from BASF (especially shaped particles of high compressibility). All these materials can be classified as modified excipients as their chemical nature remains unchanged. Modified excipients usually meet compendial specifications but often with stricter limits.

7.1.1.2 Coprocessed Excipients Sometimes, one excipient is simply not enough and combinations are necessary in order to achieve particular features of a drug formulation. Thus, several well-established materials can be combined in such a way that they create new and/or superior physical properties; the result is that they interact synergistically. However, in this case, the performance of the excipient formulation cannot be achieved by simple physical mixing. These so-called "coprocessed" excipients consist of two or more compendial excipients that are formulated without significant chemical change [7–9]. Thus, in most cases, the safety profile of a coprocessed excipient will match that of a corresponding physical mixture. Typical manufacturing processes are mixer granulation, fluid bed granulation, spray formulation, and microencapsulation. Coprocessed excipients typically facilitate the manufacturing processes of drug products. Their use can reduce manufacturing costs due to the higher process efficiency achieved and reduced testing and documentation requirements by reducing the number of excipients. Sometimes, they even allow the

TABLE 7.1 Main Categories and Examples of New Excipients

Category	Examples
Modified excipients	Tween™ 80 HP (Croda) Polyplasdone® Ultra (ISP) Kollidon VA 64 Fine (BASF) Swelstar™ Mx1 (Asahi Kasei) GalenIQ™ 721 (Palatinit)
Coprocessed excipients	Spectrablend™ HS (Sensient) Prosolv® ODT (JRS) Ludiflash® (BASF) Aquarius® (Ashland) StarCap 1500® (Colorcon)
Novel excipients	Kollicoat® IR (BASF) Soluplus® (BASF) Kollicoat® Smartseal 30 D (BASF) Captisol® (CyDex)

number of manufacturing steps in the production of a dosage form to be reduced. They are commonly used for direct compression and coating applications, thereby easing and speeding up drug developments and subsequent production [10].

7.1.1.3 Novel Excipient ICH Guideline M4Q defines an excipient as being novel if it is used for the first time in a drug product or in a new administration route. Therefore, all excipients with a completely or partially new chemical structure that was not known or used before are novel and require thorough characterization with the focus on physicochemistry (including impurities and stability) and safety. These excipients must be fully toxicologically characterized since they represent material that had not been applied to humans before so that safety must be proven [11]. Biotechnologically or biologically derived substances such as albumin (Recombumin®) or transferrin (CellPrime® Transferrin AF) are also considered to be novel excipients, even though the substances are well known.

The development and use of novel excipients are triggered by formulation challenges that cannot be overcome with established materials or formulation technologies. The most prominent example is the bioavailability improvement of poorly soluble or poorly permeable drugs.

Table 7.1 lists main categories of new excipients and gives examples for commercially available products.

7.2 DEVELOPMENT OF NOVEL EXCIPIENTS

The most challenging task facing these three categories is the development of novel excipients that can be considered to be new chemical entities (NCE) as these could take at least 6–7 years and incur very high costs. Here, we have to distinguish between two kinds of developments: one where a new product is a derivative or a successor of an already known excipient and the other where the product is completely new with no predecessor. Examples of the first kind are the developments of hydroxypropyl beta-cyclodextrin and sulfobutyl beta-cyclodextrin (CAPTISOL®), which are derivatives of the already known and approved beta-cyclodextrin [12]. This material, however, has some drawbacks, particularly in certain administration routes, such as low aqueous solubility and toxic effects when given parenterally. By introducing or changing functional groups, such drawbacks can be eliminated and new properties created. The substituted beta-cyclodextrin derivatives exhibit a much higher solubility in water, thus preventing nephrotoxic effects caused by precipitation in the kidneys when the material is administered parenterally. Furthermore, the new derivatives are characterized by a higher solubilization capacity for a variety of drugs compared to the unchanged beta-cyclodextrin [13]. The second kind, the development of a novel excipient where no basic structure exists, has to start with a screening phase to find the most suitable one. Table 7.2 indicates types of new excipients and the respective development times.

In general, the main phases of novel excipient development are screening, product and process optimization, scale-up, toxicological studies, and documentation. The extent of toxicological studies is more or less the same as for a new active ingredient. Similarities exist also with regard to the manner in which the development is performed, due to the fact that many units are involved, for example, polymer laboratory, production, process engineering, regulatory affairs, marketing, and toxicology, all of which should collaborate on a project team basis.

Besides the application-related performance of a novel excipient, other aspects such as environmental issues, safety of manufacturing, c-GMP production, product safety, and costs are very crucial.

The dilemma of all developers of novel excipients is the fact that customers require monographed excipients or at least examples of drugs containing the novel excipients in the relevant markets in order to minimize the risk of new drug developments. But,

TABLE 7.2 Types of Excipient Developments

New Chemical Entity (NCE)	Modified Excipient	Coprocessed Excipient
Not yet known	Modification of the physical form	Composed out of approved pharma ingredients
↓	↓	↓
Long development (6–7 years)	Medium development time (2–4 years)	Medium development time (3–5 years)
Examples		
Soluplus (Polyvinylcaprolactam–polyvinyl acetate–polyethylene glycol graft copolymer)	Kollidon VA 64 Fine (Fine copovidone)	Ludiflash (Mannitol-Kollicoat® SR 30 D-crospovidone)

how can excipient developers obtain such approvals when pharmaceutical companies are very reluctant to use novel excipients? This is similar to the chicken and egg conundrum.

It also means that the launch of a novel excipient does not signify the end of the project. Getting monographs of new excipients into the pharmacopoeias and obtaining drug approvals in the relevant markets are also required. Approximately 3–4 years can be assumed for the time required by pharmaceutical companies to develop and register new drugs with a novel excipient. These long development times strongly decrease the profitability of novel excipients, making such developments less attractive.

These facts – long development times, high risks, high costs, long launch phases, and long payback periods – are the background of A.T. Florence's famous rhetorical question:

Where are the new excipients, where are the new solubilizers, sustained release excipients … ?

By posing this question some time ago, he was criticizing the lack of novel excipients in pharma; and the situation has not significantly changed today. However, some have appeared on the market recently and these will be illustrated in more detail.

A brief summary is given in Figure 7.1, which illustrates the most important stages of development: product development, toxicological testing, scale-up of production, documents for regulatory filings, and application data. Only then can the new polymer be launched. However, even after the launch of the material, numerous activities have to be carried out to obtain drug approvals and get monographs in the various pharmacopoeias.

Since the development of a novel excipient is a rather complex process, it should be run by a project team coordinating all the various activities. Many of these activities interfere with others or have consequences for others; they need to be discussed and decisions made. Typical parties involved in the project and therefore part of the team are mentioned in Figure 7.2.

Generally, there is considerable similarity between the development of a new active and the development of a new excipient.

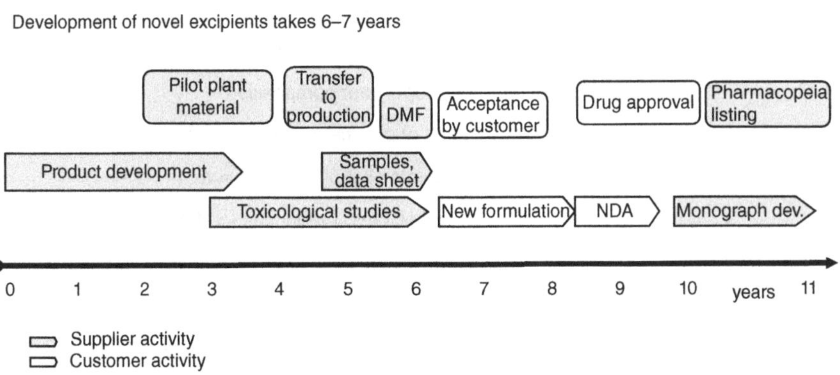

Figure 7.1 Development chart of new excipients.

Figure 7.2 Development organization and structure.

7.2.1 Product Development

After one has evaluated the idea and clearly defined the project goal including the target profile, project development generally starts with a phase called product development, the first part of which involves the screening of various polymer classes and polymerization techniques in order to find the most suitable polymer class. Subsequent milestones are optimization of the polymer composition and the polymerization process, scale-up, and transfer to production. These steps and milestones in product development are illustrated in Figure 7.3.

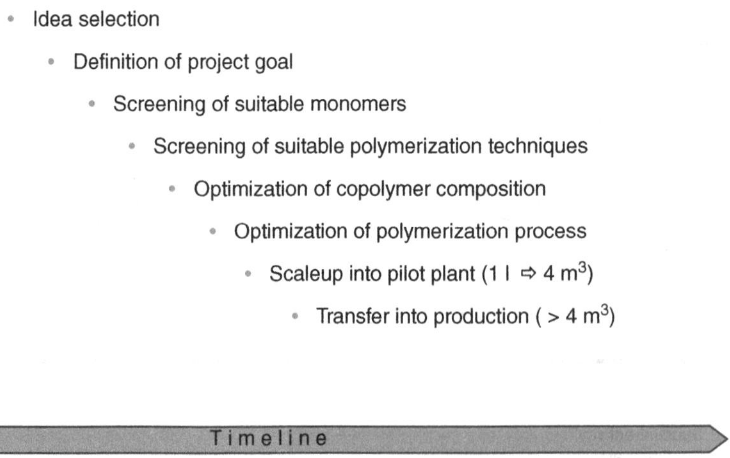

Figure 7.3 Steps and milestones in product development.

Of course, the target profile strongly influences which monomers and monomer combinations should be employed in the screening process. For example, if the resulting polymer should be water soluble, monomers that are water soluble at a certain concentration are required; otherwise, the polymer will not dissolve in water at all. Typical polymer classes are, for instance, polymethacrylates, polyacrylates, polyvinyllactames, polyvinylalcohols, polyesters, polyurethanes, polyureas, polyamides, and polystyrenes. Not only the monomers themselves, but also the polymerization technique exerts a strong influence on polymer properties. If various techniques are applicable for certain monomers, the differences in properties of the resulting products should be determined.

In this phase, numerous vastly different polymer structures need to be tested in order to find the most suitable one (Figure 7.4).

Having explored this field, optimization within the polymer class takes place. Here, usually, the ratio of comonomers, molecular weight, and, if applicable, the degree of grafting or lengths of various building blocks are systematically varied and evaluated.

During this stage of development, again numerous polymers have to be synthesized and tested for their physicochemical and application-related properties. It is of the utmost importance to determine whether the final candidate meets the original target profile. In addition, other characteristics of the polymer must be investigated in order to avoid significant drawbacks in this regard.

Only a very intense collaboration between chemists and pharmacists, or in other words between synthesis and application testing allows for streamlined and rapid optimization. This process can be schematically illustrated as the pathway to the top of a hill where many changes in direction must be made to reach the summit.

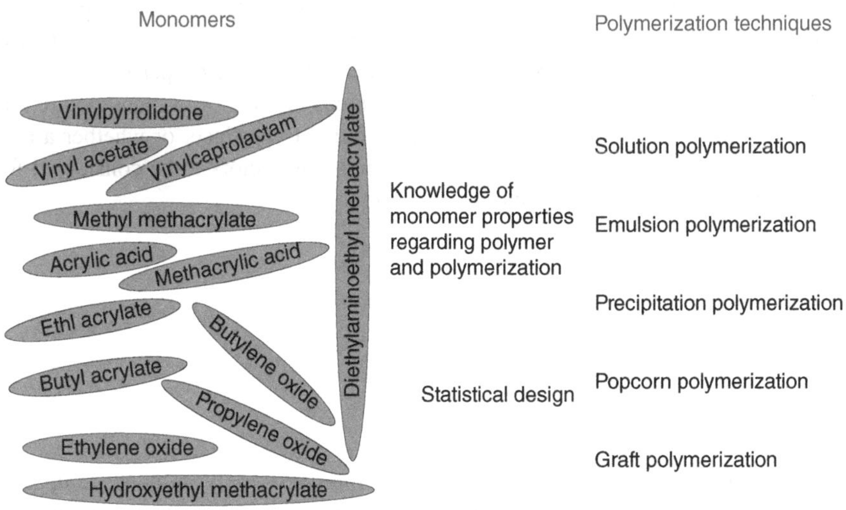

Figure 7.4 Monomers and polymerization techniques.

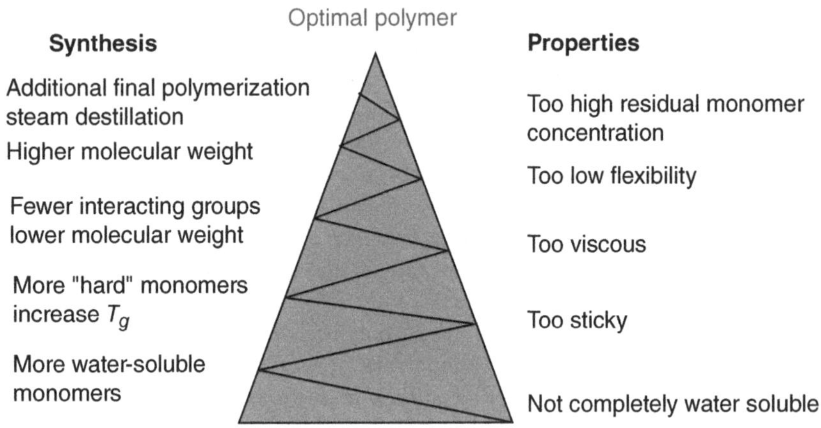

Figure 7.5 Optimization helix.

Numerous countermeasures must be taken to strengthen the required features and reduce the unwanted ones (Figure 7.5).

7.2.2 Scale-Up and Transfer to Production

Having found the ideal polymer candidate, scale-up is then initiated. This requires an increase in batch size from typically 1 l to 4 m^3. Of course, one cannot go from 1 l directly to the cubic meter scale; there are several steps in between. At each scale, the polymer characteristics must be checked to determine whether they have changed significantly or not. If there are changes, countermeasures must be taken to achieve the originally targeted profile. Here, comprehensive knowledge of polymer processes and polymer properties are of great help. It is of some advantage to know what process settings must be changed in order to achieve a certain polymer property.

The next step is then to establish whether an existing plant can manufacture this material, whether a production line can be modified to produce, or whether a new polymer production line has to be installed. Important factors to be considered for the decision on the production line are as follows:

- GMP conditions
- Idle capacity
- Suitability of plant/line
- Costs for adaptation of plant/line
- Manufacturing costs.

It is important to select suitable production lines very early in the scale-up process in order to have the opportunity to tailor polymer synthesis to the production prerequisites.

As already mentioned, polymer properties should not change during this transfer process. Very experienced chemists and engineers are a prerequisite for a smooth process.

During the final stage of this development process, validation of the excipient manufacturing must be performed. This validation shows that the material can be produced in a reproducible and reliable manner.

Packaging is the final step in excipient manufacture and it plays an important role. During development, various tests must be performed to select the most suitable primary packing material, the so-called inliner. The decision has to be made, for example, whether a common foil such as a PE can be used or whether a denser variety is required or even whether packaging in aluminum is the best choice. Hygroscopic materials require a denser packaging material in order to prevent the uptake of water. Only tests at various temperatures and humidities can reveal what is needed for a certain shelf life of the material. Even though stress testing is carried out, these run for some time; they should thus be initiated during scaling up of polymer synthesis. The decision on the secondary packaging material and type, for example, carton or drum, is mainly made by marketing and production. It depends on what capabilities production has and what is most appropriate for the intended use of the excipient.

7.2.3 Analytical Characterization

Product development and scale-up are always accompanied by an in-depth investigation and evaluation of starting materials, impurities, and by-products. Polymer synthesis should be carried out in such a way that by-product and residual monomers are kept at a very low level. Sometimes, this cannot be guaranteed by polymer synthesis itself and other purification steps must be added in order to get rid of undesired or reactive impurities.

The maximum level of impurities is usually set based on the requirements of the pharmacopoeia or – in case these are not mentioned there – from toxicological evaluation.

Not only the impurities, additives, or degradants that can be formed during storage must be specified but also the parameters or properties of the polymer itself: identity tests and other tests that characterize the polymer (e.g., quantification of functional groups or molecular weight) are needed. The overall principle is to establish what is required to exactly characterize a new pharmaceutical excipient and to make sure that it can fulfill its designated task in a drug formulation [14, 15]. Should similar excipients be described already in the pharmacopoeia, those parameters can be used; otherwise, new parameters must be established.

The full range of specification parameters is based on the pharmacopoeia, technical capabilities, and the relevance for the targeted application, as illustrated in Figure 7.6.

If test methods are available in various pharmacopoeias, these should be used first. However, if new parameters are needed, this usually requires the development of new methods. For example, if a new monomer is used or a new impurity occurs, an HPLC or GC method needs to be developed, of course, including its validation. It should

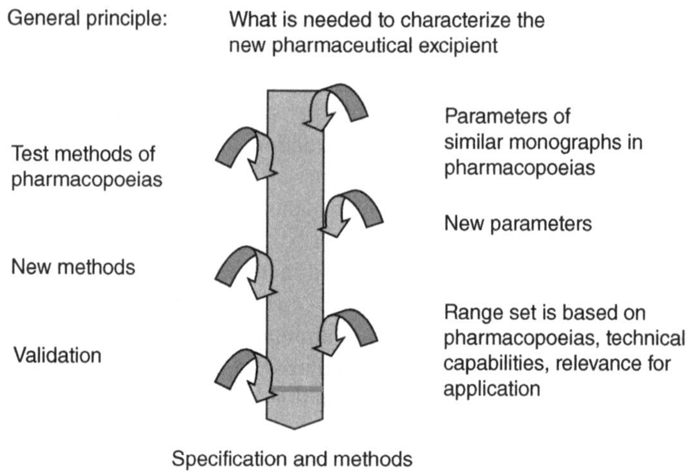

Figure 7.6 Analytical characterization of a new excipient.

always be the target to develop analytical methods that are fit for the intended purpose and easy to implement by excipient users [16].

7.2.4 Stability Studies

The stability of excipients varies and it is important to provide evidence that the excipient will continue to meet the specification throughout the recommended retest period. Packaging and storage conditions can have a major impact on stability and their suitability should be demonstrated.

Stability studies as outlined in ICH Q1A (R2) [17] are usually not required, but a systematic evaluation based on literature data and actual test results for the respective packaging is a minimum recommendation for all excipients [18].

A more thorough investigation is necessary for all classes of new excipients. It has to be demonstrated that their physicochemical characteristics remain stable. Bracketing and matrixing studies are particularly suitable for modified or simple types of coprocessed excipients.

A more comprehensive study design is needed for complex coprocessed excipients in order to demonstrate the absence of chemical change due to interaction of the components during storage. It may be necessary to develop new analytical methods that are fit for this purpose.

Last but not least, the stability evaluation of novel and highly functional excipients is a major task during the development phase. Degradation pathways have to be identified, and the impact of temperature, oxygen, and moisture and the influence of the selected packaging have to be determined. Furthermore, analytical test methods for stability-indicating parameters have to be in place before the stability study can be initiated. A study design as is mandatory for new drug substances including long-term and accelerated storage conditions and with the inclusion of follow-up batches can be appropriate for novel excipients. In all cases, it is possible to define a preliminary retest period based on the initial test results and to extend it step by step as more supporting data become available.

7.2.5 Toxicological Studies

After polymer structure, composition, and its relevant properties have been decided, toxicological testing can start. This is usually performed according to the FDA guidance notes on "Nonclinical studies for the safety evaluations of pharmaceutical excipients," Rockville, MD, 2005 [19]. Further information can be found in the USP general information chapter <1074> Excipient Biological Safety Evaluation Guidelines. Figure 7.7 illustrates the toxicological studies for novel excipients.

Principally, the toxicity study should be performed with a product sample of representative quality with impurity levels as they are expected in the commercial grade. The toxicological test program is typically started with acute toxicity and irritation studies followed by genotoxicity studies. The larger chronic and subchronic studies also require dose-finding studies of a shorter period prior to this in order to find out the most appropriate doses.

Radioactively labeled polymer is used to perform the ADME studies, which is the abbreviation for absorption, distribution, metabolism, and elimination. For this,

Toxicological Studies for Novel Excipients according to FDA Guidance for Industry (May 2005)

- Acute studies
  Acute toxicity (rat)
  Acute dermal toxicity (rat)
  Acute dermal irritation
  Acute eye irritation
  Skin sensitization
- Genotoxicity
  Ames test
  Chromosome aberration test
  Mouse lymphoma test
  Mouse micronuclei test
- Absorption, distribution, metabolism, elimination
  ADME test with radioactively labeled polymer
- Reproductive toxicology
  Fertility and early embryonic development (rat)
  Teratogenicity (rat and rabbit)
  Prenatal and postnatal toxicity (rat)
- Subchronical and chronical toxicology
  (incl. clinical pathology, histopathology, and toxicokinetic analysis)
  1 month (rat; dose finding)
  1 month (dog; dose finding)
  3 months (rat; optional)
  6 months (rat)
  9 months (dog)
- Additional studies for other routes of application
  (pulmonary, injectible, topical)

- Additional
  (not mandatory)
- Ecotoxicological tests
  Aquatic toxicity
  Algae growth inhibition test, acute daphnia immobilisation test, acute fish toxicity, short-term respiration test
  Environmental fate
  Biodegradability test

Figure 7.7 Toxicological studies required for pharmaceutical excipients for oral applications.

polymer synthesis must be downscaled to a very small scale and it must be ensured that the radioactively labeled polymer characteristics are the same as those of the non-labeled one. The radioactively labeled polymer is rather unstable and should therefore be used within a short period of time. Polymers of larger molecular weight such as polyvinylalcohol–polyethylene glycol graft copolymer (Kollicoat IR) at 45,000 Da are usually not absorbed and are excreted completely with the feces. This is one of the reasons why polymers usually do not show toxicological reactions in the body when applied orally.

Ecotoxicological studies such as aquatic toxicity and environmental fate are not mandatory for human use, but they indicate whether there is a risk for an environmental hazard coming from the polymer and its production. A company should know what happens when a material is released to the environment, for instance, by an accident in manufacturing. These studies are part of the care package of responsibility/sustainability for a new excipient.

7.2.6 Drug Master File (DMF) and Certificate of Suitability (CEP)

The established regulatory procedures for excipients are based on prior pharma use, the reference to food additive/GRAS status, compliance with a pharmacopoeia monograph or, in Europe, the use of the European Pharmacopoeia (Ph. Eur.) Certificate of Suitability (CEP). This practice is suitable for modified grades of well-established excipients. However, complex coprocessed excipients and novel excipients, in particular, cannot use this procedure [20].

Some countries such as the United States, Canada, and Japan allow quality and safety information about an excipient to be submitted by the manufacturer in the form of a drug master file (DMF). In general, a DMF is a submission of information about a component of a drug product (e.g., drug substance, excipient, or packaging materials) to the health authorities. The DMF allows the authority to review this information in support of a third party's submission without disclosing it to the third party. In the United States, Type IV and V DMFs are applicable to excipients: Information on chemistry, manufacturing, and controls (CMC) is submitted to the US FDA in form of a Type IV DMF. Nonclinical overview material and safety evaluation data can be either included as a separate volume in the Type IV DMF or filed independently as a Type V DMF. It is important to state that there is no legal or regulatory requirement to file a DMF. Quality and safety information can be either included in the application (filed by the excipient user) or in a DMF (filed by the excipient manufacturer). Furthermore, CMC information for a compendial excipient (where the quality is covered by a USP-NF monograph) is usually not reviewed by the FDA. Consequently, standard grades of monographed excipients do not require a DMF [21].

For complex coprocessed and novel excipients, the DMF system offers the important advantage that sensitive information about the manufacture and safety evaluation can be disclosed to health authorities but kept confidential from other parties.

Structure and content of an excipient master file should follow the Common Technical Document (CTD) as outlined in ICH M 4 [22]. For novel excipients, full CMC details with references to supporting safety data should be provided according to the drug substance format (Module 2: Quality Overall Summary 2.3 S) & Module 3 Quality, Body of Data, 3.2.S). In the case of a novel synthetic polymer, the Type IV DMF should include the following information as shown in Table 7.3.

It is good practice to disclose scientific information to the applicant in the form of a "regulatory information file." This document should contain sufficient information to enable the applicant to evaluate the suitability of the excipient specification for his/her control procedures. It is emphasized that this information is considered confidential and may only be shared under a confidentiality agreement. The content of such a document could, for example, be similar to the so-called "applicant's part" or "open part" of an Active Substance Master File (ASMF) as it is used for active ingredients in Europe [23].

Each type of excipient may require a slightly different approach when establishing the design of the DMF to facilitate the review process (e.g., biological excipients present special challenges in their characterization).

The framework described in "The IPEC-Americas® Excipient Master File Guide" follows the structure and the requirements of ICH M 4, highlights aspects that are specific to excipients, and explains how quality and safety information should be presented [24].

A DMF is neither approved nor disapproved. A DMF is only reviewed within the context of a third party's application, where the content of the DMF is being referenced. Nevertheless, the DMF holder has the obligation to keep the DMF up to date and to notify sponsors or applicants who have referenced the DMF about any pertinent changes – if appropriate even before implementation. Current information on US DMFs is available on the FDA *www.fda.gov* website. In 2011, Type IV DMFs (excipients) accounted for 11% of all active DMFs.

In other countries where the master file system does not exist or is not open to excipients (Europe), details about CMC and safety of a novel excipient can only be submitted by the pharmaceutical manufacturer as part of the regulatory dossier for a new drug product. This puts an even greater burden on both parties (excipient innovator and pharmaceutical manufacturer) and intensifies the regulatory dilemma (see Section 2.8).

7.2.7 Launch

After having successfully completed the entire development of the new excipient, one can prepare the launch. Here, one should have a clear launch strategy and a comprehensive launch concept, since just offering the new excipient to users is not sufficient. The launch concept describes in detail what documents and actions are necessary and in what time period they should be compiled and carried out, respectively. Basically, decisions on pricing, advertisements, and how to approach customers must be made. Is the internal sales force and technical service group in place and capable of handling the launch of the new excipient or should it be outsourced to a distributor? Excipient suppliers with their own sales and technical service teams usually have a

TABLE 7.3 Content of a Type IV Drug Master File and of Proposed Content for a "Regulatory Information File"

	Overview of Type IV DMF Contents	Regulatory Information File
3.2.S.1	General information	X
3.2.S.1.1	Nomenclature Chemical name, compendial name, CAS-number	X
3.2.S.1.2	Structure Structural formula, stereo chemistry, chemical formula, molecular weight	X
3.2.S.1.3	General properties Physicochemical properties and performance or functionality-related characteristics	X
3.2.S.2	Manufacture	X
3.2.S.2.1	Manufacturer(s) Name and address	X
3.2.S.2.2	Description of manufacturing process and controls • Flowchart (including molecular formulae, weights, yield ranges, starting materials, intermediates, reagents, operating conditions, and solvents) • Sequential procedural narrative of the manufacturing process (quantities of raw materials, solvents, catalysts reflecting representative batch scale for commercial manufacture) • Identification of critical steps, process controls, equipment, and operating conditions (temperature, pressure, pH, time) • If applicable, alternative processes and identification and justification of reprocessing steps	Only flowchart and short description
3.2.S.2.3	Control of materials Materials used in manufacture, for example, raw materials, starting materials, solvents, reagents, and catalysts	–
3.2.S.2.4	Control of critical steps and intermediates	Only as far as this information is relevant to the applicant
3.2.S.2.5	Process validation and/or evaluation	–
3.2.S.2.6	Manufacturing process development Significant changes made to the process and/or site of the excipient used in nonclinical trials, pilot, scale-up, or production batches	–

TABLE 7.3 (*Continued*)

	Overview of Type IV DMF Contents	Regulatory Information File
3.2.S.3	Characterization	X
3.2.S.3.1	Elucidation of structure and other characteristics and appropriate analytical tests to substantiate proof of structure	X
3.2.S.3.2	Impurities	X
3.2.S.4	Control of the excipient	X
3.2.S.4.1	Specification	X
3.2.S.4.2	Analytical procedures	X
3.2.S.4.3	Validation of analytical procedures	X
3.2.S.4.4	Batch analysis	X
3.2.S.4.5	Justification of specification	X
3.2.S.5	Reference standards or materials	X
3.2.S.6	Container closure system	X
	Identification, specification, including critical dimensions, with drawings of primary and secondary packaging	
3.2.S.7	Stability	X
3.2.S.7.1	Stability summary and conclusion	X
3.2.S.7.2	Postapproval stability protocol and stability commitment	X
3.2.S.7.3	Stability data • Long-term, forced degradation and stress conditions • Analytical procedures and validation of these procedures	X

clear advantage, since these employees have a much deeper knowledge of the new excipient and its properties, capabilities, and limitations.

From an application point of view, it is crucial to have technical information available displaying all relevant data in a comprehensive manner, for example, chemical and physicochemical characteristics, regulatory information, application-relevant data, and in particular examples of the intended applications with model drugs. A formulator in a pharmaceutical company should be able to use the excipient in a proper way based on the information contained in the technical information leaflet.

The new excipient must become known and popular, all of which take time. In addition, a pharmaceutical company needs several years to develop a drug formulation with it. During this development period, only small quantities are supplied for laboratory scale and pilot plant scale purposes. If the drug is approved by the authorities – a procedure that also takes some time – increasing sales can be expected. However, this usually happens more than 12 years after having started the development process.

7.2.8 The Regulatory Dilemma

Pharmaceutical excipients are not approved as they are but only in drug formulations [25]. Thus, the approval is given to the drug formulation and it can then be concluded that the excipient contained in the given dose and the targeted route of administration is also "approved." The FDA lists all the excipients, their routes of administration, and maximum doses in their "Inactive Ingredient Database (IIG)" *www.fda.gov* [26].

Generally, pharmaceutical formulators want to have approved excipients because this does not pose any additional risk to their drug developments. The most preferred option of course is to have an excipient with a monograph in the pharmacopoeias.

However, when launching a novel excipient, neither a monograph nor approved drugs containing this excipient exist. Pharmaceutical formulators are therefore somewhat reluctant to use the novel excipient. This then delays excipient sales and makes the overall excipient development project for the supplier less profitable and attractive. From the patient's perspective, it often appears that not the best DDS are being developed but the ones with the lowest regulatory risk. In order to make significant progress in this field, the hurdles described must be overcome [27]. In this regard, it is absolutely crucial that the supplier enjoys a high degree of reliability and reputation and does the whole development and documentation according to state of the art and the regulations of the authorities. The excipient documentation handed over to the customer should be complete and should fulfill all current requirements. Of course, it is of tremendous benefit and provides a high degree of trust if the supplier has a good track record in that they can show that they have already developed several novel excipients that have been accepted by the authorities in new drug applications. Figure 7.8 illustrates the relationship between excipient supplier, pharmaceutical company, and health authority.

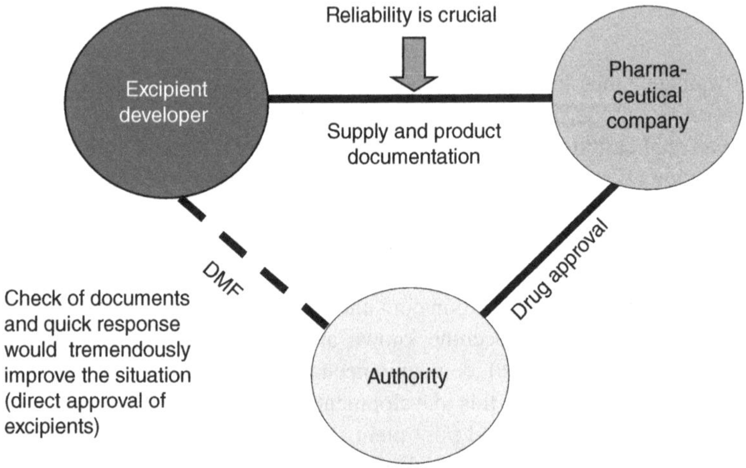

Figure 7.8 Relationship between excipient supplier, pharmaceutical company, and health authority.

There could be another, even better solution for this dilemma. Excipients could have their own approval process and should not be approved only in combination with drugs. This individual approval would have various benefits because every pharma company would then know that the use of a novel excipient at a certain dose range is safe. This process would facilitate the use of novel excipients and also strongly support innovation in excipients by enabling novel excipients to penetrate the market more quickly. Subsequently, suppliers would also be more motivated and invest more money in new developments. As a result, we would see more modern DDS with unique properties being developed.

All parties would benefit from such an approach: the pharma company, the excipient supplier and, finally, the patient by receiving a unique medicine.

Significant efforts of the industry (The International Pharmaceutical Excipients Council (IPEC)) toward such a harmonized and consistent approach for quality standards, safety assessment, and excipient information packages for regulatory filings are ongoing.

7.2.9 Development Examples: Polyvinylalcohol–Polyethylene Glycol Graft Copolymer (Kollicoat® IR)

The development of polyvinylalcohol–polyethylene glycol graft copolymer was initiated in 1997 after intensive investigation of the excipient market. This revealed that there was a need for a new immediate release polymer with better properties than the existing market standards. It was clearly defined that the new polymer should have a much lower viscosity and a much higher elongation at break than the standard HPMC and should not need any plasticizer. After a 2-year screening period of polymer classes and testing of more than 50 different polymer classes, polyvinylalcohol–polyethylene glycol graft copolymers were found. The ratio of both molecular parts and the molecular weight was optimized to fulfill the set targets. Numerous loops between synthesis and application testing had been necessary to come to the final structure. At the beginning of this period only the most relevant properties such as viscosity, elongation at break, and dissolution of films in 0.1 N HCl and phosphate buffer pH 6.8 had been tested but the more the optimization progressed, the more other properties were included since any significant drawbacks of the new polymer should be avoided. These further tests consisted of sprayability, appearance, coating tests under various conditions, gloss, moisture permeability, oxygen permeability, and so on.

The synthesis and final structure of the new polymer, Kollicoat IR (polyvinylalcohol–polyethylene glycol graft copolymer), are shown in Figures 7.9 and 7.10. It can be considered as an internally plasticized polyvinyl alcohol, which means that the plasticizer – the PEG moiety – is covalently bound to the molecule and thus cannot migrate or be evaporated. The high number of hydrophilic functional groups enables high solubility and quick dissolution in all aqueous media to be achieved, independent of pH.

The molecule is synthesized by grafting vinyl acetate onto polyethylene glycol in a first step. In a second step, under the impact of a strong alkaline catalyst, the

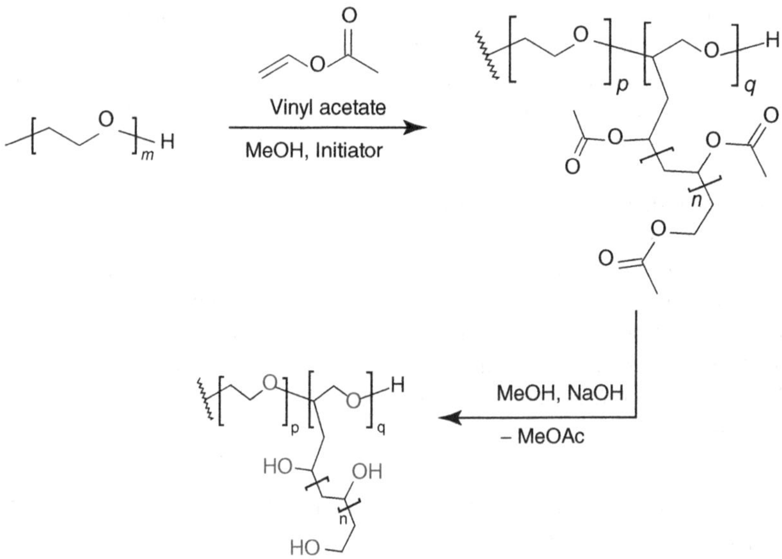

polyvinyl alcohol–polyethylene glycol
graft copolymer

Figure 7.9 Synthesis of Kollicoat IR.

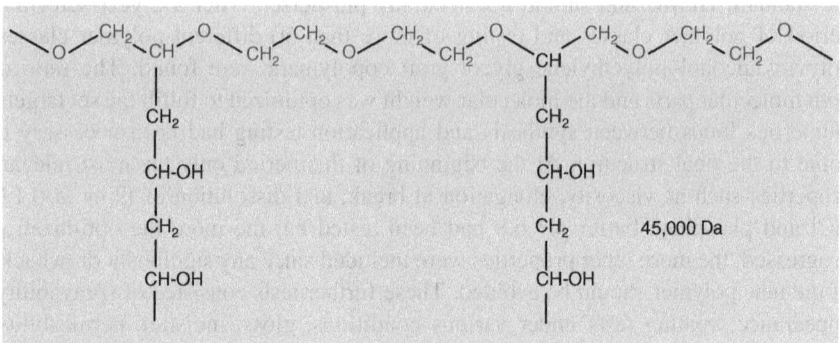

polyvinyl alcohol–polyethylene glycol graft copolymer

PVA-units:	75%
PEG-units:	25%
Solubility in water:	>50%

Figure 7.10 Structure of Kollicoat IR.

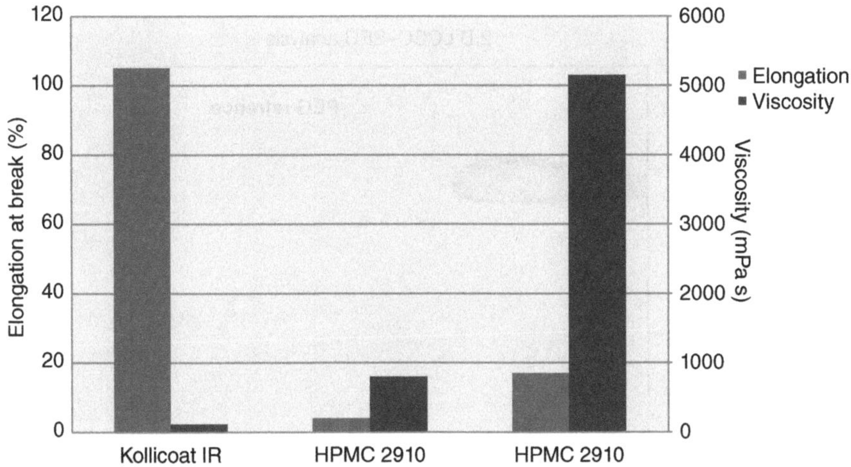

Figure 7.11 Comparison of most relevant properties of immediate release coatings: Kollicoat IR versus market standard.

acetate group is transferred to an alcohol and subsequently polyvinyl alcohol side chains are formed. After solvent exchange against water, the aqueous polymer solution is spray-dried to a powder. All steps are monitored and controlled by appropriate in-process controls and they are carried out under GMP conditions.

The differences compared to the market standard are shown in Figure 7.11.

Viscosity determined as a 20% aqueous solution was very low (approx. 120 mPa s) and flexibility expressed as elongation at break very high (approx. 100%). Both properties are usually related to the molecular weight of the polymer but in an oppositional way. This means the higher the molecular weight, the higher the viscosity and the higher the elongation at break. Thus, a compromise had to be found resulting in a molecular weight of approx. 45,000 Da [28, 29]. However, the grafted structure strongly improves both parameters and outperforms other polymer structures.

Thus, the targets set at the beginning of the development were achieved. Due to its favorable physicochemical properties, Kollicoat™ IR is also suitable as a carrier in drug-loaded film strips [30]. Its use as a hydrophilic pore former in combination with sustained release coating agents [31] as well as a wet binder [32] or carrier in solid dispersions [33, 34] is also described in the literature.

The final product was investigated for free plasticizer, that is, free PEG using a special analytical method, LCCC–SEC combination (Figure 7.12). No free PEG could be detected, which proved that the material can be designated plasticizer-free. This shows that all PEG chains had been grafted with at least one side-chain of polyvinyl alcohol.

Other parameters describing the new excipient more from a compendial point of view are derived from either polyethylene glycol or polyvinyl alcohol. Thus, the monomers ethylene oxide and vinyl acetate, the by-products dioxane and acetate and the physicochemical characteristics pH, viscosity and loss on drying need to be determined.

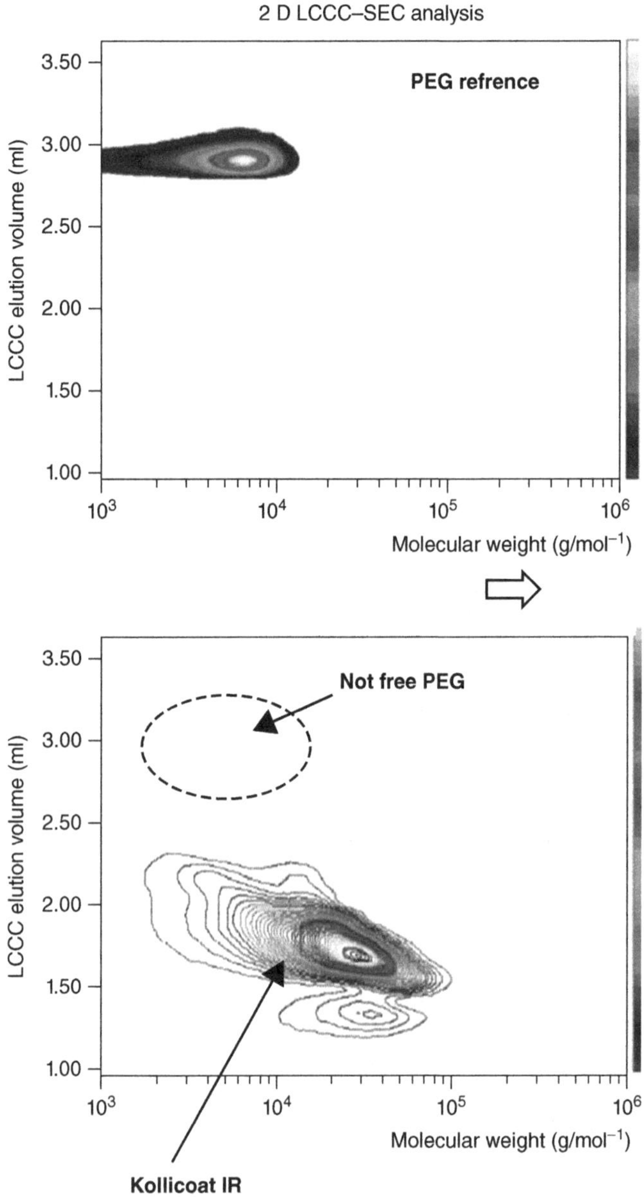

Figure 7.12 LCCC–SEC analysis of Kollicoat IR.

The toxicological tests were performed according to the existing guidelines and all results were without any sign of toxicity. This result was expected due to the fact that the polymer was not absorbed from the GI tract as shown by ADME studies. When polymers have higher molecular weight – even when they are completely water soluble – it is highly unlikely that they are absorbed via the intestinal mucosa.

In summary, it can be stated that, during the development of the polyvinyl alcohol–polyethylene glycol graft copolymer, the originally set physicochemical, toxicological application-related targets were achieved.

The new polymer was introduced onto the market in 2003. Due to its benefits, there was a rapid uptake by several pharma companies for use in drug formulations. The first approvals were in the EU and South America in 2006, followed by an approval by PMDA in Japan in 2007. The entire registration in major countries was completed on approval in the United States in 2008. Five years after having launched the new polymer, it had been registered in all major pharmaceutical countries in the world.

The fact that the new excipient was contained in approved drugs enabled it to be included in the various pharmacopoeias. The monograph entitled "Macrogol – poly(vinyl alcohol) grafted copolymer" was published in 2010 in the European Pharmacopoeia and under the title "Ethylene glycol and vinyl alcohol graft copolymer" in 2011 in the USP. A monograph is also included in the 2012 edition of Japanese Pharmaceutical Excipients (JPE). Kollicoat IR has been self-affirmed to be generally recognized as safe (GRAS) for use as coating for dietary supplements in the United States. The use in solid food supplements (E1209, Polyvinyl alcohol–polyethylene glycol-graft-copolymer) was authorized in Europe in 2014. Regulatory aspects of Kollicoat IR are illustrated in Figure 7.13.

Other typical examples of novel excipients are methyl methacrylate–diethylaminoethyl methacrylate (Kollicoat Smartseal 30 D), [35] which is mainly used for taste-masked formulations and protective coatings, and polyvinylcaprolactam–polyvinyl acetate–polyethylene glycol graft copolymer (Soluplus) [36],which offers unique capabilities for solid solutions with the overall target of strongly improving the bioavailability of poorly soluble drugs.

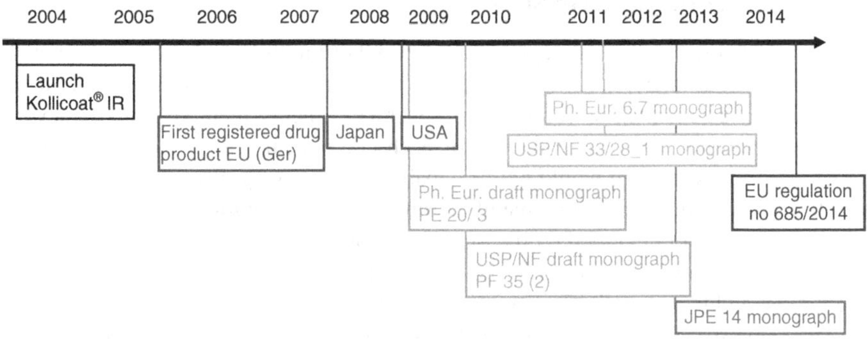

Figure 7.13 Regulatory aspects of Kollicoat IR.

7.3 DEVELOPMENT OF COPROCESSED EXCIPIENTS

7.3.1 Product Development

There are some similarities but also significant differences between the development of a novel excipient and a coprocessed excipient. After defining the target profile that covers mainly physicochemical and application characteristics, the development of a coprocessed excipient starts with a screening process to identify the ideal partners for the formulation. The first step is to find out what kind of partners and how many are necessary to achieve the target profile, that is, which excipient categories must be combined, for example, a filler, a binder, and a disintegrant. Once this has been defined, the most suitable partners are screened within each category and the best binding or disintegrating material for the given target selected.

Of course, the right manufacturing process must also be selected since different processes result in different product properties. Furthermore, materials that can be easily combined in a certain manufacturing process such as mixer granulation might create problems in other processes such as spray drying. Sometimes, the process cannot be carried out anymore or it produces a product characterized by inferior quality. Because of the strong interaction of the formulation components and the formulation process, selection of the optimal composition can only be made for a specific formulation process.

Finally, a decision must be made regarding the process and the qualitative composition. After optimizing the best concentrations for each component within the formulation and finishing the laboratory scale development, the product must be scaled up to pilot plant scale and subsequently to production. Typically, laboratory scale comprises batches between 0.5 and 5 kg, pilot plant scale between 10 and 100 kg, and production scale between 200 kg and 10 t.

Having successfully established manufacturing in the production plant, one should perform process validation in order to ensure reproducible and reliable product quality. This is a prerequisite for all kinds of excipient developments. Table 7.4 describes these stages involved in the development of coprocessed excipients.

7.3.2 Further Development Steps

Several further development steps are similar to those in the development of a novel excipient such as analytical characterization, which also needs to be validated. Since the coprocessed excipient is composed of several excipients, methods must be developed to determine the exact quantity of each component within the formulation; otherwise, release testing would not be capable of identifying discrepancies in composition and fluctuations during manufacturing. In addition, it must be shown that the components are not degraded by the manufacturing process.

Stability testing, compilation of the regulary package, e.g. DMF, technical documentation, and launch activities are also similar to those of a novel excipient.

The biggest difference can be found in toxicological testing since this is not mandatory for a coprocessed excipient, if the absence of significant chemical change can be demonstrated and if the components have already been approved and used in the contained quantities.

TABLE 7.4 Development of Coprocessed Excipients

Composition	Process
Screening for the best excipient categories	Screening for the best process
Screening for the best partner in each category	For each of the selected processes
Selection of qualitative composition and process	
Optimization of quantitative composition	
Optimization of process parameters	
Scale-up	
Transfer to production	

Because of the lack of toxicological testing, the development of a coprocessed excipient usually takes 2–4 years only.

7.3.3 Development Example: Mannitol–Polyvinyl Acetate–Crospovidone (Ludiflash®)

Ludiflash is a typical example of a coprocessed excipient since it consists of already approved, well-known excipients. These are formulated in a way that new properties are created compared to the simple physical mixture of the individual components [37].

When the project was started, there were only a few excipients on the market, but all had significant drawbacks such as poor taste, sandy mouthfeel, insufficient disintegration, poor compressibility, special machinery, and technology needed for manufacturing and one material was only available upon signing agreements and paying royalties. Based on this, the targets of the new project were easily set:

- Quick disintegration
- Excellent taste and mouthfeel
- High compressibility
- Easy to process.

It was obvious that for good taste and mouthfeel, sugar or sugar alcohols were required or other fillers without any off-taste. All, but particularly the latter, must be incorporated in very small particle sizes in order to avoid a sandy mouthfeel. A very effective binder should be used to prepare the ground for high compressibility and nonstickiness of the formulation. Finally, a superdisintegrant should enable fast disintegration of tablets compressed from the new formulation.

Thus, summarized, the formulation should consist of the following:

- A suitable filler
- An effective binder
- A superdisintegrant.

As with other excipient formulations, the final properties are defined by the ingredients and the process used for manufacturing. Thus, a screening period is scheduled to select the best fillers, the best binders, the best disintegrant and, in parallel, the best technology to formulate them. All ingredients have a certain necessary role in such a formulation but in addition performance of these ingredients is impacted by the technology. This means that one cannot optimize the formulation first and then look for the best process or vice versa. At intermediate stages, it has to be checked whether the selected ingredients and the selected process still produce the best result. At the beginning, highest emphasis was placed on finding the optimal composition starting with a filler screening, keeping the binder and disintegrant constant, and, of course, using a standard agglomeration procedure. All monographed sugars, sugar alcohols, inorganic fillers, celluloses, and starches were tested. In this regard, testing means that they were agglomerated, compressed into tablets, and the tablets then investigated for disintegration, hardness, and mouthfeel. Prior to this study, a preselection study was performed where all candidates were evaluated for taste and mouthfeel. Only those materials with good taste and mouthfeel or at least neutral ones were used for the agglomeration and tableting study. As the filler is in high concentration in the final formulation, no compromises can be made regarding taste and mouthfeel. Poor properties in this regard can never be compensated by the minor components of the formulation.

The screening process was set up as follows:

1. Preselection study by oral tasting
2. Agglomeration and tableting study for fillers
3. Agglomeration and tableting study for disintegrants
4. Agglomeration and tableting study for binders
5. Technology (process) selection.

Approximately 50% of all fillers were discarded in the preselection study and not tested further. The main study revealed mannitol as the best filler material followed by some other sugar alcohols; these were considered to be a second option.

As expected, all water-soluble binders improved compressibility and hardness but also prolonged disintegration of tablets. Surprisingly, it was found that water-insoluble and highly plastic binders produced hard tablets, which then disintegrated quickly. The material of choice was polyvinyl acetate dispersion (Kollicoat SR 30 D) with polyvinyl acetate as its main component. Its action can be explained by its delaying of dissolution of mannitol, thus allowing water to penetrate quickly and deeply into the tablet and the disintegrant to start working within the spaces of the tablet. In the case of a water-soluble binder, the incoming water dissolves the mannitol and the binder. This results in a viscous solution that blocks the pores and channels in the tablet so that the water cannot quickly reach all zones of the tablet. In this case, the tablet slowly erodes from the surface to the core zone, a process that takes much longer.

In disintegrant screening, crospovidone performed best; this can be attributed to its limited swelling behavior and the fact that these particles do not form a gel in water but remain in a particle-like structure, which is of course swollen but with a well-defined border. Almost all other disintegrants form a gel upon contact with water and thereby block further penetration of water into the deeper zones of the tablet.

Having performed all these experiments, it was quite obvious that the formulation should consist of the following:

- Mannitol
- Polyvinyl acetate
- Crospovidone.

In parallel, various agglomeration techniques were tested and one selected for further development, starting with the optimization of the ratio of components. It was found that 90% mannitol, 5% crospovidone, and 5% polyvinyl acetate produced the best results.

The SEM photo (Figure 7.14) reveals the nicely granulated, relatively porous structure of Ludiflash, which allows the particles to flow and be compressed easily, forming a strong network within the tablet.

Mannitol is known to have a very flat sorption isotherm [38] and thus it is not surprising that Ludiflash does not take up significant amounts of water at higher humidities. However, loss on drying should be kept low since higher amounts of water can reduce the effectiveness of crospovidone in accelerating disintegration. Table 7.5 shows the other characteristics of mannitol.

Figure 7.15 reveals that hard tablets can be achieved at a very low compression force while keeping the porosity high, thus allowing water to penetrate quickly into the tablet and disintegrate it [39]. A hardness of more than 40 N for a medium-sized tablet of 10 mm in diameter is necessary to withstand mechanical stress during handling, packaging, and removal from the blister.

The originally set target "easy to process" was proven by direct compression trials using various actives, in different formats, at various tableting speeds and different tableting machines. A deep knowledge of the application properties and the behavior

Figure 7.14 Particle structure of Ludiflash (SEM photo).

TABLE 7.5 Particle Characteristics of Ludiflash

Bulk density (g/ml)	0.53
Tap density (g/ml)	0.64
Particle size (D 4,3) (μm)	180
Span	3.7
Angle of repose (°)	35.7
Sorption isotherm (%)	< 2% at 80% r.h.

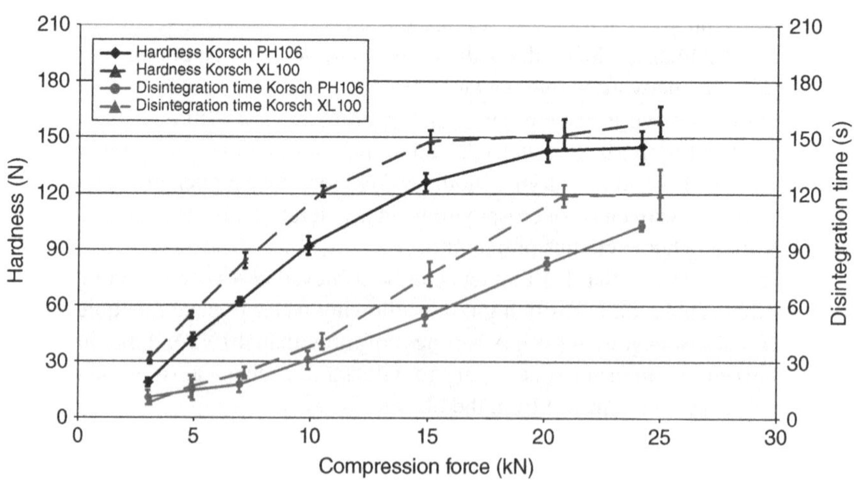

Figure 7.15 Hardness–disintegration time–compression force profile of Ludiflash (comparison of different tablet presses).

of an excipient under various conditions serves as an excellent basis for guiding customers and enables them to quickly develop a new tablet formulation.

As with all oral formulations such as oral liquids, chewables, and lozenges, orally disintegrating tablets also remain in the mouth at least for a certain period of time or have intimate contact with the mucosa, thus palatable sensations are of high importance. Therefore, oral testing of tablets was performed in order to make sure that the disintegration time in the mouth fulfills these requirements and that there is a smooth mouthfeel and excellent taste without any off-taste. In this regard, a taste panel consisting of more than 20 test persons was used in order to evaluate the developed excipient. This evaluation included slight variations of Ludiflash as well as competitive products, as it is worthwhile knowing the differences between the various products for marketing purposes.

7.4 DEVELOPMENT OF MODIFIED EXCIPIENTS

Since the development of modified excipients has a high similarity to coprocessed excipients, it is not described in detail here. However, an example is given in order to illustrate such a development.

The typical development time for such an excipient is 2–3 years.

7.4.1 Development Example: Copovidone Special Fine Grade (Kollidon® VA 64 Fine)

Kollidon® VA 64 is a well-established excipient [40] mainly used as a wet binder, dry binder, and matrix for solid solutions [41]. However, deeper investigations on the mode of action of a dry binder showed that possibilities exist to strongly increase its dry binding efficacy by changing the physical structure in a certain way. A better action in dry binding would fit the trend in the pharmaceutical industry to speed up manufacturing and reduce costs by changing from wet granulation to direct compression. Quite often the existing dry binders are not able to produce tablets of a sufficient hardness, in particular when poorly compressible actives are used in higher loading in the formulation. Furthermore, there is a trend toward smaller tablets since they are easier to swallow, thereby improving convenience and compliance. This limits the amounts of other excipients in a tablet and requires a very effective dry binder.

Based on theoretical and practical studies with different materials, studies were carried out to find out how particle size and particle structure influence compressibility, hardness, and friability.

It is quite obvious that a reduction in particle size strongly increases the contact area between binder and active particles in the tablet since volume increases as a function of diameter at exponential 3 and the projected area at exponential 2. This means that the same concentration of binder in a tablet glues particles together better, establishing a stronger network [42, 43]. Figures 7.16 and 7.17 illustrate these principles further. Thus, one of the first goals was to significantly reduce the particle size of Kollidon VA 64.

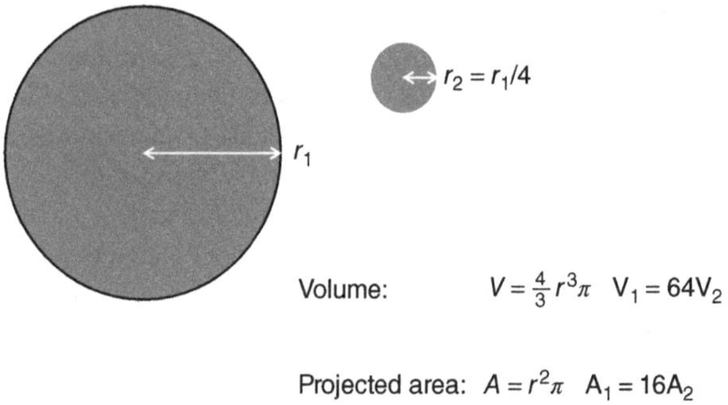

Figure 7.16 Relationship between volume and projected area of spheres.

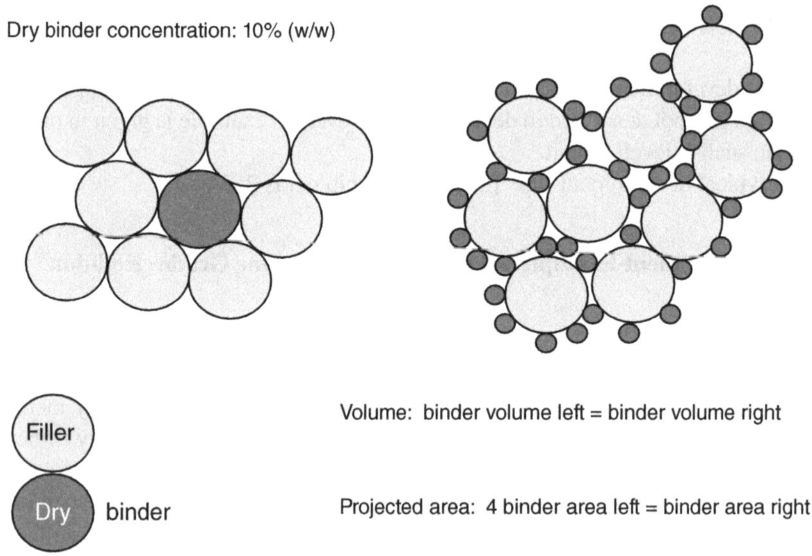

Figure 7.17 Impact of particle size on binding properties.

In a next step, it was investigated how the ratio between volume or projected area to mass can be increased. As a result, hollow particles with extremely thin shells were produced, which on being compressed cover a much larger surface, thus producing a higher degree of hardness for the same concentration in a tablet. Such particles should show enhanced deformability and plasticity, further contributing to the mechanical strength of the tablet [44].

These two main features, a very fine particle size associated with a hollow structure with a thin shell, contribute strongly to the increased binding action of Kollidon VA 64 Fine [45]. Figure 7.18 reveals the particle characteristics of Kollidon VA 64 Fine by SEM photos. The benefit of using Kollidon VA 64 Fine to enhance compactibility of the formulation is represented in Figure 7.19.

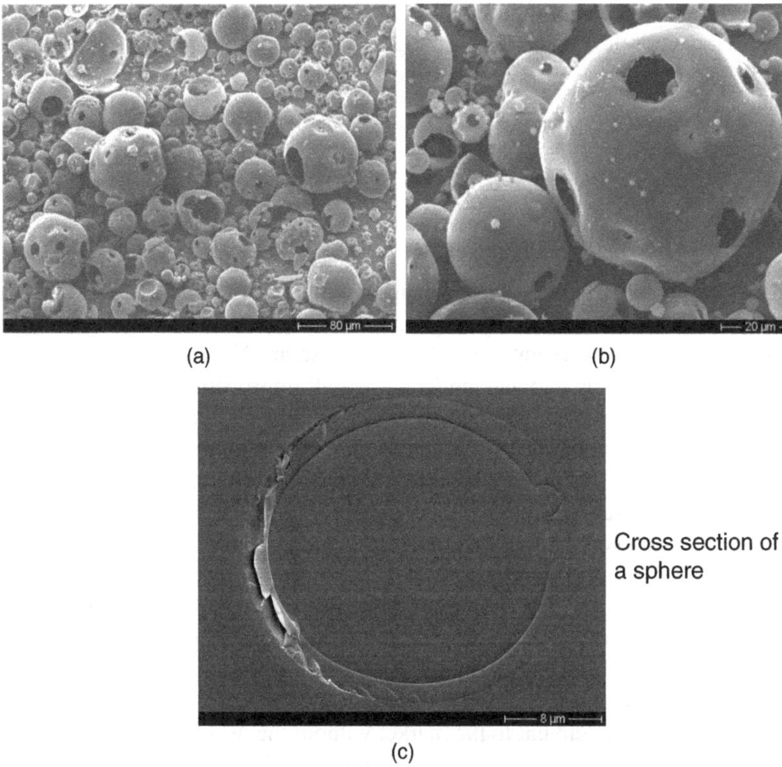

Figure 7.18 SEM photos of Kollidon VA 64 Fine.

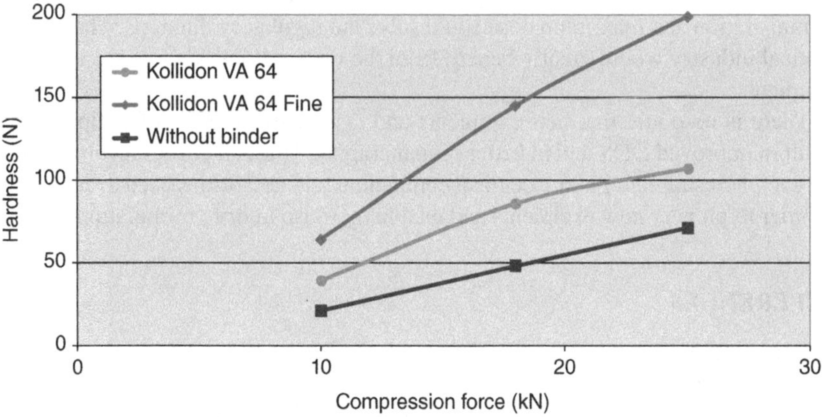

Figure 7.19 Hardness–compression force profile of vitamin C tablets.

In this development, the polymer was not changed in its chemical structure but only in terms of physical properties. Kollidon VA 64 and Kollidon VA 64 Fine meet the requirements of the copovidone monograph in the various pharmacopoeias. This can serve as an example of how properties can be adjusted in order to optimize a material for a certain application.

7.5 SUMMARY

New pharmaceutical excipients can be assigned to three different categories: Modified excipients with only slight changes compared to established ones, coprocessed excipients characterized by a synergistic combination of already established excipients, and novel excipients, which in this case are NCEs. The major target of excipient development is to gain significant overall improvement in performance or to optimize the performance in a particular application. Similar to the development of a new drug, the development of a new excipient is a complex process requiring a well-structured approach with the involvement of specialists with expertise in various fields. A screening phase for a suitable composition and manufacturing technology is followed by optimization of the compound and the process and subsequent scale-up into large scales and finally into production. Clearly defined milestones ensure that the originally set targets are achieved. This chapter illustrates with three case studies that the development of new excipients poses a real challenge. It explains the difficulties that may arise from regulatory and safety requirements, development costs, manufacturing restrictions, and long time to market.

Introducing a new excipient to the market without the necessary approval process also presents a big challenge. The pharmaceutical industry prefers excipients that have a preapproved functional role in drug products in order to avoid an additional perceived risk in drug product development. The example of Kollicoat IR shows that the transition of a "novel excipient" to a commonly accepted excipient can easily take 7–10 years. A direct approval process for new excipients respecting the route of administration and maximum dose could solve the regulatory dilemma. The pharmaceutical industry would greatly benefit from the reduced risk when using innovative excipients.

There is no doubt that new excipients and new characteristics of excipients will result in improved DDS and in better drug therapies. Thus, all parties involved meaning excipient supplier, pharmaceutical companies, and authorities need to collaborate in order to pioneer new excipients and enable their use in drug formulations.

REFERENCES

1. Pifferi, G., Santoro, P., and Pedrani, M. (**1999**) Quality and functionality of excipients. *Il Farmaco*, **54**, 1–14.
2. Jenkins, J.K. CDER New Drug Review: **2011** Update. FDA/CMS Summit. www.fda.gov.
3. Olson, M.K. (**2009**) PDUFA and initial U.S. drug launches. *Michigan Telecommunications and Technology Law Review*, **15**, 393–416.

4. Breitenbach, J. (**2002**) Melt extrusion: From process to drug delivery technology. *Eur. J. Pharm. Biopharm.*, **54** (2), 107–117.

5. Ilic, I., Kasa, P., Dreu, R. *et al.* (**2009**) The compressibility and compactability of different types of lactose. *Drug Dev. Ind. Pharm.*, **35** (10), 1271–1280.

6. Doelker, E. and Gurny, R. (**1987**) Degrees of crystallinity and polymerization of modified cellulose powders for direct tableting. *Powder Technology*, **52** (3), 207–213.

7. Nachaegari, S. and Bansal, A.K. (**2004**) Coprocessed excipients for solid dosage forms. *Pharm. Technol.*, **2004**, 52–64.

8. Saha, S. and Shahiwala, A.F. (**2009**) Multifunctional coprocessed excipients for improved tabletting performance (2009). *Expert Opinion on Drug Delivery*, **6** (2), 197–208.

9. The IPEC Excipient Composition Guide, **2009**.

10. Gupta, P. *et al.* (**2006**) Improved excipient functionality by coprocessing, in *Excipient Development for Pharmaceutical, Biotechnology and Drug Delivery Systems* (ed. Katadare A.), Informa Healthcare USA, Inc., New York.

11. DeMerlis, C.C., Goldring, J.M., Velagaleti, R. *et al.* (**2009**) Regulatory update: The IPEC novel excipient safety evaluation procedure. *Pharm. Tech.*, **33** (11), 72–82.

12. Thompson, D.O. (**2006**) Cyclodextrins – enabling excipients: A case study of the development of a new excipient – sulfobutylether ß-cyclodextrin CAPTISOL®, in *Excipient Development for Pharmaceutical, Biotechnology and Drug Delivery Systems* (ed. Katadare A.), Informa Healthcare USA, Inc., New York.

13. Brewster, M.E., Vandecruys, R., Peeters, J. *et al.* (**2008**) Comparative interaction of 2-hydroxypropyl-β-cyclodextrin and sulfobutylether-β-cyclodextrin with itraconazole: Phase-solubility behavior and stabilization of supersaturated drug solutions. *Eur. J. Pharm. Sci.*, **34**, 94–103.

14. Waterman, K.C. and Adami, R.C. (**2005**) Accelerated aging: Prediction of chemical stability of pharmaceuticals. *International Journal of Pharmaceutics*, **293**, 101–125.

15. Sheehan, C. and Amidon, G. E. (**2011**) Compendial standards and excipient performance in the QbD Era: USP excipient performance chapter <1059>. *American Pharmaceutical Review* **14**, http://www.americanpharmaceuticalreview.com/Featured-Articles/37322-Compendial-Standards-and-Excipient-Performance-in-the-QbD-Era-USP-Excipient-Performance-Chapter-1059/.

16. Schmeller, T., Guth, F., and Kolter, K. (**2011**) Characterization requirements of new excipients. Two case studies. *Chimicaoggi*, **29**, 10–12.

17. ICH Q1A (R2) Stability testing of new drug substances and products. International Conference on Harmonisation of Technical Requirements for Registration of Pharmaceuticals for Human use, http://www.ich.org.

18. The IPEC Excipient Stability Program Guide, The International Pharmaceutical Excipients Council 2010, http://ipec-europe.org.

19. http://www.fda.gov/Drugs/GuidanceComplianceRegulatoryInformation/Guidances.

20. Denton, K. (**2010**) Regulatory challenges are hindering the use of novel excipients. *Pharm. Technol.*, **22** (7), 35–36.

21. Shaw, A.B., Drug Master Files. FDA Small Business Office Webinar, November **2011**.

22. ICH M 4 The Common Technical Document. International Conference on Harmonisation of Technical Requirements for Registration of Pharmaceuticals for human use, http://www.ich.org.

23. Guideline on Active Substance Master File Procedure, European Medicines Agency, **2004**, http://www.ema.europa.eu.
24. Excipient Master File Guide. The International Pharmaceutical Excipients Council of the Americas (IPEC-Americas®), **2004**, www.ipecamericas.org.
25. Baldrick, P. (**2000**) Pharmaceutical excipient development: The need for preclinical guidance. *Regul. Tox. Pharmacol*, **32** (2), 210–218.
26. FDA Inactive Ingredient Search for Approved Drug Products. As accessed at http://www.accessdata.fda.gov/scripts/cder/iig/index.cfm.
27. Ku, S. and Velagaleti, R. (**2010**) Solutol HS 15 as a novel excipient. Identification of the need for and implementation of a US Regulatory Strategy. *Pharmaceutical Technology*, **34**, 11.
28. Kolter, K. (**2001**). Innovation in instant release film-coating, AAPS Annual Meeting and Exposition Denver 2001.
29. Kolter, K., Gotsche, M., Schneider, T. (**2001**). Physico-chemical characterization of Kollicoat IR, AAPS Annual Meeting and Exposition Denver 2001.
30. Garusch, V. and Breitkreutz, J. (**2009**) Novel analytical methods for the characterization of wafers. *European Journal of Pharmaceutics and Biopharmaceutics*, **73**, 195–201.
31. Siepmann, G. and Hoffmann, A. (**2007**) How to adjust desired drug release patterns from ethylcellulose-coated dosage forms. *Journal of Controlled Release*, **119**, 182–189.
32. Agnese, T., Cech, T. (**2010**). Comparing the wet granulation properties of PVA–PEG grafted copolymer and different PVP grades in fluid bed granulation processes applying different spray rates. 2nd Conference Innovation in Drug Delivery 2010, Aix-en-Provence, France.
33. Guns, S. (**2010**) Characterization of the copolymer poly(ethyleneglycol-g-vinylalcohol) as potential carrier in the formulation of solid dispersions. *European Journal of Pharmaceutics and Biopharmaceutics*, **74**, 239–247.
34. Guns, S. (**2011**) Comparison between hot-melt extrusion and spray drying for manufacturing solid dispersions of the graft copolymer of ethylene glycol and vinyl alcohol. *Pharmaceutical Research*, **28**, 673–682.
35. Kolter, K. (**2010**). Physico-chemical characteristics of a new aqueous polymer designed for taste-masking and moisture protection. AAPS Annual Meeting and Exposition 2010, New Orleans, USA.
36. Linn, M. and Collnot, E.M. (**2012**) Soluplus® as an effective absorption enhancer of poorly soluble drugs in vitro and in vivo. *European Journal of Pharmaceutical Sciences*, **45**, 336–343.
37. Kruse, S., Gebert, S. Meyer-Boehm, K., Maschke, A. Kolter, K. (**2007**). Development of orally disintegrating tablets based on a new excipient, AAPS Annual Meeting and Exposition 2007, San Diego.
38. Rowe, R.C., Sheskey, P.J., and Owen, S.C. (**2006**) *Handbook of Pharmaceutical Excipients*, Pharmaceutical Press, The American Pharmacists Association, London, Washington, pp. 449–453.
39. Stoltenberg, I. and Breitkreutz, J. (**2011**) Orally disintegrating mini-tablets (ODMTs) – A novel solid oral dosage form for paediatric use. *Eur. J. Pharm. Biopharm.*, **78** (3), 462–469.
40. Rowe, R.C., Sheskey, P.J., and Owen, S.C. (**2006**) *Handbook of Pharmaceutical Excipients*, Pharmaceutical Press, The American Pharmacists Association, London, Washington, pp. 201–203.

41. Repka, M.A. (**2012**) Melt extrusion: Process to product. *Expert Opinion on Drug Delivery*, **9**, 105–125.

42. Nystroem, C., Mazur, J., and Sjögren, J. (**1982**) Studies on direct compression of tablets. The influence of the particle size of a dry binder on the mechanical strength of tablets. *Int. J. Pharm.*, **10** (3), 209–218.

43. Adolfsson, A., Caramella, C., and Nystroem, C. (**1998**) The effect of milling and addition of dry binder on the interparticulate bonding mechanism in sodium chloride tablets. *Int. J. Pharm.*, **160** (2), 187–195.

44. Duberg, M. and Nystroem, C. (**1986**) Studies on direct compression of tablets XVII. Porosity–pressure curves for the characterization of volume reduction mechanism in powder compression. *Powder Tech.*, **46** (1), 67–75.

45. Kolter, K. Meyer-Boehm, K. (**2005**). Dry binding efficacy of a new copovidone grade (Kollidon VA 64 Fine), AAPS Annual Meeting and Exposition 2005, Nashville.

8

PATability OF EXCIPIENTS

CLAUDIA CORREDOR
Pharmaceutical Development, Bristol-Myers Squibb Company, New Brunswick, NJ, USA

8.1 INTRODUCTION

The concept of quality by design (QbD) has been adopted in the pharmaceutical industry since the introduction of several initiatives such as the US Food and Drug Administration (FDA)'s Pharmaceutical cGMPs for the 21st Century and the International Conference on Harmonization (ICH)'s guidelines Q8, Q9, and Q10 [1–4]. QbD is defined as "a systematic approach to development that begins with predefined objectives and emphasizes product and process understanding based on sound science and quality risk management" [1]. One key aspect of QbD is the implementation of Process Analytical Technology (PAT). PAT is defined as "A system for designing, analyzing, and controlling manufacturing through timely measurements (i.e., during processing) of critical quality and performance attributes of raw and in-process materials and processes, with the goal of ensuring final product quality." As stated in the FDA PAT guideline, "consideration should be given to the quality attributes of incoming materials and their processability for each unit operation. Establishing effective processes for managing physical attributes of raw and in-process materials requires a fundamental understanding of attributes that are critical to product quality" [5].

As a direct result of the regulatory expectations, the product risk assessment should link the excipient critical material attributes (CMA) and process parameters to the product Critical Quality Attributes (CQAs) [2]. Studies should be performed to understand the impact of excipients on the manufacturability and performance of a product [6–8]. There are, however, several challenges that pharmaceutical companies

Pharmaceutical Excipients: Properties, Functionality, and Applications in Research and Industry,
First Edition. Edited by Otilia M.Y. Koo.

face in this regard. Contrary to the active pharmaceutical ingredient (API), chemical and physical properties of excipients may not be entirely under the control of the pharmaceutical user. Many excipients are made in large chemical plants primarily designed for producing chemicals for other industries. Consequently, their chemical characteristics and certain physical properties may be primarily tailored to other markets' needs. This can become a bigger issue taking into account that excipients often constitute up to 90% by weight of the formulation. Another challenge is the lack of understanding of the influence of raw material variability on the excipient functionality (qualitative classifications describing the purposes or roles of an excipient in a drug product). Chamarty et al. illustrated this in their work on excipient functionality. In this study, soluble starch was subjected to an extra washing step during purification, producing two lots. The two lots gave identical results when subjected to a series of physical tests. However, when subjected to a compaction process, the two lots were found to be functionally different, due to differences in the surface properties, which were not detected by the physical testing [9].

The actual overreliance on excipient testing based only on compendial monographs is another challenge that should be addressed. This practice poses the risk of not identifying or measuring CMAs. The USP Excipient Performance General Chapter <1059> [10] can help identifying the excipient CMAs. It organizes excipients by functional categories, including a general description, the mechanisms by which excipients achieve their function, physical properties common to these excipients, chemical properties, and a list of USP general chapters that can be useful in the development of specific tests, procedures, and acceptance criteria to ensure that CMAs are adequately monitored and controlled. Although this chapter is an excellent reference, there is still a need for a thorough understanding of the excipient CMA to minimize the risk. This is critical to determine supplier-to-supplier variability and to assure interchangeability of sources [7, 11, 12].

In this chapter, we present recent QbD studies elucidating the effect of different excipient properties on traditional unit operations. PAT tools used to elucidate the impact of excipient variability on the performance of a product are introduced. Case studies on in-line PAT tools were used to study common CMAs, such as particle size and size distribution, density, moisture content, homogeneity, and content uniformity. Particular attention was paid to excipient variability studies in some of the most important unit operations such as blending, dry granulation, wet granulation, fluid bed drying (FBD), tablet compression, and coating. A case study for the development and validation of a quantitative in-line NIR method for magnesium stearate to monitor two blending operations (before roller compaction and tablet compression) is presented.

8.2 ELUCIDATING RAW MATERIAL VARIABILITY WITH PAT TOOLS

Several PAT tools have been used to demonstrate the impact of excipient variability on the CMAs and subsequently on product performance. These include in-line or at-line process analyzers based on NIR and Raman spectroscopies [13–19], microwave resonance [20, 21], in-line probes for particle size analysis [22], and chemical imaging [19, 23]. In-line process analyzers have several advantages. They are nondestructive, do not require sample preparation, provide real-time data due to their fast acquisition and processing times, and are often noninvasive. PAT tools offer a green alternative as they minimize waste from traditional wet chemistry. The use of these tools allows the timely analysis of excipients in a batch or continuous process based on statistical analysis of hundreds of data points (i.e., during each revolution in a bin blender or hundreds of tablets during tablet compression). They provide information to generate a design space and enable process monitoring inside/outside the design space. Real-time data collection allows the detection of unknown perturbations, which provides basis for continuous improvement [1, 5]. In the case of blending operations, NIR instruments allow in-line monitoring of the blend, without the use of thief sampling that could potentially result in improper sampling (sampling bias) [24], could expose operators to harmful materials, and generates waste that must be disposed of in an environmentally conscious manner. PAT tools are not only limited to in-line or at-line process analyzers but also include multivariate tools for design, data acquisition and analysis, process control tools, continuous improvement, and knowledge management tools [25, 26].

Perhaps the most common PAT tools are based on NIR and Raman spectroscopies. They have been used within the pharmaceutical industry for many years for raw material identification. Based on spectroscopic libraries, the spectrum of the particular lot of the excipient is compared to the spectral signature from an extensive library of the material in question. If the lot of the material being tested closely matches the spectral properties of library standards, then the subject lot is positively identified [27–29]. Recently, the suitability of NIR spectroscopy to identify potassium sorbate, sodium starch glycolate, calcium ascorbate, calcium carbonate, candelilla wax, maltodextrin monohydrate, and anhydrous lactose was determined [29]. These substances are often used in the manufacture of solid dosage forms as binders, diluents, disintegrants, or lubricants. The aim of the study was to identify the pharmaceutical excipients inside USP vials. Although raw material identification is an important (and mandatory) step in manufacturing (for the purpose of cGMP), more studies need to be conducted in order to determine the excipient CMAs in regard to the particular formulation and the process. A particular example is characterization of different polymorphs in an excipient to minimize the risk in formulation due to supplier-to-supplier variability. This was illustrated in the work of Ali et al. that proposed an FT-NIR method for the *in situ* and nondestructive identification of anhydrous and monohydrated forms of lactose [29].

There has been a significant increase in the recognition of NIR and Raman as viable tools for in-line and at-line process monitoring as well as its potential use to determine raw material variability. NIR has been the technique of choice in the majority of studies. NIR instruments have been available since the mid-1980s. Many materials in the pharmaceutical industry present characteristic absorption bands in the NIR region, providing regions in the NIR suitable to develop analytical methods. The development of qualitative and quantitative methods based on NIR has been well understood for some decades. Characteristic bands or single wavelengths

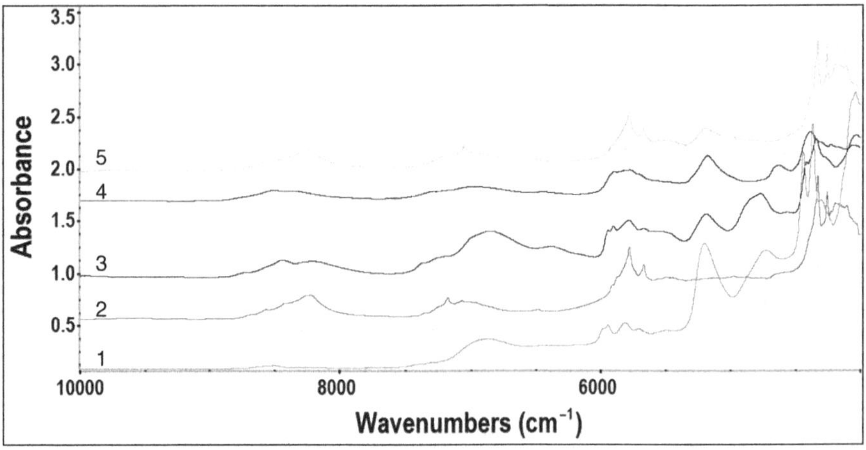

Figure 8.1 NIR spectra of L-(+) lactic acid (1), stearic acid (2), hydroxypropyl cellulose LF (3), crospovidone NF (4) and magnesium stearate (5).

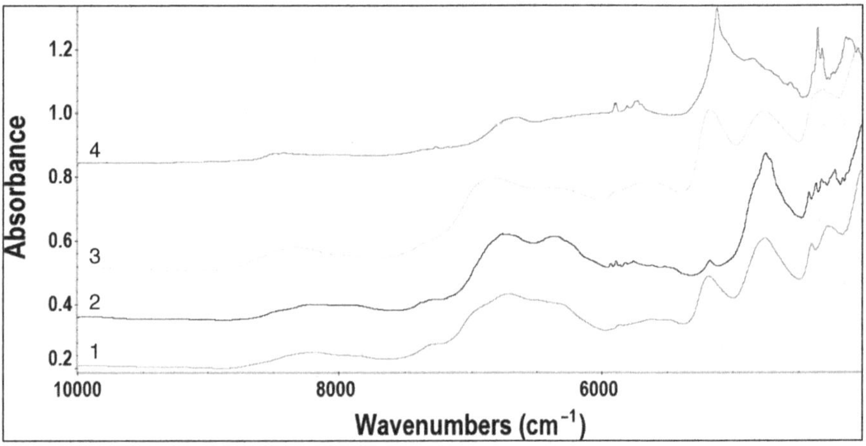

Figure 8.2 NIR spectra of MCC PH 102 (1), lactose anhydrous DC NF (2), sodium starch glycolate (3) and sodium citrate dihydrate FCC USP (4).

in the spectra have been used, together with multivariate analysis based on partial least squares (PLS), multiple linear regression (MLR), multivariate curve resolution (MCR), and so on. Figures 8.1 and 8.2 show NIR spectra of common excipients used in formulations.

There are challenges in the development of PAT methods based on NIR or Raman spectroscopy for an excipient. Very often, there is a marked overlap between the spectrum of the API and other excipients on the spectrum of the excipient of interest, impacting the selectivity and accuracy of the method. This effect can be minimized with an appropriate selection of particular wavelength regions where mostly the excipient has characteristic absorption bands with minimal overlap of peaks from the API and other components of the formulation [29]. Since overlap of the spectral signals are common (Figures 8.1 and 8.2), the development of a method often requires an appropriate sample design (to minimize cross-correlation) and the use of chemometrics (Table 8.1). Besides the possibility of poor selectivity, there is the possibility of sensitivity issues; therefore, monitoring minor components in the formulation with NIR or Raman spectroscopy can be challenging.

Some excipients in the formulation are used at very low levels and do not show strong absorption bands. Analytical selectivity and sensitivity (e.g., % w/w analyte per absorbance unit change) should be established [31] to develop robust methods and appropriate acceptance limits for qualitative or quantitative methods. A significant challenge is when nondestructive spectroscopic methods depend on insensitive or inaccurate reference methods for the development of multivariate calibration models (very often based on chromatographic techniques). Kauffman et al. demonstrated that an NIR PLS method for magnesium stearate monohydrate and dihydrate (developed using samples of magnesium stearate with known amount of monohydrate and dihydrate) was more sensitive and more accurate than the traditional TGA mass loss on drying (LOD) used frequently as reference method [33].

Raman spectroscopy has been used to a lesser extent for the analysis and quantification of excipients, mainly because they are usually poor Raman scattering materials (nonaromatic, noncrystalline materials) in comparison to APIs that are often small aromatic heterocycles. Figures 8.3 and 8.4 show Raman spectra of common excipients used in formulations. As shown in the figures, identifying strong excipient peaks in Raman spectra, even at much higher concentrations than the API, is less likely.

Another challenge is the fluorescence background from common excipients and APIs in the Raman spectrum [34, 35]. Short laser excitation wavelengths (532 nm) can induce fluorescence in organic molecules with long UV absorption. This effect is minimized with the use of laser sources at longer excitation wavelengths such as NIR excitation at 785, 830, and 1064 nm. The fluorescence background of Raman spectra excited with a 1064 nm laser excitation source was over 500 times weaker than that obtained with 785 nm wavelength excitation. Furthermore, the background was more stable in 1064 nm such that background reduction with photobleaching was minimal. While longer excitation wavelengths reduced the fluorescence background, the Raman scattering intensity was also significantly reduced (since the Raman intensity is inversely proportional to the fourth order of the excitation wavelength). One advantage of Raman is the insensitivity of the Raman spectra to water and its greater chemical selectivity compared to NIR [34, 35].

Water content in excipients can be a CMA. Moisture in APIs and excipients can impact drug product manufacturing unit operations such as granulation, conveyance, compaction, and drying. A number of studies have confirmed that moisture content in excipients influences compaction properties, tensile strength, and viscoelastic properties. Moisture significantly affects the mechanical properties of fillers such as microcrystalline cellulose (MCC). Partially crystalline MCC is able to rapidly absorb moisture, acting as a plasticizing agent and lowering its yield strength [36]. Water in the pores of MCC may act as an internal lubricant, reducing tablet density variation by providing a better transmission of the compression force [12]. Moisture also affects

TABLE 8.1 NIR PAT Tools and Conditions Used to Monitor Pharmaceutical Excipients in Blends

Excipient	Property	Region (nm)	Method	Range (Weight %)	#Factors	R^2	RMSECV (%)	Reference
Magnesium stearate	Uniformity	1100–2400	2nd der. PLS	0.3–0.7%	17	0.99	0.011	[30]
	Uniformity	1690–1800	SNV, 2nd der. PLS	1–14%		0.99	0.238	[31]
	Uniformity	1140–1830	2nd der. PLS	0–3%	5	0.98		[13]
	Uniformity	1385–1429 1677–1779 2262–2370	1st der. + SNV, PLS	0.2–1.6%	4	0.99	0.03	[14]
	Micromixing	1000–2500	SBC method	1.5%	NA	NA	NA	[23]
Lactose monohydrate	Content (% w/w)	1600–2400	PLS	16.3–51.3%	3	0.92	3.21	[32]
		1600–2400	MCR-ALS		NA	0.90	3.50	
	Micromixing	1000–2500	SBC method	34–47.5%	NA	NA	NA	[23]
Microcrystalline cellulose	Content (% w/w)	1600–2400	PLS	17.7–51.3%	3	0.96	2.27	[32]
		1600–2400	MCR-ALS		NA	0.92	3.03	
	Micromixing	1000–2500	SBC method	34–47.5%	NA	NA	NA	[23]
Crospovidone	Uniformity	1100–2400	2nd der. PLS	3–7%	14	0.99	0.092	[30]
Prosolv®	Uniformity	1100–2400	2nd der. PLS	44.5–84.5%	13	0.99	0.37	[30]

NA, nonapplicable; der., derivative; SNV, standard normal variate; PLS, partial least square; MLR, multiple linear regression; MCR-ALS, multivariate curve resolution-alternating least squares; SBC, science-based calibration (SBC) based on pure component spectral data.

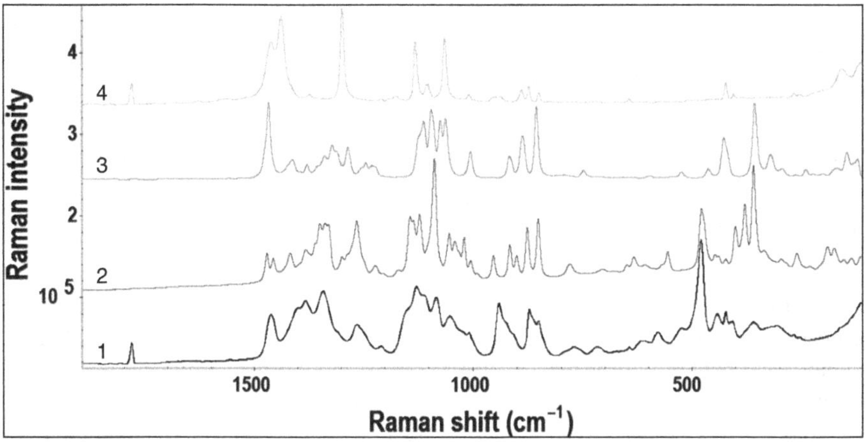

Figure 8.3 Raman spectra of corn starch (1), lactose monohydrate (2), xylitol (3) and magnesium stearate (4).

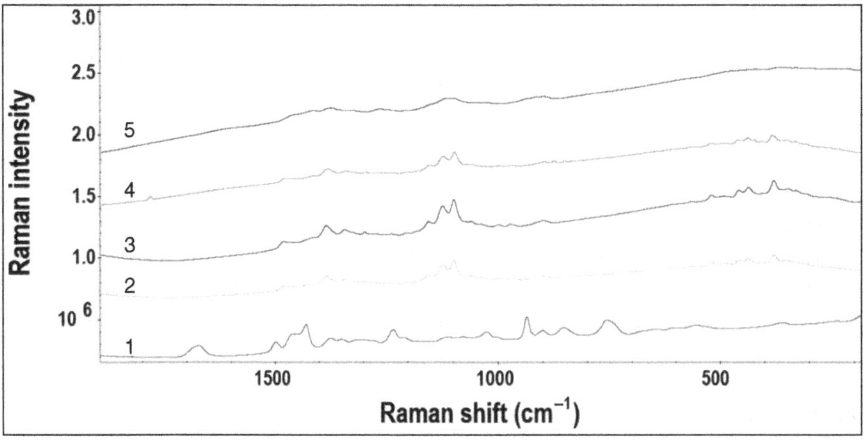

Figure 8.4 Raman spectra of crospovidone (1), MCC PH 102 (2), MCC PH 200 (3), MCC PH 101 (4) and croscarmellose sodium (5).

the mechanical properties of lactose. At ambient conditions, β-lactose anhydrous is converted into α-lactose monohydrate in the presence of moisture.

NIR spectroscopy is well suited for the measurement of moisture because water shows strong NIR absorption bands; most prominent are the first overtone OH stretch at around 6800–7100 cm^{-1} (1470–1408 nm) and the combination band at around 5100–5300 cm^{-1} (1960–1890 nm). In-line NIR has been used not only to determine water content but also for the simultaneous determination of other quality attributes during drying such as drying endpoint, API assay, residual solvents, granule particle size, fluid bed pellet coating, and to visualize the different stages of the FBD process [20, 37, 38].

Despite the challenges in the implementation of PAT tools, there is a growing interest in determining excipient variability and many studies have been recently published. The following section summarizes some of these studies by unit operation.

8.3 PAT FOR EXCIPIENTS: CASE STUDIES BY UNIT OPERATIONS

8.3.1 Blending

Blending is a key unit operation in the manufacturing of tablets. Blend homogeneity is directly linked to the CQA of content uniformity that is related to product safety and efficacy. An adequate control strategy needs to be in place to assure API blend homogeneity and content uniformity. In-line real-time monitoring of blending processes (blending trajectory, uniformity, and end point) most commonly rely on qualitative or quantitative methods based on the API response (with the use of chemometrics) [39, 40].

However, the potential effect of excipients material properties on blending, flow, and cohesiveness has been recognized. Recently, there have been a number of studies investigating blending homogeneity not only of the API but also of key excipients in the formulation [12–14, 23, 30, 31, 41]. Table 8.1 lists some of the conditions used in studies of common excipients such as magnesium stearate, lactose, and MCC by NIR.

Magnesium stearate is one of the most commonly used lubricants. Different techniques have been used to study the magnesium stearate blending process and its effect on slowing dissolution time [41]. The amount and homogeneity of the lubricant can affect downstream processability. Underlubrication can cause sticking during roller compaction or tablet compression, while overlubrication might cause a delay in dissolution or a decrease in compactibility. It has been demonstrated that increased blending time with the lubricant has negative effects on the tablet hardness and compactibility, since the thickness of the magnesium stearate layer depends on the blending time and the intensity of blending [16]. Hence, the distribution of the lubricant for a given type and amount, the type of blender, blender scale and rotation speed, and blending time need to be determined.

NIR PAT instruments have been used in the study of other excipients during blending such as lactose monohydrate, crospovidone, and MCC (Table 8.1). In general, a positioning sensor in the NIR instrument triggers the collection of the spectra every revolution of the bin. Reflectance spectra are collected through a bin lid provided with a sapphire window. Data acquisition is triggered when the powder is in contact with the window. NIR data for calibration batches is often collected in-line (or a combination of in-line and off-line data). Different locations have been used to install the NIR. The significance of the sensor location on the blender has been studied [31]. To ensure specificity, excipient levels should be varied randomly relative to the active concentration (to ensure noncolinearity). The use of a very high number of factors in a method (i.e., 17 factors for magnesium stearate [30]) suggests a lack of specificity, poor selection of an appropriate wavelength region, or poor calibration sample design (cross-correlation with API or other excipients). The number of latent variables or factors in a PLS model should be selected so that the covariance between the spectral data and the reference values is well described and the model should not overfit because of noise [32].

NIR chemical imaging (NIR-CI) tools have been used to study in-line the micromixing blend behavior of a formulation containing acetaminophen, MCC (Avicel PH 101) lactose monohydrate, amorphous fumed silica (Cab-O-SilM-5P), and magnesium stearate. Osorio et al. [23] used the imMix® system (Middleton Research, Middleton, WI), with a camera positioned to scan the blend through a window at the bottom of the rotating blender. A computer-controlled motor rotates the blender with the NIR camera programmed to scan the blender window on specific rotations. Hyperspectral data was collected throughout the blending process, and composition maps (spatial dispersion) were created for all blend ingredients. The imMix® system proved to be useful in monitoring the spatial distribution and aggregate sizes of the API and excipients in the blends.

8.3.2 Dry Granulation

Roller compaction is a continuous dry granulation process whereby a homogeneous dry powder blend containing the API and excipients are compacted using two counter-rotating rollers producing densified sheets or "ribbons" of material that are then subsequently milled to form granules of the desired particle size distribution. It requires both feeding (as powder is fed into the rolls prior to compaction) and processing (compaction) conditions to be carefully controlled. Process parameters such as roll pressure, feed screw speed, and roll speed, and material attributes such as morphology, particle size, and moisture have an impact on the ribbon density (and hence solid fraction), ribbon moisture, granule particle size, flow, and compressibility [21, 42, 43].

Despite the apparent simplicity of the functioning principle of the roller compaction, its mechanism is not well understood. Many efforts have been undertaken to understand the influence of process parameters and material attributes on the CQAs of the dosage form. The type and amount of excipients in the formulation (particularly the lubricant and glidant) can have a significant effect on the tablet properties. Compared to direct compression and wet granulation, excipients experience increased shear during the roller compaction and milling steps. The effect of the increased shear on minor components of the formulation (such as magnesium stearate and sodium lauryl sulfate SLS) was studied by Pandey et al. They observed a high reduction in the tensile strength of roller compacted batches compared to wet granulated batches of the same formulation. The higher observed reduction in compaction (tabletability and compactibility) was attributed to additional shear imparted [6]. He et al. confirmed that for an MCC-based formulation, an overlubrication effect caused by magnesium stearate was observed during roller compaction and especially by the subsequent milling step. When ribbons were compacted to a relatively high solid fraction (>0.6), they were harder to mill and required more time in the mill, causing a more intimate mixture of excipients with magnesium stearate, and thus subject to overlubrication.

Ribbon density, moisture content, tensile strength, and Young's modulus have been monitored by PAT techniques [21, 42, 44–46]. NIR has been most commonly used for the in-line determination of ribbon density. The roller compacted ribbon does not have a uniform density but exhibits a three-dimensional density (solid fraction) distribution. The density is lower at the edges and higher at the center of the ribbon. Several approaches have been evaluated for real-time monitoring of ribbon density with NIR. One of the approaches uses the slope of best-fit line through an NIR spectra's baseline to correlate with the strength or density of ribbons [36, 45]. A second qualitative approach was developed with a principal component analysis (PCA) of spectra taken in-line during the production of ribbons. The third approach involved multivariate PLS calibration models using caliper, pycnometer, and in-line laser as reference methods. All the approaches showed high accuracy of prediction [45].

Besides NIR, microwave resonance has been used for the determination of roller-compacted ribbon density and moisture. The moisture content of granules significantly affects their flowability, cohesivity, and compressibility. The presence of moisture can lead to the formation of both liquid and solid bridges between particles. In most cases, this leads to increased cohesion and friction, which reduce the flowability of the material [21]. Roller-compacted ribbons of MCC in the density range of 0.675–1.216 g/cc and the moisture content range of 2.1–5.5% were tested. Over this moisture content range, significant changes to MCC's flowability and yield strength were observed. The root mean squared error of prediction (RMSEP) for moisture content using microwave sensing (0.065%) was approximately half of the error seen using NIR analysis (0.105%).

Modeling is also part of the PAT tools available for the study of roller compaction. Several models have been developed to describe the process, particularly Johanson's rolling theory for granular solids and the modified approach to Johanson's theory presented by Reynolds et al. The use of instrumented rolls containing force transducers to determine density distributions has also been reported. Nesarikar et al. illustrated the development of statistical models using placebo preblend to express ribbon density as a function of maximum normal stress and gap [47]. Soh et al. [46] identified raw material properties critical to the modeling of granule and ribbon properties as part of the optimization of roller compaction (RC) processes. The utility of roll gap (RG) and NIR signal, specifically, the spectral slope, as process critical control parameters (PCCPs) was evaluated. Raw material tabletability, particle size, size distribution span, and tapped density were found to be the most important factors for building robust predictive models. RG and NIR spectral slope in combination with RC operating parameters yielded models with good predictability for RC responses. Results from this study supported the suitability of RG and NIR spectral slope as PCCPs in roller compaction, specifically, through ribbon density monitoring [46].

8.3.3 High Shear Wet Granulation

High shear wet granulation (HSWG) is extensively used in the pharmaceutical industry. In HSWG, powder particles are granulated using agitation and a liquid binder. Granulation improves the flowability of powders, reduces dustiness and segregation, and improves content uniformity during tablet compression. HSWG is capable of producing granules that are of small sizes (typically less than 1 mm). Characteristic features of a typical high shear granulator include a stainless steel granulator bowl and a central impeller blade, used to agitate the powder and promote densification. A chopper blade is usually located off-center to help break apart agglomerates or promote the growth of smaller particles. Typical critical attributes in HSWG are wet mass consistency, granule particle shape/size distribution, granule bulk density and porosity, moisture content, drug content uniformity/polymorphism, and granule strength and friability. These critical attributes are directly affected by granulation process parameters such as water addition rate, impeller speed, and end point of the granulation process [17, 48].

Excipient attributes such as their particle size, particle size distribution, particle shape, moisture content, cohesiveness, static charge, wettability, and stickiness can affect the process [12]. The binder solution variables influencing granule quality attributes are the type of solvent, type of binder, binder concentration, binder viscosity, type of surfactant, and surfactant concentration. The ability to monitor these properties provides added knowledge that leads to more robust manufacturing process and sustainable product quality.

PAT tools (modeling and probes) supporting in-line measurements of CQAs in HSWG have been recently reviewed [17, 48, 49]. Existing techniques include the use of torque measurement, power consumption, capacitance, microwave, NIR and Raman spectroscopies, in-line particle size measurements, stress and vibration measurements, and acoustic emissions. In-line particle size measurements such as focused beam reflectance measurement (FBRM) and spatial filtering technique (SFT) have been used to track real-time changes in particle size and distribution in the process. Although in-line data collected during the HSWG process from multiple probes providing different types of signals have been correlated to product performance, each particular probe presents challenges and limitations, some of them related to harsh process conditions. For instance, the impeller torque is scale dependent and not always sensitive enough to characterize the granulation process. The power consumption is not reliable since the wear and tear of mixer and motor may cause power fluctuations. Materials may be sticky and adhesive to begin with or get sticky during granulation, causing probe fouling. Probe position relative to the impeller and the chopper influence results and in some cases cause increased probe fouling. Although some of those challenges have been addressed (e.g., FBRM probe incorporating a scraping mechanism to maintain a clean probe window ensuring consistent measurements has been used), there is still a need for improvement of PAT applications for HSWG.

8.3.4 Fluid Bed Granulation and Fluid Bed Drying

Granulation is a key manufacturing step in the production of tablets. The resulting granule particle size critically influences powder flow rate, blend uniformity, and tablet properties such as crushing strength, average mass, and friability. Traditionally, the pharmaceutical fluid bed granulation (FBG) process is controlled by monitoring a few process parameters (e.g., process air flow, temperature, humidity). Fluid Bed Drying (FBD) can also be controlled by monitoring inlet and outlet temperatures.

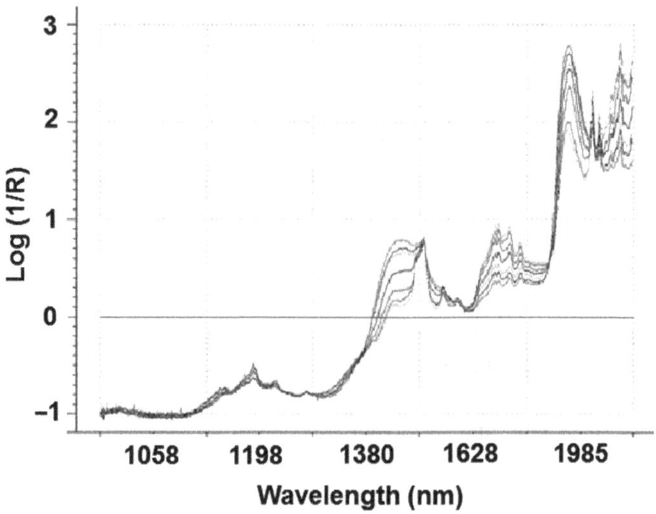

Figure 8.5 NIR spectra of calibration blends during fluid bed drying showing the main absorption bands at 1470–1408 nm and 1960–1890 nm. *See color plate section for color representation of this figure.*

However, these are considered indirect measurements that may be inaccurate or may not account for changes in excipient properties or external disturbances. To monitor the process effectively and reliably in real time, continuous in-line measurements of key product properties such as moisture and particle size distribution are of great importance in FBG and FBD [37, 50–53].

Monitoring moisture during FBG and FBD is important in terms of product quality, and numerous methods such as NIR and acoustic emission have been developed and studied [37, 50–53]. NIR spectroscopy is well suited for the measurement of moisture because water shows strong NIR absorption bands at around 6800–7100 cm^{-1} (1470–1408 nm) and 5100–5300 cm^{-1} (1960–1890 nm). Figure 8.5 shows the main absorption bands of water in the NIR.

During FBD, wet granules from the wet granulation process are transferred to a fluid bed dryer. FBD is usually a short process, taking ~18 minutes. Figure 8.6 shows typical drying profiles for three different batches of a wet granulated product. The endpoint determination is critical for the quality of the product. In this particular example, moisture levels between 2% and 3% were required at the end point. An in-line NIR method was developed with excellent accuracy at this level of moisture. The NIR probe was placed against a quartz sight-glass window of the fluid bed dryer at the same height of the system-integrated sampling probe. Wet granules are often found to stick to sight glass, rendering analysis through the glass problematic; however, this product was found to pose no such problems. During drying, granules

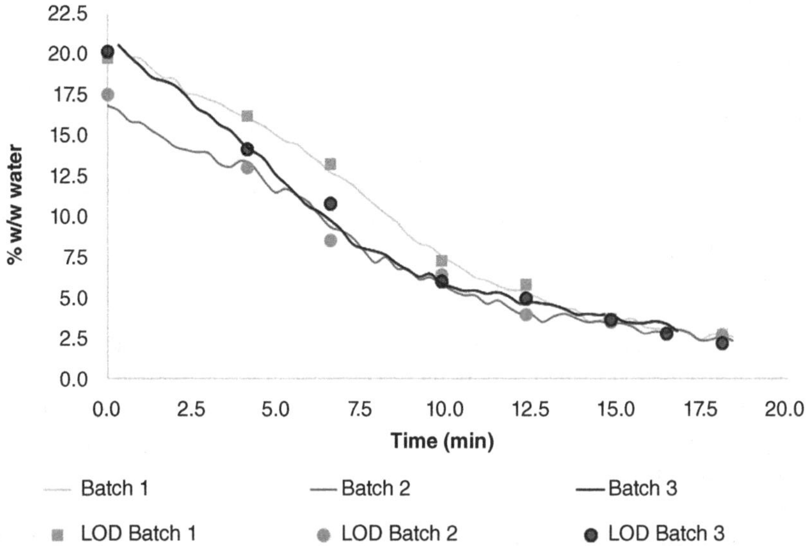

Figure 8.6 % w/w water versus time for three different batches of a wet granulated product and LOD reference values. *See color plate section for color representation of this figure.*

were sampled every 3–5 minutes. The primary reference value for the water content of the sampled granules was determined by LOD. Spectra collected during the drying cycle were transformed to second derivative spectra followed by Savitzky–Golay smoothing prior to PLS regression analysis.

NIR spectroscopy has also been used for the timely monitoring of particle size. Real-time methods for measuring particle size in fluid bed granulators have been developed, including an imaging probe, spatial filtering velocimetry (SFV), acoustic emission, NIR, FBRM, and particle image velocimetry (PIV). Burggraeve et al. [50] reviewed the published work in the field of PAT for FBG to monitor and control the process, as well as to improve process understanding. The implementation of PAT to monitor particle size and moisture during FBG processes is very important. NIR spectroscopy has been extensively used for the timely monitoring of both properties during FBG. Since NIR is a noninvasive technique that does not require sample preparation and provides real-time data, it has been selected as a tool of choice for drying control during FBD. Recently, the spatial filtering technology (SFT) was described as a new in-line particle size analysis method for FBG. The advantage of this method is that no calibration of the sensor with the actual product is necessary. The sensor position in a fluid bed granulator and insertion depth and rotation angle have a relevant influence on the determined particle size and reliability of in-line particle size measurement. The best probe position was determined experimentally [22].

8.3.5 Tablet Compression

A large fraction of pharmaceutical products in the market and development are prepared as tablets and consequently significant emphasis has been placed on understanding the manufacturing process and the CQAs pertaining to efficacy and performance of such solid oral products. These include individual tablet assay (average by potency and variation by content uniformity), moisture, and dissolution, which have been extensively explored by NIR, Raman, microwave resonance, and imaging methods [18–20, 34, 39, 54–56].

Content uniformity assessment has been traditionally based on the distribution of the drug, neglecting other components (i.e., excipients) in the powder blends or tablets. However, there is a growing interest to examine alternative analytical techniques capable of analyzing not only the API but also the distribution of excipients [6, 18, 35, 57]. This is because excipients could play an important functionality role in the performance of the formulation. Table 8.2 lists some of the conditions used to monitor excipients in tablets by NIR.

NIR chemical imaging (NIR-CI) is one emerging technology used for this purpose. NIR-CI adds spatial distribution information to the spectral information by combining traditional NIR spectroscopy with digital imaging. Ravn et al. visualized the spatial distribution of an API, MCC (Prosolv SMCC®), and lactose monohydrate in tablets [57].

NIR chemical imaging (NIR-CI) in combination with chemometric methods such as classical least square (CLS) and MCR were used to provide quantitative and spatial information if the API, magnesium stearate, MCC, lactose and talc in tablets [58]. Raman imaging has been successfully applied to characterize lubricant distribution in tablets. Lakio et al. determined the distribution of magnesium stearate in MCC tablets using Raman imaging, despite low concentration of the lubricant [16].

8.4 CASE STUDY: MAGNESIUM STEARATE BLEND UNIFORMITY BY NIR

Blending homogeneity is often determined by monitoring the concentration of the API. In recent years, a number of studies have recognized the importance of determining the homogeneity of excipients during blending (as described in Section 8.3.1). The amount and distribution of lubricant in a blend will influence the powder properties during roller compaction (to avoid sticking of the blend to the rolls) and tablet compression (to avoid sticking of the blend to the tablet tooling). Since prolonged periods of shear can result in overlubrication and extended dissolution times, an accurate determination of blending end point is important without increasing the blending time.

This case study illustrates the development of an NIR fit-for-purpose PLS method to determine the concentration (%w/w) and %Relative Standard deviation (RSD) of magnesium stearate in two blending operations (preblend and final blend) in the manufacture of a solid dosage form. The preblend (PB) step consists of the blend of API

TABLE 8.2 NIR PAT Tools and Conditions to Monitor in Tablets

Excipient	Property	Region or Signature Wavelength	Method	Range (Weight %)	#Factors	R^2	RMSECV (%)	Reference
Magnesium stearate	Content (% w/w)	980–1900 nm	SNV, 1st derivative, PLS	0.5–1.5%	4	0.947	0.084	[18]
		980–1400 nm			5	0.868	0.131	
		1400–1900 nm			5	0.939	0.089	
		1408 nm	MLR		NA	0.867	0.129	
Lactose monohydrate	Content (% w/w)	980–1900 nm	SNV, 1st derivative, PLS	78–80%	6	0.942	0.797	[18]
		980–1400 nm			5	0.831	1.322	
		1400–1900 nm			6	0.864	1.157	
		1538 nm	MLR		NA	0.615	1.636	
		2210–2350 nm 1850–2000 nm	1st derivative, PLS	63.6–79.4%		0.99	0.62	[57]
Microcrystalline cellulose	Content (% w/w)	980–1900 nm	SNV, 1st derivative, PLS	20–30%	6	0.927	0.904	[18]
		980–1400 nm			5	0.853	1.289	
		1400–1900 nm			5	0.880	1.188	
		1489 nm	MLR		NA	0.740	1.551	
		2230–2450 nm	1st derivative, PLS	14–26%		0.98	0.70	[57]

NA, nonapplicable; SNV, standard normal variate; PLS, partial least square; MLR, multiple linear regression.

and main excipients before roller compaction, containing 0.8% w/w of lubricant. The final blend (FB) step consists of the blend of the granules after roller compaction with additional 0.8% w/w of lubricant (1.6% w/w final concentration). During PB and FB, the bin is rotated for 7 minutes at 15 rpm for a total of 105 revolutions. In the following section, the procedures and considerations into developing the PAT method are described in detail. Since the concentration and %RSD at the end of blending showed good correlation to process CCP and product CQAs, the method was validated (following ICH Q2 guidelines) and used during technical transfer and as part of the robustness program.

8.4.1 Instrument Conditions

The instrument conditions are reported in Table 8.3. Spectra are recorded through a bin lid sapphire window with the spectrometer placed outside the bin lid and powder in the bin is on the inner surface of the sapphire window. There is no contact between the NIR instrument and powder/granules inside the bin. A microelectromechanical system (MEMS)-based NIR spectrometer (Antaris Target® blend analyzer, Thermo Scientific, Madison, WI) equipped with two NIR tunable laser sources (covering the range of 7400–5550 cm^{-1}) and an InGaAs detector was selected.

8.4.2 Calibration Standards

Calibration blends were designed to enable the development of a PLS model correlating the NIR spectral and the gravimetric reference values. Calibration standards were prepared at 2.0 kg scale (using the same manufacturing procedure as production batches). The magnesium stearate concentration in the calibration samples expanded the range of 0.6–2.7% w/w. The calibration set consisted of 32 calibration samples. A calibration test set of 16 samples was used for optimization and selection of model parameters. NIR data for calibration batches was collected in-line.

TABLE 8.3 NIR Instrument Parameters

Parameter	Value
Detection mode	Reflectance
Spectral range	7400–5650 cm^{-1}
Resolution	8 cm^{-1}
Number of sample scans per revolution	8
Number of background scans	8
Data preprocessing	SNV (7401–5557 cm^{1}), 2nd derivative. Smoothing: Savitzky–Golay Data Point: 19 Polynomial Order: 2
Wavelength range of loadings	5980–5740 cm^{1}
Analysis type	PLS
Number of factors	3

8.4.3 Modeling Approach and Variable Selection

The concentration of magnesium stearate was determined by a chemometric model, which uses a precollected set of calibration spectra that are correlated by PLS regression analysis to the gravimetric concentration. During method development, different spectral pretreatments and wavelength regions were used. Magnesium stearate has strong absorption bands in the regions of 7220–6980 cm^{-1} and 5820–5641 cm^{-1} (Figure 8.1). Spectral pretreatments were used to remove or reduce the multiplicative scatter effects resulting from particle size variations. Figure 8.7a shows representative spectra of calibration samples after preprocessing. Figure 8.7b shows the calibration curve (theoretical vs predicted by NIR). The RMSEP for independent blends ranged from 3.0% to 12.5% for PB4 and 5% to 8% in FB.

8.4.4 Validation

In addition to the traditional chemometric measures of validity of an NIR multivariate method, validation criteria found in the EMA [59] and ICH Q2 [60] guidelines were followed. The method validity was challenged using independent validation samples with respect to specificity, linearity, accuracy, and precision, as shown in Table 8.4.

The intent of the calibration model developed was to meet the needs of each blending step, particularly the determination of the blending end point and blending homogeneity. Method development and validation were done in the context of fit for purpose. After evaluating the method performance for PB4 and FB blends the acceptance criteria of the PB4 method were broadened in terms of accuracy and precision due to the difficulty of determining very low magnesium stearate concentrations (0.6–0.8%). All tests passed the set acceptance criteria.

A System Suitability Test (SST) was developed and validated based on the calculation of the uncertainty term. For this purpose, calibration test set and an independent validation test set were used. The uncertainty values were calculated using the Thermo® TQ Analyst software and the equation showed in Bu et al [61].

8.4.5 Batch Monitoring

The NIR method for magnesium stearate was used during the development and scale-up of a pharmaceutical product. Figure 8.8 shows typical magnesium stearate blending profiles for two development batches. In PB, the profiles plateau at around 30 revolutions. The predicted concentration was close to the target concentration of 0.8%. In FB, the profiles plateau at around 25 revolutions. At the end of the blending cycle, the predicted concentration was close to the target concentration of 1.6%.

Plots of the relative standard deviation (RSD) of magnesium stearate in PB and FB are shown in Figure 8.9. The %RSD plots can also be used as criteria for blending homogeneity. In both PB and FB, the %RSD profiles plateau at around 40 revolutions.

The NIR magnesium stearate method was also used during lump investigations. Lumps were found in the PB of a particular batch. The lumps were visible on the

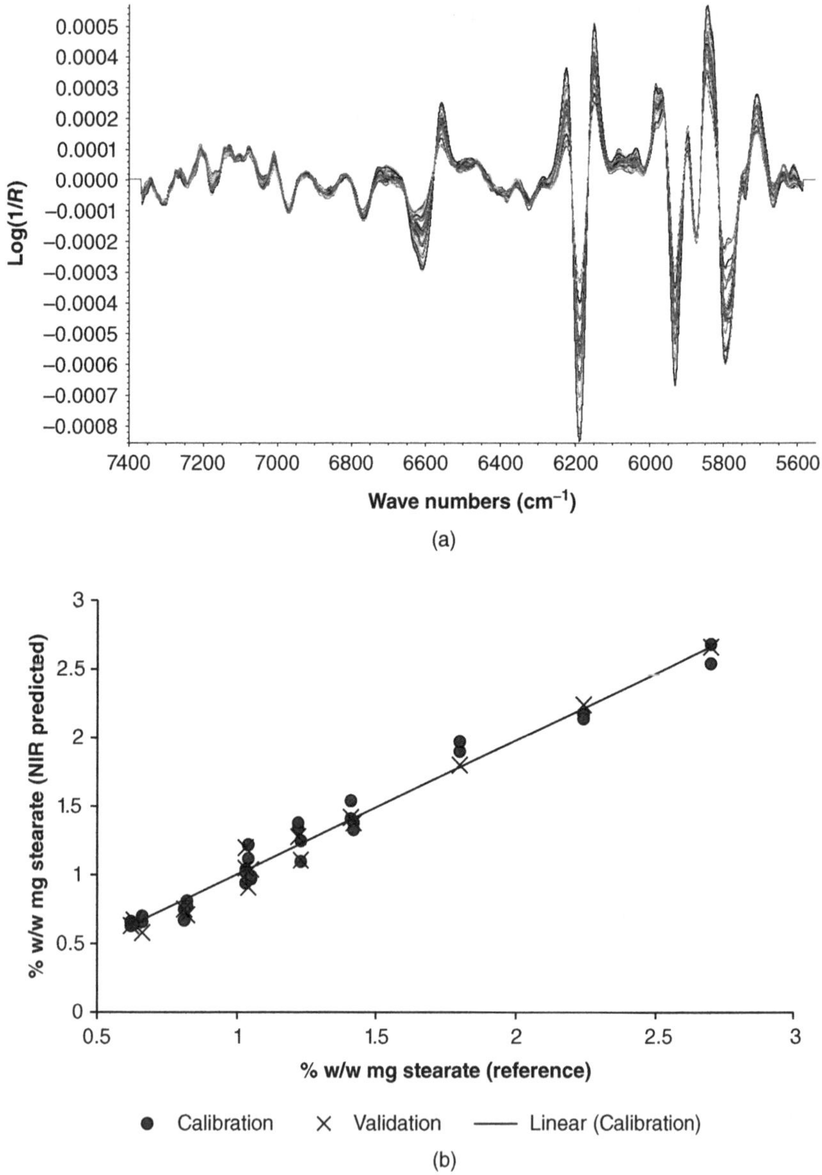

Figure 8.7 (a) Representative preprocessed calibration spectra covering the range for the model generation (b) Calibration curve showing the regression of % w/w magnesium stearate values (gravimetric weight) to the NIR-predicted values.

TABLE 8.4 NIR Method Validation Parameters and Results

Parameter	Result	Acceptance Criteria
Selectivity	Uncertainty PB4: 0.22 Uncertainty FB: 0.20	Uncertainty < 0.25
Linearity	R^2: 0.996; Slope: 0.90	$R^2 \geq 0.95$ Slope: 0.89–1.05
SEP	SEP PB4: 12.5% SEP FB: 4.2%	SEP PB4 < 15% SEP FB < 8%
Accuracy[a]	Theoretical: 0.60% w/w Predicted: 0.49% w/w (81.67%)	0.49–0.72% w/w (80–120%)
	Theoretical: 1.37% w/w Predicted: 1.29 % w/w (94.16%)	0.49–0.72% w/w (90–110%)
	Theoretical: 2.73% w/w Predicted:2.83 % w/w (103.66%)	0.49–0.72% w/w (80–120%)
Precision – Scan Repeatability	PB %RSD < 3.7% FB %RSD < 1.6%	PB %RSD ≤ 8% FB %RSD < 5%

[a] Average of three replicates.

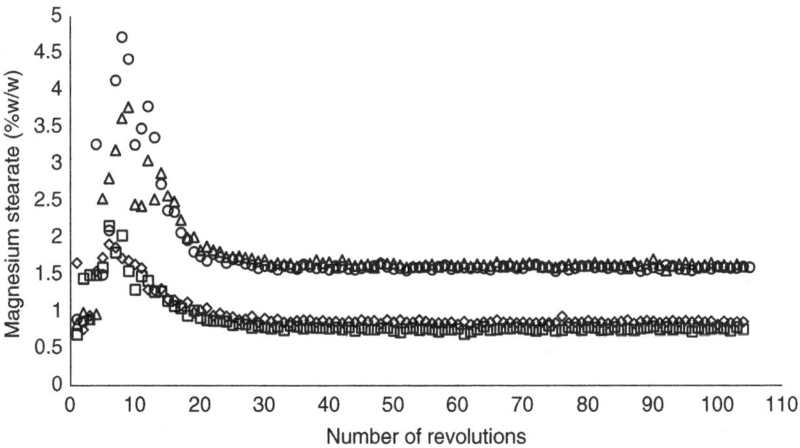

Figure 8.8 Typical blending profile of magnesium stearate in PB (□: Batch 1, ◇: Batch 2) and FB (Δ: Batch 1, ○: Batch 2). *See color plate section for color representation of this figure.*

top of the blend after completing PB. Lumps were removed and analyzed by NIR. Figure 8.10 shows the spectrum of the lumps (in red), pure magnesium stearate (in purple), and PB (in blue). The NIR spectra of the lump corresponded mainly to pure magnesium stearate.

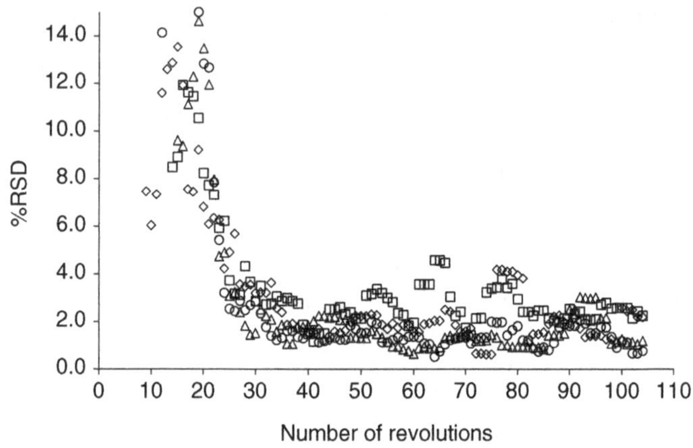

Figure 8.9 %RSD vs. time for magnesium stearate in PB (□: Batch 1, ◇: Batch 2) and FB (Δ: Batch 1, ○: Batch 2).

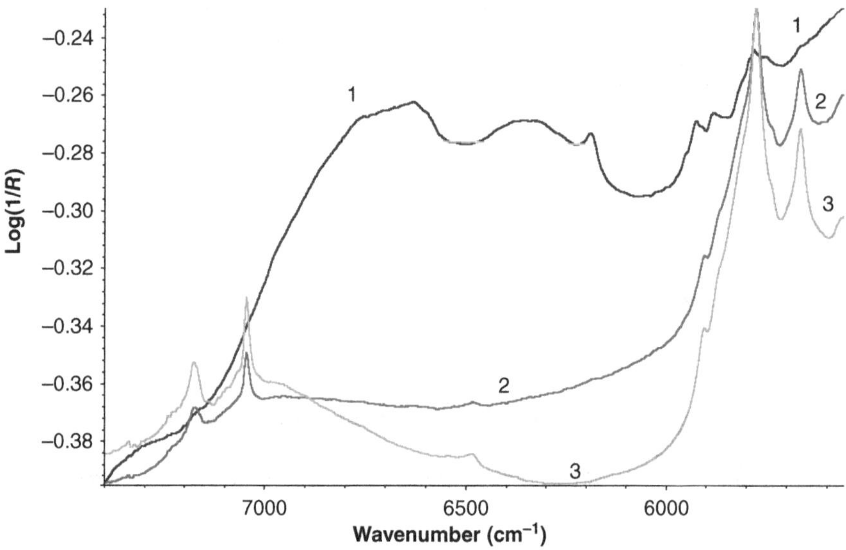

Figure 8.10 NIR spectrum of FB (1) pure magnesium stearate (2), and lump found in FB (3).

The NIR method developed for magnesium stearate provided information about the blending process and blending end point from the perspective of an excipient at lower concentrations (0.8% in PB4 and 1.6% in FB), compared to the API (at 20% w/w). The method provided magnesium stearate data that was used as part of the Process Robustness program implemented at the manufacturing site for the purpose of technical transfer. NIR was also used to conduct investigations related to the blending step.

8.5 CONCLUSION

Development of robust formulations under the QbD paradigm requires a much better understanding of the functional effect that excipients have on a process and the final product quality. For the proper selection and use of excipients, consideration should be given to the CMA of incoming materials and their impact on product processability for each unit operation. In this chapter, several examples of QbD approaches applied to the study of excipient variability in different unit operations were presented. PAT tools based on NIR and Raman spectroscopies and particle size probes used to monitor CMA were introduced. A case study of the development of an in-line NIR method to monitor blending of magnesium stearate was presented. A similar development approach can be taken for other excipients, providing blending information not only for the API, but for other components in the formulation. Understanding the variability of the physical properties in excipients is critical to understand performance tests that can address functionality.

ACKNOWLEDGMENTS

The authors would like to thank Douglas Both and Pankaj Shah for helpful discussions.

REFERENCES

1. FDA/International Conference on Harmonisation (ICH) Q8: Guidance for Industry Pharmaceutical Development. 2009.
2. FDA/International Conference on Harmonisation (ICH) Q9: Quality Risk Management. 2005.
3. FDA/International Conference on Harmonisation (ICH) Q10: Pharmaceutical Quality System. 2008.
4. FDA Pharmaceutical cGMPs for the 21st century-A risk-based approach-Final report. 2004.
5. FDA, Guidance for industry PAT- A framework for innovative pharmaceutical development, manufacturing, and quality assurance. 2004.
6. Pandey, P. *et al.* (**2014**) Excipient-process interactions and their impact on tablet compaction and film coating. *Journal of Pharmaceutical Sciences*, **103** (11), 3666–3674.
7. Saluja, V. and B.S. Sekhon, The regulation of pharmaceutical excipients. *Journal of Excipients and Food Chemicals*, **2013**. **4**(3): p. 95–106, 12 pp.
8. Yu, L.X. *et al.* (**2014**) Understanding Pharmaceutical Quality by Design. *AAPS J.*, **16** (4), 771–783.
9. Chamarthy, S., Pinal, R., and Carvajal, M. (**2009**) Elucidating raw material variability—importance of surface properties and functionality in pharmaceutical powders. *AAPS PharmSciTech*, **10** (3), 780–788.

10. USP32-NF37, Excipient Performance General Chapter <1059>.
11. Carlin, B. (**2012**) Quality risk management of compliant excipients. *Journal of Excipients and Food Chemistry*, **3** (4), 11.
12. Thoorens, G., et al., Microcrystalline cellulose, a direct compression binder in a quality by design environment-A review. *International Journal of Pharmaceutics*. (Amsterdam, Neth.), **2014**. **473**(1–2): p. 64–72.
13. Duong, N.H. *et al.* (**2003**) A homogeneity study using NIR spectroscopy: Tracking magnesium stearate in Bohle bin-blender. *Drug Development and Industrial Pharmacy*, **29** (6), 679–87.
14. Nakagawa, H., et al., Real-time monitoring of lubrication properties of magnesium stearate using NIR spectrometer and thermal effusivity sensor. *International Journal of Pharmaceutics*. (Amsterdam, Neth.), **2013**. **441**(1–2): p. 402–413.
15. De Beer, T., Burggraeve, A., Fonteyne, M. *et al.* (**2011**) Near infrared and Raman spectroscopy for the in-process monitoring of pharmaceutical production processes. *International Journal of Pharmaceutics*, **417** (1–2), 32–47.
16. Lakio, S. *et al.* (**2013**) Challenges in detecting magnesium stearate distribution in tablets. *AAPS PharmSciTech*, **14** (1), 435–444.
17. Hansuld, E.M. and L. Briens, A review of monitoring methods for pharmaceutical wet granulation. *International Journal of Pharmaceutics*. (Amsterdam, Neth.), **2014**. **472**(1–2): p. 192–201.
18. Karande, A.D., Heng, P.W.S., and Liew, C.V. (**2010**) In-line quantification of micronized drug and excipients in tablets by near infrared (NIR) spectroscopy: Real time monitoring of tabletting process. *International Journal of Pharmaceutics*, **396** (1–2), 63–74.
19. Puchert, T. *et al.* (**2011**) A multivariate approach for the statistical evaluation of near-infrared chemical images using Symmetry Parameter Image Analysis (SPIA). *European Journal of Pharmaceutics and Biopharmaceutics*, **78** (1), 117–124.
20. Corredor, C.C., Bu, D., and Both, D. (**2011**) Comparison of near infrared and microwave resonance sensors for at-line moisture determination in powders and tablets. *Analytica Chimica Acta*, **696** (1–2), 84–93.
21. Austin, J. *et al.* (**2013**) The use of near-infrared and microwave resonance sensing to monitor a continuous roller compaction process. *Journal of Pharmaceutical Sciences*, **102** (6), 1895–1904.
22. Rossteuscher-Carl, K., et al., In-line monitoring of particle size in a fluid bed granulator: Investigations concerning positioning and configuration of the sensor. *International Journal of Pharmaceutics*. (Amsterdam, Neth.), **2014**. **466**(1–2): p. 31–37.
23. Osorio, J.G., et al., Characterization of pharmaceutical powder blends using in situ near-infrared chemical imaging. *Chemical Engineering Science*, **2014**. **108**: p. 244–257.
24. Momose, W. *et al.* (**2011**) Process analytical technology applied for end-point detection of pharmaceutical blending by combining two calibration-free methods: Simultaneously monitoring specific near-infrared peak intensity and moving block standard deviation. *Powder Technology*, **210** (2), 122–131.
25. Kushner, J. (**2013**) Utilizing quantitative certificate of analysis data to assess the amount of excipient lot-to-lot variability sampled during drug product development. *Pharmaceutical Development and Technology*, **18** (2), 333–342.
26. Garcia-Munoz, S. (**2014**) Two novel methods to analyze the combined effect of multiple raw-materials and processing conditions on the product's final attributes: JRPLS and TPLS. *Chemometrics and Intelligent Laboratory Systems*, **133**, 49–62.

27. Blanco, M. and M.A. Romero, Near-infrared libraries in the pharmaceutical industry: A solution for identity confirmation. *Analyst* (Cambridge, U. K.), **2001**. **126**(12): p. 2212–2217.
28. Alcala, M. *et al.* (**2013**) Qualitative and quantitative pharmaceutical analysis with a novel hand-held miniature near infrared spectrometer. *Journal of Near Infrared Spectroscopy*, **21** (6), 445–457.
29. Ali, H.R.H., Edwards, H.G.M., and Scowen, I.J. (**2009**) Noninvasive in situ identification and band assignments of some pharmaceutical excipients inside USP vials with FT-near-infrared spectroscopy. Spectrochim. *Acta, Part A*, **72A** (4), 890–896.
30. Li, W. and Worosila, G.D. (**2005**) Quantitation of active pharmaceutical ingredients and excipients in powder blends using designed multivariate calibration models by near-infrared spectroscopy. *International Journal of Pharmaceutics*, **295** (1–2), 213–219.
31. El Hagrasy, A.S., Chang, S.-Y., and Kiang, S. (**2006**) Evaluation of risk and benefit in the implementation of near-infrared spectroscopy for monitoring of lubricant mixing. *Pharmaceutical Development and Technology*, **11** (3), 303–312.
32. Igne, B., et al., Modeling strategies for pharmaceutical blend monitoring and end-point determination by near-infrared spectroscopy. *International Journal of Pharmaceutics*. (Amsterdam, Neth.), **2014**. **473**(1-2): p. 219–231.
33. Kauffman, J.F. *et al.* (**2008**) Near infrared spectroscopy of magnesium stearate hydrates and multivariate calibration of pseudopolymorph composition. *Journal of Pharmaceutical Sciences*, **97** (7), 2757–2767.
34. Townshend, N. *et al.* (**2012**) Comparison of the determination of a low-concentration active ingredient in pharmaceutical tablets by backscatter and transmission Raman spectrometry. *Analytical Chemistry*, **84** (11), 4671–4676.
35. Sasic, S. *et al.* (**2013**) Raman chemical mapping of magnesium stearate delivered by a punch-face lubrication system on the surface of placebo and active tablets. *Applied Spectroscopy*, **67** (9), 1073–1079.
36. Gupta, A. *et al.* (**2005**) Influence of ambient moisture on the compaction behavior of microcrystalline cellulose powder undergoing uni-axial compression and roller-compaction: A comparative study using near-infrared spectroscopy. *Journal of Pharmaceutical Sciences*, **94** (10), 2301–2313.
37. Corredor, C.C. *et al.* (**2010**) Monitoring of blending uniformity: Form conversion and fluid bed drying by near infrared and Raman spectroscopy. *American Pharmaceutical Review*, **13** (1), 66–72.
38. Peinado, A., Hammond, J., and Scott, A. (**2011**) Development, validation and transfer of a near infrared method to determine in-line the end point of a fluidised drying process for commercial production batches of an approved oral solid dose pharmaceutical product. *Journal of Pharmaceutical and Biomedical Analysis*, **54** (1), 13–20.
39. Corredor, C.C., Wan, B., and McGeorge, G. (**2012**) Pharmaceutical analysis and NIR-A review. *European Pharmaceutical Review*, **17** (1), 7–10.
40. Corredor, C. (**2015**) Analytical method quality by design for an on-Line near infrared method to monitor blend potency and uniformity. *Journal of Pharmaceutical Innovation*, **10** (1), 47–55.
41. Abe, H. and Otsuka, M. (**2012**) Effects of lubricant-mixing time on prolongation of dissolution time and its prediction by measuring near infrared spectra from tablets. *Drug Development and Industrial Pharmacy*, **38** (4), 412–419.

42. McAuliffe, M.A.P. *et al.* (**2015**) The use of PAT and off-line methods for monitoring of roller compacted ribbon and granule properties with a view to continuous processing. *Organic Process Research and Development*, **19** (1), 158–166.
43. Feng, T. *et al.* (**2008**) Investigation of the variability of NIR in-line monitoring of roller compaction process by using Fast Fourier Transform (FFT) analysis. *AAPS PharmSciTech*, **9** (2), 419–24.
44. Kona, R. *et al.* (**2015**) Quality-by-design III: Application of near-infrared spectroscopy to monitor roller compaction in-process and product quality attributes of immediate release tablets. *AAPS PharmSciTech*, **16** (1), 202–216.
45. Acevedo, D. *et al.* (**2012**) Evaluation of three approaches for real-time monitoring of roller compaction with near-infrared spectroscopy. *AAPS PharmSciTech*, **13** (3), 1005–1012.
46. Soh, J.L.P. *et al.* (**2007**) Importance of raw material attributes for modeling ribbon and granule properties in roller compaction: Multivariate analysis on roll gap and NIR spectral slope as process critical control parameters. *Journal of Pharmaceutical Innovation*, **2** (3–4), 106–124.
47. Nesarikar, V.V., et al., Roller compaction process development and scale up using Johanson model calibrated with instrumented roll data. *International Journal of Pharmaceutics*. (Amsterdam, Neth.), **2012**. **436**(1–2): p. 486–507.
48. Kumar, A. *et al.* (**2013**) Model-based analysis of high shear wet granulation from batch to continuous processes in pharmaceutical production - A critical review. *European Journal of Pharmaceutics and Biopharmaceutics*, **85** (3PB), 814–832.
49. Huang, J. *et al.* (**2010**) A PAT approach to improve process understanding of high shear wet granulation through in-line particle measurement using FBRM C35. *Journal of Pharmaceutical Sciences*, **99** (7), 3205–3212.
50. Burggraeve, A. *et al.* (**2013**) Process analytical tools for monitoring, understanding, and control of pharmaceutical fluidized bed granulation: A review. *European Journal of Pharmaceutics and Biopharmaceutics*, **83** (1), 2–15.
51. Soppela, I. *et al.* (**2014**) On-line monitoring of fluid bed granulation by photometric imaging. *European Journal of Pharmaceutics and Biopharmaceutics*, **88** (3), 879–885.
52. Nieuwmeyer, F. *et al.* (**2007**) Granule characterization during fluid bed drying by development of a near infrared method to determine water content and median granule size. *Pharmaceutical Research*, **24** (10), 1854–1861.
53. Morris, K.R. *et al.* (**2000**) Accelerated fluid bed drying using NIR monitoring and phenomenological modeling. *Drug Development and Industrial Pharmacy*, **26** (9), 985–988.
54. Boiret, M., Meunier, L., and Ginot, Y.M. (**2011**) Tablet potency of Tianeptine in coated tablets by near infrared spectroscopy: Model optimisation, calibration transfer and confidence intervals. *Journal of Pharmaceutical and Biomedical Analysis*, **54** (3), 510–516.
55. Pelletier, M.J., Larkin, P., and Santangelo, M. (**2012**) Transmission Fourier transform Raman spectroscopy of pharmaceutical tablet cores. *Applied Spectroscopy*, **66** (4), 451–457.
56. Said, M.M. *et al.* (**2011**) Near-infrared spectroscopy (NIRS) and chemometric analysis of Malaysian and UK paracetamol tablets: A spectral database study. *International Journal of Pharmaceutics*, **415** (1–2), 102–109.
57. Ravn, C., Skibsted, E., and Bro, R. (**2008**) Near-infrared chemical imaging (NIR-CI) on pharmaceutical solid dosage forms-comparing common calibration approaches. *Journal of Pharmaceutical and Biomedical Analysis*, **48** (3), 554–561.

58. Amigo, J.M. and Ravn, C. (**2009**) Direct quantification and distribution assessment of major and minor components in pharmaceutical tablets by NIR-chemical imaging. *European Journal of Pharmaceutical Sciences*, **37** (2), 76–82.
59. (EMEA), E.M.A., Guideline on the use of near infrared spectroscopy by the pharmaceutical industry and the data requirements for new submissions and variations. 27 January 2014.
60. International Conference on Harmonisation ICH Q2 (R1). Validation of analytical procedures: text and methodology. 2005.
61. Bu, D., B. Wan, and G. McGeorge, (**2013**) A discussion on the use of prediction uncertainty estimation of NIR data in partial least squares for quantitative pharmaceutical tablet assay methods. *Chemometrics and Intelligent Laboratory Systems*, **120**, 84–91.

INDEX

abbreviated new drug application (ANDA), 234
see also new drug application (NDA)
acetaminophen–gelatin system, 62
acid-modifying excipient, 19
active pharmaceutical ingredient (API), 1, 11, 52, 199, 304
active substance master file (ASMF), 281
additives, 204
adsorption, 35–36
adsorption-induced aggregation, 178
advanced NMR techniques, 18–20
adverse immunogenic side effects, 146
AFM *see* atomic force microscopy (AFM)
aggregation-competent species, 181
aldehydes, 27–29, 41–42
alkali cellulose, 35
American National Standards Institute (ANSI), 261
amino acids, as excipients, 170–171
amorphous fumed silica (Cab-O-SilM-5P), 311
amorphous magnesium stearate, 70
amorphous solid dispersions (ASDs), 101
development of, 106–107
SDs, 107
stabilization of, 106
analytical testing, 10
angiotensin-converting enzyme (ACE) inhibitor, 40
anhydrous lactose, 54
anhydrous lactose NF, 217
animal-derived products, 177
antiagglomeration agents, 78
antibody-mediated immune responses, 146
anticaking agents, 78
antimicrobial preservatives, 180–181
antioxidants, 78–79
API *see* active pharmaceutical ingredient (API)
arginine, 170–171
ASDs *see* amorphous solid dispersions (ASDs)
asparagine–glycine sequence, 149
atomic force microscopy (AFM)
DCP, 20
dicalcium phosphate dihydrate, 22
low-field NMR analysis, 21
solid-solid interface, 20
autoxidation, 176

base-catalyzed ester hydrolysis, 182
binder–binder contact, 62

Pharmaceutical Excipients: Properties, Functionality, and Applications in Research and Industry,
First Edition. Edited by Otilia M.Y. Koo.

binders
performance, 61–65
types
natural polymers, 57–58
sugars, 60–61
synthetic polymers, 58–60
bioavailability, 101 *see also* oral bioavailability
biopharmaceutical classification system (BCS), 101–102
biotechnology products, excipients in
buffer agents, 185
bulking agents, 185–186
common classes of, 163–164
amino acids, 170–171
buffer agents, 164–167
polymers and proteins, 177–180
preservatives, 180–182
salts, 171–173
sugars and polyols, 167–170
surfactants, 173–177
tonicity modifiers, 167
formulation development, 146–148
lyophilization, 183–184
preservatives, 188
proteins degradation pathways
chemical degradations, 148–152
physical degradations, 152–163
solid dosage forms of, 182–183
stabilizers, 186–187
surfactants, 187–188
blend homogeneity, 310
blending
magnesium stearate, 310
NIR-CI, 311
bond percolation threshold, 121
browning/Maillard-type reaction, 54
browning reaction, 29, 32
buffer agents
chemical and physical stabilities, 164
protein solution phase stability, 166
sodium and potassium phosphate salts, 167
solid dosage forms, 185
buffer-catalyzed deamidation, 149
bulking agents, lyophilized formulations, 185–186

carbon-13 NMR spectrum, 18
CCS *see* croscarmellose sodium (CCS)
cellulosic excipients, 26
cellulosic polymers, 121
certificate of suitability (CEP), 280–281
chaotropes, 172
charged polymers, 178
chemical and physical properties, pharmaceutical excipients
categorization of, 2–3
processing methods, 6
universal tests, 5
USP monographs, 2
chemical degradations
deamidation, 148–149
glycation, 152
oxidation, 149–151
peptide backbone cleavage, 151–152
chemical imaging
chemometric processing tool, 11
drug product stability, 11–12
FTIR images, 14
NIR-CI techniques, 11
pH modifiers, 11
Raman mapping, 15
chemical synthesis, 26
chemistry, manufacturing, and controls (CMC), 225
chromotropic acid (CTA), 41
Claisen–Schmidt condensation reaction, 32
CMAs *see* critical material attributes (CMAs)
coating polymers, 73–74
coating-related excipients
coating polymers, 74–75
other additives, 75
plasticizers, 75
cold denaturation, 183
colloidal instability, 156–158
color additives and flavors, 242–243
colorants, 75–76
Colorcon, 259
common technical document (CTD), 244, 246, 258, 259
compendia
food
Codex Alimentarius, 215
FCC, 215
JPE, 215
compendial (pharmacopoeial) monograph specification, 206
compendial test methods
monograph revisions, 7–9
pharmacopoeial harmonization, 7
composition, additives and processing aids
elemental impurities, 264
visible particles, 264
concentration-dependent inhibitors, 178
concomitant components, 202
conformational instability
aggregation rates of, 155

protein aggregation, 154
rhGCSF, 154
rhIFN-g, 154
controlled porosity osmotic pumps (CPOP), 129–130
controlled release solid dispersions (CRSDs), 103
conventional glucose assays, 41
conventional oral solid dosage forms, in excipients
anticaking agents, 78
antioxidants, 78–79
binders, 57–65
coating-related excipients, 73–75
colorants, 75–76
coprocessed excipients, 79–83
diluents/fillers, 52–57
disintegrants, 64–69
lubricants, 69–73
pH modifiers, 76–78
copovidone, 295–298
coprocessed excipients, 79–83, 270–271
atypical actives, 265
development steps, 290–291
product development, 290
cosolvents, 113
critical material attributes (CMAs), 97, 303, 304
critical micelle concentrations (CMCs), 175
critical process parameters, 101
critical quality attributes (CQA), 101, 303
croscarmellose sodium
adsorption, 35–36
solid-state reactions, 36
specific interaction, 36
croscarmellose sodium (CCS), 65–66
crospovidone, 34, 65, 293
crystalline solid dispersions (CSDs), 102
C-terminal aspartic acid, 152
customer and sales specification, 206
cyclodextrin (CD), 163, 179

DDS *see* drug delivery systems (DDS)
deamidation, chemical degradations, 148–149
degree of polymerization (DP), 55
destabilizing effect, 181
dibasic calcium phosphate (DCP), 53
dibasic calcium phosphate dihydrate (DCP-D), 205, 217, 224
diethylenetriaminepentaacetic acid (DTPA), 180
diethylstilbestrol preparations, 37–38
differential scanning calorimeter (DSC), 39, 107
diluents/fillers
performance, 54–57
types
DCP, 53
lactose, 53–54
mannitol, 54
MCC, 52–53
direct compression, 135
disaccharides sucrose and trehalose, 186–187
disintegrants
performance, 66–69
types
CCS, 65–66
crospovidone, 65
SSG, 65
starch, 64–65
disodium edetate (EDTA), 180
DMF *see* drug master file (DMF)
drug delivery systems (DDS), 115, 269
drug-excipient interactions, 24–26
drug incompatibility, 25
drug master file (DMF), 280–281
confidential document, 245
IPEC excipient master file guide format, 247
PR-MF, 246
type-V, FDA-accepted reference information, 245
drug nanocrystals, 114 *see also* nanocrystals
drug product
marginal stability, 147
stability, 26
dry binder, 295–296
dry granulation, 311–312

economically motivated adulteration (EMA), 213, 232–233
EDTA *see* disodium edetate (EDTA)
electrostatic screening, 172
electrostatic stabilization, 116
embryo and fetal development study, 253
enteric coating, 74
Environmental Protection Agency (EPA), 255
ethylcellulose (EC), 59–60
European pharmacopoeia (EP), 98, 209–210
excipient information package (EIP), 247
excipient–protein interactions, 145, 164
excipients
additives, 204, 218–220
alternate test methods, 212–213
API, 199
biotechnology products, 163–164
amino acids, 170–171
buffer agents, 164–167
polymers and proteins, 177–180
preservatives, 180–182
salts, 171–173
sugars and polyols, 167–170
surfactants, 173–177
tonicity modifiers, 167

excipients (*Continued*)
compatibility studies
croscarmellose sodium, 35
lactose, 32–33
magnesium stearate, 38–40
MCC, 33
silicon dioxide, 37–38
sodium starch glycolate, 36
starch, 36–37
stearic acid, 38
compendial test methods
monograph revisions, 7–9
pharmacopoeial harmonization, 7
composition, 235
IPEC excipient, guide, 203, 220
processing aids and additives, 203–204
conventional oral solid dosage forms
anticaking agents, 78
antioxidants, 78–79
binders, 57–63
coating-related excipients, 73–75
colorants, 75–76
coprocessed excipients, 79–83
diluents/fillers, 52–57
disintegrants, 64–69
lubricants, 69–73
pH modifiers, 76–78
development of
coprocessed, 290–295
DDS, 269
modified, 295–298
novel, 271–289
types, 270–271
drug-excipient interactions, 24–26
drug master file (DMF), 280–281
functionality, 98
harmonization, 237
EMA, 232–233
excipient, 226–232
International Conference on, 225–226
process, 231–232
QbD, 233–235
impurities
aldehydes, 27–29
inorganic, 222–224
metals, 29
monochloroacetate, 30
organic acids, 29–30, 222–224
peroxides, 26–27
reducing sugars, 27
residual solvents, 222, 224
life cycle, 200–202
novel characterization techniques
advanced NMR techniques, 18–20
AFM, 20–23
chemical imaging, 10–17
PAT, 23–24
PATability of
magnesium stearate blend uniformity, 316–322
QbD, 303
raw material variability, 304–310
unit operations, 310–316
performance
chemical properties, combination, 204–207, 224
pharmacopeia monograph, 204
QbD, 205
pharmacopeias/pharmacopoeias
European, Japanese and United States, 209–214
International Pharmacopeia, 214
monographs, 216–225
National pharmacopeias, 214
processing aids, 203–204
regulatory information, 281–282
color additives and flavors, 242–243
composition, additives and processing aids, 263–265
drug master files, 245–247
drug product applications, 243–245
IPEC, safety evaluation procedure, 243, 257–259
new developments affecting excipients, United States, 248–249
OTC monograph products, 242
safety evaluation, 249–256
supporting, 247–248
tiered approach, testing, 251
total excipient control system, 259–263
specifications
compendia, 206
customer, 206
formal test methods, 205–209
GMP, 205
in-process specification, 206
'QTTP', 207
raw material specification, 206, 207
release, 206
sales, 206
setting of, 207–209
trace reactive excipient impurities
aldehydes, 41–42
organic acids, 42
peroxides, 42
reducing sugar, 41
variability, 101, 235–237
visible particles, 264
expert working group (EWG), 254

FDA *see* Food and Drug Administration (FDA)
fertility study, 253
filler-binders, 52
fine particle ethylcellulose (FPEC), 60
Flavor and Extract Manufacturers Association (FEMA), 243
fluid bed drying (FBD), 304, 313–315
fluid bed granulation (FBG), 313–315
flux regulators, 130–131
focused beam reflectance measurement (FBRM), 313
Food and Agriculture Organization (FAO), 215
Food and Drug Administration (FDA), 241
Food Chemicals Codex (FCC), 215
Food, Drug and Cosmetic Act of 1938, 242
formaldehyde
  formic acid formation, 28
  impurity, reactions with, 35
formulation development, of biotechnology products, 146–148
fourth generation SDs, 102–103, 106, 107
FRCs *see* functionality-related characteristics (FRCs)
freeze drying, 135
Freundlich/Langmuir adsorption isotherms, 36
FRT *see* functionality-related testing (FRT)
functionality, of excipients, 98
functionality-related attributes
  flux regulators, 130
  plasticizers, 130–131
  pore forming agents, 129–130
  semipermeable membranes, 128
  wicking agents, 129
functionality-related characteristics (FRCs)
  anhydrous lactose, 100
  excipients, characteristics of, 98, 217
  FRT, 99
  hydrophobicity, 101
  in ODTs, 136
  and pharmacopoeial harmonization, 99–100, 217
functionality-related testing (FRT), 98

gas chromatography (GC), 41
gelatin, 199
generally recognized as safe (GRAS), 243
general protein aggregation pathway, 153
genotoxicity, *in vivo* test
glycation, chemical degradations, 152
good laboratory practice (GLP), 249, 252
good manufacturing practice (GMP), 205
guanidine hydrochloride, 171

harmonization, 225–237
  EMA, 232–233
  excipient, 226–232
  International Conference on
    excipients, PDG harmonization process, 229, 230
    ICH guidelines, 227–228
    PDG, 226–230
  process, 231–232
  QbD, 233–235
Hatch–Waxman extension, 201
heavy metal impurities, 10
high-molecular-weight polymers, 199
high shear wet granulation (HSWG), 312–313
Hofmeister effect, 172
HPC *see* hydroxypropyl cellulose (HPC)
human equivalent dose (HED), 259
human serum albumin (HSA), 163
hydrochlorothiazide (HCTZ), 27, 58
hydrolysis and catalytic effects, 33, 149
hydrolytic degradation, 39–40
hydroperoxide (HPO), 42
hydrophilic matrices, 120–121, 124
hydrophilic matrix-based MR systems, 121
hydrophilic polymers, 123, 178
hydrophobicity, 101
hydrophobic matrices, 121, 123
5-hydroxymethyl-2-furfuraldehyde (HMF), 28
hydroxypropyl cellulose (HPC), 34, 59
hydroxypropyl methylcellulose (HPMC), 60, 74, 101
hygroscopicity, 136
hygroscopic materials, 277

ibuprofen-starch granulation, 23
ICH *see* International Conference on Harmonization (ICH)
ICH safety testing guidelines, 250, 251
immunogenicity, 146, 149
impurities
  inorganic impurities, 222–223
  organic impurities, 223
inactive ingredient database (IID), 242, 284
in situ chemical imaging *see* chemical imaging
insoluble excipients, 67, 75
interface-induced denaturation and aggregation, 162
interfacial instability, 159–163
International Conference on Harmonization (ICH), 200, 225–226, 303
International Pharmaceutical Excipients Council (IPEC), 99, 230, 241
  EIP, 247
  new excipient safety evaluation procedure

International Pharmaceutical Excipients Council (IPEC), (*Continued*)
IPEC-Americas Procedure, 257–258
Novel Excipient Evaluation Committee Review of Surelease, 258–259
organizations, 243
International Pharmaceutical Excipients Council of the Americas (IPEC-Americas), 245
intravenous immune globulin (IVIG), 146
Investigational New Drug (IND) applications, 243
IPEC *see* International Pharmaceutical Excipients Council (IPEC)
IPEC-Americas Safety Committee, 257
IPEC-Pharmaceutical Quality Group (PQG), 260
irreversible inactivation, 149

Japanese pharmaceutical excipients (JPE), 215

Kelvin–Ostwald–Freundlich equation, 115
Kollicoat IR, 285
LCCC–SEC analysis of, 287, 288
vs. market standards, 287
regulatory aspects of, 289
sructure of, 286
synthesis of, 286
Kollidon VA 62, 295–298
binding properties, particle size, 296
SEM photos, 297
volume and projected area of spheres, 296
kosmotropes, 172

lactate dehydrogenase (LDH), 178
lactose, 28, 32–33, 53–54
Claisen–Schmidt condensation reaction, 32
hydrolysis and catalytic effects of, 33
lactose anhydrous, 217, 223, 224
lactose-dicalcium phosphate formulation, 66
LBFs *see* lipid-based formulations (LBFs)
LFCS *see* lipid formulation classification system (LFCS)
linoleic acid methylester (LME), 38
lipid-based excipients
mixed glycerides and derivatives, 113
vegetable oils, 110–112
lipid-based formulations (LBFs)
definition of, 107–109
excipients used in, 110–113
FRCs of, 113–114
lipid formulation classification system (LFCS), 109–110
lipid matrices, 121–123
lipid-soluble antioxidants, 113
liquid–liquid phase segregation, 158
liquid–liquid phase separation, 158
liquid protein drug, 183
LME *see* linoleic acid methylester (LME)
long-chain triglycerides (LCTs), 110
lubricants, 69
performance, 71–73
types
magnesium stearate, 70–71
stearic acid, 70
Ludiflash
agglomeration techniques, testing, 293
hardness–disintegration time–compression force profile, 294
particle structure of, 294
superdisintegrant, 292
Lumry–Eyring framework, 153
lyophilization, 183–184

magnesium stearate, 70–71
amine groups, 40
chemical structure of, 38–39
hydrolytic degradation at basic pH, 39–40
metal ion-mediated degradation, 40
oxidation, 40
potential interactions, 39
magnesium stearate blend uniformity
batch monitoring, 319–322
calibration standards, 318
instrument conditions, 318
modeling approach, variable selection, 319
validation, 319
Maillard reaction, 33, 152
mannitol, 54, 293
master file system (MF), 246
matrix-coated pellets, 133
matrix MR systems, 120
excipients used in, 122
FRCs of excipients, 123–125
hydrophilic matrices, 120–121
hydrophobic matrices, 121
lipid matrices, 121–123
types of, 120
MCC *see* microcrystalline cellulose (MCC)
medium-chain triglycerides (MCTs), 110
metal-catalyzed oxidation, 180
metal ion-mediated degradation, 40
metals, pharmaceutical excipients, 30
methionine, chemical stability of, 150
methylcellulose (MC), 58–59, 121
microcrystalline cellulose (MCC), 219, 220, 225
diluents, 52–53
formaldehyde impurity, reactions with, 35
HPC, 34
moisture content of, 33

NF, 218
povidone and crospovidone, 33–34
mixed glycerides and derivatives, 113
modified excipients, 270, 295–298
modified release (MR) dosage forms *see* oral modified release dosage forms
moisture scavenging, 37
molar phase composition, 20
molecule therapeutics, 146
monochloroacetate, 30
monograph revisions, USP, 7–9
monomers and polymerization techniques, 275
multidose, single-container formulations, 180
multiparticulate drug delivery systems
excipients used in, 128, 133
pellets, 133
vs. single-unit dosage forms, 134
multiple internal reflectance infrared (MIR), 39
multiple linear regression (MLR), 306
multivariate curve resolution (MCR), 11, 306

nanocrystals
advantages of, 115
definition of, 114
excipients used in, 117
FRCs of, 117
preparation methods, 115
stabilization, 115–116, 118
National Formulary (NF), 210
native protein conformation, 146
natural polymers, 57–58
NDDS *see* novel drug delivery systems (NDDS)
neutral crowders, 171
new chemical entities (NCE), 271
new drug application (NDA), 234, 242
new excipient, 269–298
New Excipient Evaluation Committee (NEEC), 257
NF *see* National Formulary (NF)
NIR chemical imaging (NIR-CI), 11, 311
NIR spectroscopy, 314, 316
nominal components, 202
noncovalent interactions, 146
nonfunctional coat, 73
nonionic surfactants, 175, 176
nonnative protein aggregation, 153
novel characterization techniques
advanced NMR techniques, 18–20
AFM, 20–23
chemical imaging, 10–17
drug formulation, 9
PAT, 23–24
novel drug delivery systems (NDDS)
analytical characterization, 277–278
DMF and CEP, 280–281
excipients used in, 100–102
Kollicoat IR, 285
launch, 281–283
NCE, 271
product development, 274–276
regulatory dilemma, 284–285
scale-up and transfer, to production, 276–277
stability studies, 278–279
toxicological studies, 279–280
Novel excipient, 271–289
Noyes–Whitney equation, 115

ODTs *see* orodispersible tablets (ODTs)
opalescence, 158
optimization helix, 276
oral bioavailability, 101
oral dosage forms
lipid-based systems
definition of, 107–109
excipients used in, 110–113
FRCs of, 113–114
LFCS, 109–110
modified release dosage forms
classification of, 118–120
definition of, 117–118
matrix-based systems, 120–125
multiparticulate systems, 131–133
oral osmotically driven systems, 126–131
nanocrystals
advantages of, 115
definition of, 114
excipients used in, 117
FRCs of, 117
preparation methods, 115
stabilization, 115–116
ODTs
definition of, 133–135
excipients used in, 135–136
FRCs, 136–139
preparation methods, 135
pharmaceutical excipients
definition of, 97–98
FRCs and FRTs, 98–99
functionality of, 98
NDDS, 100–102
pharmacopoeial harmonization, 99–100
SDs, 102–103
ASDs, 107
excipients used in, 103–107
preparation methods, 103
orally disintegrating tablets (ODT), 255
oral modified release dosage forms
classification of, 118–120

oral modified release dosage forms (*Continued*)
definition of, 117–118
matrix-based systems, 120–125
multiparticulate systems, 131–133
oral osmotically driven systems
characteristic feature of, 126
as device concepts, 126
functionality-related attributes
flux regulators, 130
plasticizers, 130–131
pore forming agents, 129–130
semipermeable membranes, 128
wicking agents, 129
osmotic agents, 126–127
organic acids, 29–30, 42
orodispersible tablets (ODTs)
definition of, 133–135
excipients used in, 135–139
FRCs, 136–139
preparation methods, 135
rapid dispersion in, 136
Ostwald ripening, 116
OTC monograph products, 242

parabens, 182
partial least squares (PLS), 11, 306
PAT *see* process analytical technology (PAT)
PCA *see* principal component analysis (PCA)
peptide and protein pharmaceuticals, 145
peptide backbone cleavage, 151–152
percolation threshold, 121
permitted daily exposure (PDE), 254
peroxides, 26–27, 43
Pharmaceutical and Medical Device Agency (PMDA), 246
pharmaceutical excipients
definition of, 97–98
FRCs and FRTs, 98–99
functionality of, 98
NDDS, 100–102
pharmacopoeial harmonization, 99–100
pharmaceutical formulation, 236
pharmaceutical product variability, 237
Pharmacopeial Discussion Group (PDG), 7, 8, 200, 226, 229–230
pharmacopoeial harmonization, 7, 99–100
pharmacopeias/pharmacopoeias
European, Japanese and United States
alternate test methods, 212–213
general chapters, 211–212
general notices, 211
reference substances, 212
USP-NF Modernization, 213–214
International, 214
monographs
assay, 220
excipient sources, 217–219
identification, 219–220
impurities, 220–224
USP-NF excipient, 216
National, 214
tests, 224
phase instability, 158–159
pH modifiers, 76–78
photolytic degradation, 151
photooxidation, 151
physical degradations, 152
colloidal instability, 156–158
conformational instability, 153–155
general protein aggregation pathway, 153
interfacial instability, 159–163
phase instability, 158–159
plasticizers, 75, 130–131
PLS *see* partial least squares (PLS)
poloxamers, 177, 178
polyethylene glycol (PEG), 60, 150, 204
polyethyleneimine (PEI), 178
polyethylene oxide–polypropylene oxide–polyethylene oxide (PEO–PPO–PEO), 177
polymer-based stabilizers, 177–180
polymeric compounds, 179
polymerization 274–276
polysorbates, 176
polyvinyl acetate, 293
Polyvinyl alcohol-polyethylene glycol graft copolymer, 286, 289
polyvinylpyrrolidone (PVP), 58
pore forming agents, 129–130
potential interactions, 39
povidone (polyvinyl alcohol, PVA), 30, 33–34, 58, 217, 218, 224
preferential hydration, 169
pregelatinized starch, 57–58
preservatives, 180–182, 188
principal component analysis (PCA), 11
process analytical technology (PAT), 23–24, 98, 208, 303
product master file (PR-MF), 246
Product Regulatory Datasheet (PRD), 247
protein adsorption, 160–161
protein aggregation, 147
colloidal stability, 156
Lumry–Eyring framework of, 153
protein-based stabilizers, 177–180
protein-based therapeutics, 145
protein colloidal stability, 157
protein degradation pathways

chemical degradations, 148–152
physical degradations, 152–163
physiochemical properties, 151
protein immobilization, 157
protein instability, 146
protein–interface interactions, 163
protein–macromolecule interactions, 178
protein pharmaceuticals, 147
protein–protein interactions, 156–158
protein–protein repulsive interactions, 159
protein–protein self-interaction, 178
protein–solvent interactions, 156
protein-stabilizing cosolvents, 168
protein therapeutics, 147
Pure Food and Drug Act of 1906, 242

quality by design (QbD), 73, 98, 235–237, 262, 265
quality target product profile (QTPP), 207

Raman chemical imaging, 12, 14, 16
Raman spectra, 309
raw material variability
NIR and Raman spectroscopies, 304, 305
NIR PAT tools, conditions, 308
Raman spectra, 309
reactive amino acid, 150
reactive excipient impurities, 26
reactive oxygen species (ROS), 150
real-time data collection, 305
recombinant human granulocyte colony stimulating factor (rhGCSF), 154
recombinant human interferon-g (rhIFN-g), 154
recombinant human interleukin-1 receptor antagonist (rhIL-1ra), 181
recombinant plasminogen activator (rPA), 170
redox active metal ions, 150
reducing sugar impurities, 27, 41
regulatory dilemma, 284–285
relative standard deviation (RSD), 319
release specification, 206, 207
reservoir-coated pellets, 133
residual peroxides, 150, 176
residual processing aids, 202
residual solvents *see* excipients; impurities residual solvents
rhGCSF *see* recombinant human granulocyte colony stimulating factor (rhGCSF)
rhIFN-g *see* recombinant human interferon-g (rhIFN-g)
roller compaction, 311, 312
root mean squared error of prediction (RMSEP), 312

safety evaluation
biotechnology-derived pharmaceuticals ICH S6 (R1), 253
chronic and carcinogenicity studies, 252
developmental toxicity ICH S5 (R2), 253
excipient safety and toxicity, 250, 279
future trends, 255–256
genotoxicity testing, 252
GLP regulations and guideline, 249
immunotoxicology studies ICH S8, 253
impurities, safety considerations, 254–255
OECD guidelines, chemical testing, 254
safety pharmacology studies ICH S7A and S7B, 253
subchronic studies, 252
salt
protein formulations, 171–173
protein interactions, 172
Savitzky–Golay smoothing, 315
SDS *see* sodium dodecyl sulfate (SDS)
SDs *see* solid dispersions (SDs)
second-generation SDs, 102–103, 106–107
self-emulsifying drug delivery systems (SEDDS), 110
self-microemulsifying drug delivery systems (SMEDDS), 110
semipermeable membranes, 128
silicon dioxide
diethylstilbestrol preparations, 37
LME, 38
moisture scavenging, 37
site percolation threshold, 121
sodium dodecyl sulfate (SDS), 173
sodium starch glycolate (SSG), 36, 65
solid dispersions (SDs), 102–103
ASDs, 107
classification of, 102
excipients used in, 103–107
preparation methods, 103
solid dosage forms, of biopharmaceuticals
buffer agents, 185
bulking agents, 185–186
formulation design and excipient selection, 182
freezing and drying processes, 183
preservatives, 188
stabilizers, 186–187
stresses encountered during lyophilization, 183–184
surfactants, 187–188
solid–liquid interface, 162
solid-state NMR (ssNMR), 18
solid-state reactions, 36
solubility-limited oral bioavailability, 101
spatial filtering technique (SFT), 315

spray-dried lactose, 54–55
spray drying, 135
SSG *see* sodium starch glycolate (SSG)
ssNMR *see* solid-state NMR (ssNMR)
stability proteins, 163
stability studies, 278–279
stabilization, 116
stabilizers, 186–187
stabilizing cosolvents, 169
starch, 36–37, 57, 64–65
stearic acid, 38, 70, 72
steric stabilization, 116
sublimation, 135
sugars, 60–61
  macromolecular function and cell viability, 167
  osmolytes, 168
  preferential hydration, 169
  protein–solvent system, 168
  protein-stabilizing cosolvents, 168
  protein-stabilizing excipients, 167
surface-enhanced Raman chemical imaging, 12
    see also Raman chemical imaging
surfactants/surface-active agents, 187–188
  ionic surfactants, 173
  poloxamers, 177
  polysorbates, 176
  protein adsorption, 174
  protein formulations, 177
  protein solutions, 173
  solid–liquid interface, 174
  stabilization, 175
  structures of, 173–174
  surfactant–protein complexes, 175
  water-insoluble, 113
  water-soluble, 113
synergistic effect, 80
synthetic polymers, 58–60

tablet compression, 316
tablet core/coating, 76
tablet molding, 135
TEC *see* total excipient control (TEC)
technically unavoidable particles (TUP), 264
therapeutic proteins, 160
thermal gravimetric analysis (TGA), 39
thermodynamic stability, 146
third-generation SDs, 107
tonicity modifiers, 167, 171
total excipient control (TEC)
  Federal Food, Drug, and Cosmetic Act, 259
  IPEC's excipient control resources, TEC,
      261–262
  third-party auditing and certification, 261
  validation *vs.* process capability, 262–263
toxicity, 8
toxicological studies, 279–280
trace reactive excipient impurities
  aldehydes, 41–42
  organic acids, 42
  peroxides, 42
  reducing sugar, 41
transition-metal-mediated oxidation, 29

United States pharmacopeia (USP), 7, 209, 210
United States Pharmacopeia-National Formulary
    (USP-NF), 97
  additives, 218–220
  alternate test methods, 212–213
  modernization, 213–214
  monographs, 216–217
US Food and Drug Administration (FDA), 303

vegetable oils, 110–112
vitamin C tablets, hardness–compression force
    profile, 297

water-insoluble surfactants, 113
water-proofing effect, 71
water-soluble polymers, 179
water-soluble surfactants, 113
water structure breakers, 172
wet chemistry colorimetric test method, 8
wicking agents, 129

X-ray microtomography techniques, 15

Printed in the USA/Agawam, MA
September 16, 2026

910661.167